Levels of Perception

Springer
New York
Berlin
Heidelberg
Hong Kong
London
Milan
Paris
Tokyo

Laurence Harris
Michael Jenkin
Editors

Levels of Perception

With 175 Illustrations

**INCLUDES
CD-ROM**

Springer

Laurence Harris
Department of Psychology
York University
4700 Keele Street
Toronto, Ontario M3J 1P3
Canada
harris@yorku.ca

Michael Jenkin
Department of Computer Science
York University
4700 Keele Street
Toronto, Ontario M3J 1P3
Canada
jenkin@cs.yorku.ca

Library of Congress Cataloging-in-Publication Data
Levels of perception / editors, Laurence Harris, Michael Jenkin.
    p. cm.
    Includes bibliographical references and index.
    ISBN 0-387-95525-9 (alk. paper)
    1. Visual perception.   I. Harris, Laurence, 1953–   II. Jenkin, Michael, 1959–
QP475 .L48 2002
    152.14—dc21   2002070739

ISBN 0-387-95525-9   Printed on acid-free paper.

Printed in the United States of America.

9 8 7 6 5 4 3 2 1   SPIN 10883028

Typesetting: Pages created by the editors using a Springer TeX macro package.

www.springer-ny.com

Springer-Verlag   New York Berlin Heidelberg
A member of BertelsmannSpringer Science+Business Media GmbH

For Ian

Preface

In the summer of 2001, many scientists with various connections to Dr. Ian Porteous Howard met in Toronto to acknowledge Ian's influence on their lives and work. Given Ian's broad interests, choosing a topic for this conference and Festschrift was more challenging than it might have been. We settled on the theme of *Levels of Perception* to embrace essentially all possible topics but with the intention of emphasizing not just the science, but also people's approaches to them. In the title we wish to capture the fact that a given perceptual process can be approached experimentally and conceptually at many levels. Some aspects of vision might be constrained at the level of the retina, for example by lateral inhibition or chemical events triggered by light. If information is arranged in a certain way in the optic nerve sending visual information to the brain, then perceptual phenomena could be explained with reference to this lowest of levels. But visual processing depends surprisingly little on the retina. Our perception is so far removed from the distorted, inverted, blood-vessel-besmirched, motion-blurred, curved images on the retina that even if a perceptual phenomenon does have a robust low-level correlate, adequate explanations need to address multiple levels. Furthermore the visual system does not exist in isolation. What we think of as visual phenomena must always be associated with motor correlates, even if it is only holding the eyes (and the rest of the body) in one position for a while. Our goal in this book is to convey the importance of considering perception as a multilevel process.

The CD-ROM that accompanies this book contains colour imagery and video clips associated with various chapters and the conference itself. The CD-ROM is presented in HTML format, and is viewable with any standard browser (e.g., Netscape Navigator or Microsoft Internet Explorer). To view the videos on the CD-ROM you will need Quicktime, which is available free from Apple. To view the CD-ROM, point your browser at the file **index.htm** on the CD-ROM.

This book is in appreciation of the contributions of Ian P. Howard. He continues to be an inspiration to many. We would like to thank Teresa Manini who ran the conference, and our wives for their enduring support.

York University, Ontario, Canada
Summer 2002

Laurence Harris
Michael Jenkin

Contents

7   The Making of a Direction Sensing System for the Howard Eggmobile   127

Hiroshi Ono, Linda Lillakas, and Alistair P. Mapp

8   Levels of Processing in the Size-Distance Paradox   149

Helen E. Ross

9   The Level of Attention: Mediating Between the Stimulus and Perception   169

Jeremy M. Wolfe

10 Single Cells to Cellular Networks

Robert F. Hess

III   Eye Movements and Perception

11 Levels of Fixation

Richard V. Abadi, Richard Clement, and Emma Gowen

Contributors

Richard V. Abadi
UMIST
Department of Optometry and Neuroscience
P.O. Box 88
Manchester, M60 1QD, England
email: richard.abadi@umist.ac.uk

Dora E. Angelaki
Deptment of Anatomy and Neurobiology
Washington University School of Medicine
and Department of Research, Central Institute for the Deaf
St. Louis, MO, 63110, USA
email: anglelaki@thalamus.wustl.edu

Stuart Anstis
University of California San Diego
9500 Gilman Drive
La Jolla, CA, 92093, USA
email: sanstis@ucsd.edu

Karl Beykirch
Department of Neurology
University of Tübingen
72076 Tübingen, Germany
email: karl.beykirch@uni-tuebingen.de

Randolph Blake
Vanderbilt Vision Research Center
Vanderbilt University
Nashville, TN, 37240, USA
email: randolph.blake@vanderbilt.edu

Barbara Blakeslee
Department of Psychology
North Dakota State University
Fargo, ND, 58105-5075, USA
email: barbara_blakeslee@ndsu.nodak.edu

Richard Clement
Visual Sciences Unit
Institute of Child Health
University College London
30 Guilford Street
London, WCIN 1EH, England

Kathleen E. Cullen
Aerospace Medical Research Unit
McGill University
3655 Prom. Sir William Osler
Montreal, Quebec, H3G 1Y6, Canada
email: cullen@med.mcgill.ca

J. David Dickman
Department of Anatomy and Neurobiology
Washington University School of Medicine
and Department of Research, Central Institute for the Deaf
St. Louis, MO, 63110, USA
email: ddickman@cid.wustl.edu

Elias Economou
Department of Psychology
Rutgers University
101 Warren Street
Newark, NJ, 07102, USA
email: elias@psychology.rutgers.edu

Michael Fetter
Neurologie II /SHT
Klinikum Karlsbad-Langensteinbach
Guttmannstrasse 1
76307 Karlsbad-Langensteinbach, Germany
email: michael.fetter@kkl.srh.de

Alan L. Gilchrist
Department of Psychology
Rutgers University
101 Warren Street,
Newark, NJ, 07102, USA
email: alan@psychology.rutgers.edu

Emma Gowen
UMIST
Department of Optometry and Neuroscience
P.O. Box 88
Manchester, M60 1QD, England

Laurence R. Harris
Centre for Vision Research and Department of Psychology
York University
4700 Keele Street
Toronto, Ontario, M3J 1P3, Canada
email: harris@yorku.ca

Robert F. Hess
McGill Vision Research
Department of Ophthalmology
McGill University
Montreal, Quebec, Canada
email: rhess@astra.vision.mcgill.ca

Ian Howard
Centre for Vision Research and Department of Psychology
York University
4700 Keele Street
Toronto, Ontario, M3J 1P3 Canada
email: ihoward@hpl.crestech.ca

Michael Jenkin
Centre for Vision Research and Department of Computer Science
York University
4700 Keele Street
Toronto, Ontario, M3J 1P3 Canada
email: jenkin@cs.yorku.ca

K. Kawano
Neuroscience Research Institute
National Institute of Advanced Industrial Science and Technology
Tsukubashi, Ibaraki 305-8568, Japan

Frederick A. A. Kingdom
McGill Vision Research Unit
687 Pine Avenue West, Room H4-14
Montreal, Quebec, Canada H3A 1A1
email: fred.kingdom@mcgill.ca

Linda Lillakas
Centre for Vision Research
York University
Toronto, Ontario, M3J 1P3, Canada
email: lillakas@yorku.ca

Alistair P. Mapp
Centre for Vision Research
York University
Toronto, Ontario, M3J 1P3, Canada
email: amapp@yorku.ca

Mark E. McCourt
Department of Psychology
North Dakota State University
Fargo, ND, 58105-5075, USA
email: mark.mccourt@ndsu.nodak.edu

Daniel M. Merfeld
Department of Otology and Laryngology
Harvard Medical School
Jenks Vestibular Physiology Laboratory
Massachusetts Eye and Ear Infirmary
243 Charles Street
Boston, MA, 02114, USA
email: dan_merfeld@meei.harvard.edu

F. A. Miles
Laboratory of Sensorimotor Research
The National Eye Institute
49 Convent Drive
Bethesda, MD, 20892, USA
email: fam@lsr.nei.nih.gov

Charles M. Oman
Man Vehicle Laboratory
Massachusetts Institute of Technology
Cambridge, MA, 02139, USA
email: cmo@space.mit.edu

Hiroshi Ono
Centre for Vision Research
York University
Toronto, Ontario, M3J 1P3, Canada
email: hono@yorku.ca

C. Quaia
Laboratory of Sensorimotor Research
National Eye Institute, The National Institutes of Health
Bethesda, MD, 20892, USA

Helen E. Ross
Department of Psychology
University of Stirling
Stirling, FK9 4LA, Scotland
email: h.e.ross@stir.ac.uk

Jefferson E. Roy
Aerospace Medical Research Unit
McGill University
3655 Prom. Sir William Osler
Montreal, Quebec, H3G 1Y6, Canada

Clifton M. Schor
University of California at Berkeley
School of Optometry
Berkeley, CA, 94720, USA
email: schor@socrates.berkeley.edu

Heimo Steffen
The Johns Hopkins University School of Medicine
Baltimore, MD, 21287, USA

Martin J. Steinbach
Deparment of Psychology
York University
Toronto, Ontario, M3J 1P3, Canada
email: mjs@yorku.ca

Pierre A. Sylvestre
Aerospace Medical Research Unit
McGill University
3655 Prom. Sir William Osler
Montreal, Quebec, H3G 1Y6, Canada

A. Takemura
Neuroscience Research Institute
National Institute of Advanced Industrial Science and Technology
Tsukubashi, Ibaraki 305-8568, Japan

Mark F. Walker
The Johns Hopkins University School of Medicine
Baltimore, MD, 21287, USA

Jeremy M. Wolfe
Center for Ophthalmic Research
Brigham and Women's Hospital and Harvard Medical School
221 Longwood Avenue
Boston, MA, 02115, USA
email: wolfe@search.bwh.harvard.edu

David S. Zee
The Johns Hopkins University School of Medicine
Baltimore, MD, 21287, USA
email: dzee@dizzy.med.jhu.edu

Lionel H. Zupan
Department of Otology and Laryngology
Harvard Medical School
Jenks Vestibular Physiology Laboratory
Massachusetts Eye and Ear Infirmary
243 Charles Street
Boston, MA, 02114, USA
email: lionel_zupan@meei.harvard.edu

1

Ian P. Howard and Levels of Perception

Laurence R. Harris, Ian P. Howard, and Michael Jenkin

In the summer of 2001, many scientists with various connections to Dr. Ian Porteous Howard met in Toronto to acknowledge Ian's influence on their lives and work. Many of the attendees had direct connections with Ian such as his graduate students and postdoctoral fellows. Others had more indirect reasons to want to be there. For example Ian has influenced many people through his well-thumbed books on visual perception. Ian is a man of diverse interests and has made significant contributions to many areas including size-weight judgements, interocular transfer, colour, stereopsis, perception of disparities and binocular vision, eye movement control, perhaps especially torsion and its connection to binocular vision, and the perception of self-motion and orientation. Ian has always been interested in the history of the scientific investigation of these problems also, delighting in models of ancient stereoscopes and the like, and becoming a champion of Al Hazen. Ian points out (Howard, 1996) that Al Hazen had already considered many of the topics that most of us are working on now – and arrogantly consider as "modern questions" – a millennium ago.

Given Ian's broad interests, choosing a topic for this conference and Festschrift was more challenging than it might have been. We settled on the theme of *Levels of Perception* to embrace essentially all possible topics but with the intention of emphasizing not just the science, but also people's approaches to them. In the title we wish to capture the fact that a given perceptual process can be approached experimentally and conceptually at many levels. Some aspects of vision might be constrained by events occurring at the level of the retina, for example by lateral inhibition or chemical events triggered by light. If information is arranged in a certain way in the optic nerve sending visual information to the brain, then it seems reasonable to expect that we could explain perceptual phenoma with reference to this lowest of levels. But visual processing depends surprisingly little on the retina. Our perception is so far removed from the distorted, inverted, blood-vessel-besmirched, motion-blurred, curved images present on the retina that even if a perceptual phenomenon does have a robust low-level correlate, adequate explanations need to address multiple levels. Furthermore the visual system does not exist in isolation. What we think of as visual phenomena must always be associated with motor correlates, even if it is only holding the eyes (and of course

FIGURE 1.1. A portrait of Ian Howard by Nick Wade. Ian's eyes are on the circumference of a circle that represents the Vieth-Müller circle and also part of an embedded eye.

the rest of the body) in one position for a while. Consider the phenomenon of induced motion. Induced motion is when a stationary target appears to be moving by virtue of movement of the background. Induced motion can be analysed at the retinal level in terms of relative changes of luminance defining movement at different retinal locations, at a cortical level, at a cognitive level, or as a phenomenon arising primarily from the eye movement system only incidentally having any visual effect at all. How might these analyses fit together? Or are they like Kuhnian paradigms which can never pass useful information from one to the other? The concept of levels also makes us think of the flow of information between levels, which leads to a consideration of the roles of top-down and bottom-up flow, sometimes thought of as feed-back and feed-forward processes, respectively.

1.1   Ian's Contribution to Science

Ian Howard has made some seminal contributions to the vestibular field. Interestingly, an important contribution that he has made is in what he has <u>not</u> done! In his books *Human Spatial Orientation* and *Human Visual Orientation*, Ian included extensive italicized sections in which he described experiments that needed to be done: lacunae of knowledge. His research in human spatial orientation is still well funded and ongoing. Dr Howard is much sought after as a consultant to the U.S. space program and Canadian industry such as CAE and the Defence and Civil Institute for Environmental Medicine (DCIEM). In 1998, Ian won a place to carry out some research on the Neurolab Mission of the Space Shuttle. In preparation for this, at the age of 70, he had fun on the KC135 "vomit comet" which provides brief periods of zero gravity (see Figure 1.2. This work is funded by NSBRI

(NASA's National Sciences and Brain Research Institute) in collaboration with Chuck Oman of MIT.

In 1983, Ian was awarded the York University Walter L. Gordon Research Fellowship. In 1985 his solid reputation as a scientific researcher was recognised by his being awarded a York University Distinguished Research Professorship that makes him, as he delights in saying, a DRIP.

The year 2002 finds him in good health and actively involved in research on several well-funded projects in visual science at the exemplary age of 74. No account of Ian's life would be complete without mentioning Antonie (Toni) Howard, Ian's wife since 1956. Toni was a refugee from Hitler's Germany and had been a chemist before getting married. For many years she has helped in all aspects of Ian's scientific work. They have three children, Ruth born in 1957 and the twins Neil and Martin born in 1959. The whole family was present at the celebration, but we forgot to take a group photograph. Ian has agreed to contribute a short autobiography.

1.2   Ian on Ian

I was born in Rochdale, Lancashire, in the North of England in 1927. My father was a foundry worker from Warrington and my mother had been a weaver in the Rochdale cotton mills. They met as young Fabian socialists and cyclists before the First World War. Soon after the war my father became a full-time Union Organiser and local politician. During the Second World War my two sisters and I were evacuated to a beautiful village in the English Lake District and I attended Ulverston Grammar School. I left school at the age of 16 and worked for several years in industrial chemical laboratories, while studying Chemistry and Biology at night school. I entered Manchester University in 1948 to study Chemistry and Biology, but changed to Psychology and Physiology during my second year and obtained a B.Sc. in Psychology in 1952.

In 1952, I went to Durham University in the Northeast of England as a Research Assistant in a Psychology department that had just been founded. In 1953 I was appointed Lecturer. In those days people could gain an academic appointment without a Ph.D. Durham is a beautiful, small cathedral city. I lived in part of a thirteenth-century monastery that I renovated. I married Antonie (Toni) Eber in 1957. Toni had been a Jewish refugee from Nazi Germany. It was not long before we had three children, Ruth, and twins Neil and Martin. During the 14 years in Durham I conducted research in perceptual ambiguity, visual-motor coordination, and eye movements. Brian Templeton and Brian Craske were my graduate students.

In 1965, I was invited to be a visiting Associate Professor in the department of Psychology at New York University. It was a dramatic change for the whole family, coming from a small cathedral town in the north of England, to live on the fifteenth floor of an apartment building in Greenwich Village, New York. My

FIGURE 1.2. Ian floating in microgravity

introduction to the American academic scene was also an impressive experience. During that year I obtained a Ph.D. from Durham University. Frederick Smith was my nominal supervisor but in fact I had no supervisor since nobody in Durham worked in perception. The book *Human Spatial Orientation*, written with W. B. Templeton, appeared in the same year.

While I was in New York I received a phone call from Kurt Danziger to ask whether I would like to apply for a position at York University. I had no idea where that was, but I came up to have a look and decided to move. The thing that attracted me was the possibility of building up a group of people interested in visual perception. There were several psychologists at the Glendon campus when I arrived, including Howard Flock in visual perception. I was the first member of the department to be housed on the York campus. I was promoted to full Professor in 1967, and between 1968 and 1971 I was Chairman of the department (Figure 1.4). During this time I recruited Brian Templeton, Hiroshi Ono, Martin Steinbach, Peter Kaiser, and Len Theodor. Stuart Anstis arrived in 1972. This nucleus of visual scientists formed the York Vision Group. Later, Keith Grasse, Martin Regan, and Laurence Harris arrived and we founded the Centre for Vision Research in 1992.

Administration kept me out of the laboratory during the early years at York. In 1972, I spent a sabbatical year in England at the University of Sussex where I made an aborted attempt to write an introductory textbook on perception. Between 1977 and 1982 I was preoccupied writing *Human Visual Orientation* (Figure 1.3), part of which was written in 1980 while I was on sabbatical leave at the Smith-Kettlewell Institute in San Francisco. In cooperation with Masao Ohmi I then began research into several aspect of spatial orientation, including vection, induced motion, and torsional and vergence eye movements. My postdoctoral fellows during this period were Esther Gonzalez, Tom Heckmann, William Simpson,

FIGURE 1.3. Ian's 1982 book featured another portrait by Nick Wade.

and Li Sun. My graduate students were Bob Cheung, Christine Marton, Chieko Murasugi, Gang Hu, and Jim Zacher.

In 1988, Martin Regan and I founded the Human Performance Laboratory of the Institute for Space and Terrestrial Sciences. This was an Ontario Centre of Excellence with its headquarters at York University. It was later renamed the Centre for Research in Earth and Space Technologies (CRESTech).

In 1993, I retired from my teaching appointment at York University to become a Senior Scientist with CRESTech. At about the same time I became interested in stereoscopic vision, and in 1995 Brian J. Rogers of Oxford and I published the book *Binocular Vision and Stereopsis*. I conducted research on several aspects of stereoscopic vision with Brian Rogers, visiting scientist Byron Pierce, postdoctoral fellows Alan Ho, Masahiro Ishii, Hirohiko Kaneko, Masayuki Sato, and Kazumichi Matsumiya, and graduate students Rob Allison, Jingyu Dong, and Xueping Fang. A two-volume book on *Seeing in Depth* appeared in 2002. I have continued to work on human spatial orientation with grants from NASA, the Canadian Space Agency, and the Defence and Civil Institute of Environmental Medicine (DCIEM). At present I am aided by research assistants James Zacher and Heather Jenkin and by postdoctoral fellows Richard Dyde and Phil Duke.

My other interests include belief systems of all kinds. I am a confirmed atheist and sceptic about all things supernatural and most things political. I enjoy reading history, listening to classical music, walking, especially in the English countryside, woodworking and sculpting, and composing games. Toni and I have seven grandchildren to spoil and play with.

I am very grateful to Laurence Harris, Michael Jenkin, and others who organised the conference on *Levels of Perception* on the occasion of my retirement from CRESTech. I was impressed by the high quality of all the talks and am pleased that they are being published in this book. I was delighted to see most of my past

FIGURE 1.4. Ian P. Howard as head of the Department of Psychology at York University in 1968.

postdoctoral fellows and many of my past graduate students at the conference. Toni and I will treasure the memory of that occasion that my old colleague Masao Ohmi and his wife Heroe recorded for us in a photograph album. I will continue academic work for as long as I feel up to it.

1.3   Levels of Perception

The book is divided into parts corresponding to topics in which Ian has had an interest. One thing that attracts Ian more than anything is the presence of a lively and constructive debate in an area. This is well exemplified by an active debate concerning the origin of "lightness and brightness" effects. Brightness is perceived luminance, while lightness is perceived reflectance – a property of the surface of objects. The debate concerns whether perceived variations in either or both of these perceptions can be explained by low-level, bottom-up processes such as lateral inhibition in the retina, or whether higher-level cognitive interpretations such as determining the direction of light source and identifying shadows are required. Several of the main protagonists involved in this debate, which is carried out in the pages of the world's professional scientific vision journals, came along to present their points of view at the conference. Their chapters form Part I of this book. Not all the participants contributed chapters, but all were involved in the "discussion section" which can be found expanded on the enclosed CD-ROM.

In Part II, which we have entitled simply "Levels of Perception," various aspects of perception are considered with discussion of the different levels at which they might occur. Anstis discusses motion perception and considers at which level some of his clever illusions might arise. Blake discusses at what level binocular rivalry might occur. Ono, Lillakas, and Mapp discuss the role of perceptual geometry in the entertaining context of the Howard Eggmobile, a device that Ian built in 1978, with which he won a televised *crazy inventors* race (into which he fitted just perfectly). Anstis, Blake, and Ono et al. have also provided demonstrations that can be found on the accompanying CD-ROM. Ross considers explanations at various levels for the distortions of perceptual size that are connected to errors in distance perception. Distortions of distance can potentially arise from very low levels, such as accommodation errors, right through to high-level cognitive factors such as the expected shape of a room. Whether the processing of things that grab your attention should be considered low or high level is the subject of Wolfe's "the Level of Attention." Hess finishes Part II by reviewing how the outputs of early cortical processing are put together to build up the next level of visual analysis.

Part III looks at "Eye Movements and Perception." Ian has had a strong interest in this topic and the inevitable, interactive connection between eye movement control and perception. Abadi, Clement, and Gowen discuss levels of fixation. Schor considers the factors that contribute to the near response, sometimes called the triple response, in which vergence, accommodation, and pupil size are adjusted as a family when looking at close objects. The control of vergence eye movements is discussed by Takemura, Kawano, Quaia, and Miles. Steinbach reviews the controversial debate about how we know where our eyes are pointing and compares the use of corollary discharge, or outflow information, with in-flowing proprioceptive sensory information.

Part IV extends the eye movement work to include orientation and self-motion perception. The processing of self-motion can often be measured or assessed by looking at the compensatory eye movements associated with the movement. Maintaining and knowing about the orientation of the body relative to the world is perhaps the most fundamental perceptual task. Ian made many contributions to this task, including his well known and sadly out of print book entitled *Human Visual Orientation* (1982). This book was an extensively revised version of *Human Spatial Orientation*, which he wrote with Brian Templeton (1966).

Chapters in Part IV consider what the so-called vestibulo-ocular reflexes can tell us about the neural coding of angular (Harris et al.) and linear (Angelaki and Dickman) motion. Recent exciting discoveries at the level of the vestibular nucleus are reviewed by Cullen, Roy, and Sylvestre showing that active and passive movements are handled differently, even at this very early level of processing. Harris, Beykirch, and Fetter draw a parallel between the influence of such factors on eye movement processing and the postmodern movement in other aspects of human thought and endeavour which emphasizes the role of context in general. Merfeld and Zupan discuss the internal representation of the body and "physical quantities." Oman, who is currently actively collaborating with Ian and who was involved with Ian's recent Neurolab work, uses the microgravity environment of space to investigate the role of different frames of reference on the perception of orientation. Walker, Steffen, and Zee describe how the cerebellum is involved in the control of torsional eye movement and how damage to the cerebellum can disrupt this system.

FIGURE 1.5. After an hour-long presentation in which he summarized his life work and into which he wove fascinating biographical details, Ian Howard received a standing ovation.

References

Howard, I. P. (1982). *Human Visual Orientation*, New York: Wiley.

Howard, I. P. (1996). Al Hazen's neglected discoveries of visual phenomena. *Perception* 25: 1203-1217.

Howard, I. P. (2002). *Seeing in Depth*, Vol. I, Toronto: Porteous Press.

Howard, I. P. and Rogers, B. (1995). *Binocular Vision and Stereopsis*, Oxford: Oxford University Press.

Howard, I. P. and Rogers, B. (2002). *Seeing in Depth*, Vol. II, Toronto: Porteous Press.

Howard, I. P. and Templeton, B. (1966). *Human Spatial Orientation*, New York: Wiley.

Part I

Brightness and Lightness

2

Dualistic Versus Monistic Accounts of Lightness Perception

Alan L. Gilchrist and Elias Economou

Modern work on lightness began with the twin assumptions of raw sensations transformed by cognitive interpretation. Together, these assumptions are unfalsifiable, and leave us hopelessly trapped in the mind-body problem. We argue that Hering and Helmholtz, far from theoretical opposites, share essentially the same theory. The only radical departure was taken by Gestalt theory, which posits a single material process that results in visual experience. Following the failure of low-level models, Kingdom now proposes to combine the Hering and Helmholtz accounts. But anchoring theory, an extended, more concrete form of Gestalt theory, shows that the earlier dualism is obsolete. We claim that, in a series of experiments testing variations on the basic simultaneous contrast display, low- level contrast-based accounts of lightness either fail to make clear predictions or else make the wrong predictions. On the other hand, anchoring theory makes specific predictions that are generally supported by the data.

2.1   Introduction

From the outset of the modern era of vision research, a one-to-one correspondence between local stimulation and visual experience was assumed. Vision, in short, was faithful to the retinal image. We will refer to this assumption as the doctrine of local determination (rather than the more historically accurate, but now confusing term 'constancy hypothesis'). When Helmholtz and Hering became interested in lightness, it was precisely the violation of this doctrine that caught their attention. Lightness, they noted, is far more faithful to the distal stimulus than to the proximal, to the object than to the image.

Two phenomena, simultaneous contrast and lightness constancy, have historically received special attention, in part because both illustrate violations of the doctrine of local determination. In simultaneous lightness contrast, targets that produce equal local stimulation are perceived as different in lightness, while in lightness constancy, targets that produce different degrees of local stimulation are nevertheless seen as equal in lightness.

To accommodate the object-oriented nature of perception without sacrificing the doctrine of local determination, Helmholtz and Hering adopted a duality that persists until today. Helmholtz (1866) concluded that there are two visual responses to the stimulus: (1) raw sensations faithful to the proximal stimulus, and (2) percepts (products of a cognitive interpretation) that are faithful to the distal stimulus. Although Hering (1874) noted several logical problems with the doctrine of sensations,[1] he himself could not escape the duality. Though it has been claimed (Boring, 1942) that he rejected the doctrine of sensations, Hering emphasized peripheral mechanisms that directly influence perception but are separate from the effects of learning and memory.

Although Helmholtz and Hering are usually regarded as theoretical opposites, the difference between them is merely one of emphasis. Both embraced the concept of high and low levels, Helmholtz emphasizing the former, Hering the latter. Yet Helmholtz's theory included the sensation as a low-level process and Hering's theory included memory color as a high-level process. The only radical departure was taken by Gestalt theory (Koffka, 1935; Gelb, 1929; Kardos, 1934). The Gestalt theorists flatly rejected the notion of sensations and, by implication at least, the cognitive processes required to transform sensations into percepts. Indeed, the Gestaltists rejected the very concept of levels. They argued that when light strikes the retina, it sets in motion a chain of physical (if biochemical) actions and reactions that culminate in visual experience. It is a single process and we experience only the outcome of the process, not the early events.

Gestalt theory, marginalized by events surrounding WWII, was not taken seriously again until the emergence of the computer. In the meantime, work on lightness was dominated by theories derived from Hering. Both simultaneous contrast and lightness constancy were attributed to lateral inhibition acting retinotopically. Simultaneous contrast lent itself readily to this account. Targets of equal gray produce equal degrees of stimulation the retina, but the white region surrounding one target inhibits neural activity in that target, making it appear darker gray. A similar logic was applied to lightness constancy. A higher illumination level increases the neural excitation produced by a given target. But this increase is neutralized, either exactly (Cornsweet, 1970) or approximately (Jameson and Hurvich, 1964) by a corresponding increase in inhibition coming from a brightened surround.

In the 1970's, evidence emerged (Gilchrist, 1979, 1980) that lightness depends strongly on depth perception.[2] This presented a powerful challenge to the low-level account because retinal processes precede depth processing. Additional work (Gilchrist et al., 1983; Gilchrist, 1988) showed that, when simultaneous contrast and lightness constancy are presented in a comparable fashion, with essentially equivalent retinal images, the effects produced by the constancy display are about six times stronger than those produced by the contrast display. These displays are represented in Figure 3.1 of the accompanying chapter by Kingdom.

While these developments convinced most workers that lightness constancy

---

[1] We experience lightness as "out there," whereas sensations are thought to be within the organism. Black has the same phenomenal status as white and yet, by traditional accounts, must be the absence of sensation.

[2] Earlier work by Kardos (1934) establishing this fact was ignored.

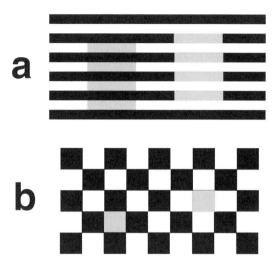

FIGURE 2.1. White's illusion and checkerboard contrast.

required a more sophisticated explanation, most were still prepared to attribute simultaneous contrast to lateral inhibition. But even this residual role for lateral inhibition has begun to erode during the past two decades. Perhaps the first important blow was struck by Michael White (1981) when he published a new stimulus, now celebrated as White's illusion, and shown in Figure 2.1a. Here the target bar that is mostly adjacent to black regions actually appears darker than those bars mostly adjacent to white, directly contradicting a contrast prediction. Todorovic (1997) composed an analogous illusion using an even more extreme aspect ratio than that of White.

Checkerboard contrast (De Valois and De Valois, 1988, p. 229), shown in Figure 2.1b, took things one step further. Though weaker than White's illusion, this pattern produces an anticontrast result even when the only regions that could induce contrast share no border at all with the target, touching each target only minimally at its corners.

Finally, this line of development was completed when reverse contrast displays were produced in three different labs (Agostini and Galmonte, 2002; Economou et al., 1998; Bressan, 2001). Here the gray target that is completely surrounded by black appears darker than the target that is completely surrounded by white. These displays provide compelling evidence that so-called contrast effects depend on organizational factors, not a structure-blind mechanism such as lateral inhibition.

Adelson's (2000) corrugated Mondrian stimulus produced a lightness difference for equiluminant targets well in excess of what would be expected based on the average luminance of the regions surrounding the targets.

Against this historical background, Kingdom (this volume) and Blakeslee and McCourt (this volume) propose to rescue a low-level account by appealing to the traditional concept of levels. Kingdom, for example, argues that many of the recent lightness demonstrations show that high-level cognitive processes are overlaid on the effects of low-level mechanisms. The very fact that Kingdom can so easily combine the theories of Hering and Helmholtz, theories that have been regarded as polar opposites, is remarkable. It supports our claim that they are rather two sides of the same coin.

Perhaps the most serious problem with Kingdom's construction is that it is not falsifiable. Results, such as that of simultaneous contrast, that go in a contrast direction are presented as evidence of low-level processes. Results that go in the opposite direction are seen as revealing the operation of high-level cognitive processes. Kingdom's edifice is not new. Its treatment of the mind-body problem seems uneconomical. Some aspects of perception are explained by the operation of the body, others by the operation of the mind.[3]

Kingdom suggests that lightness levels produced at the first level are overridden by, or combined with, those produced at higher levels. But the first level does not produce lightness levels. Most modern investigators see lateral inhibition as a key part of the mechanism by which luminance differences[4] in the retinal image are encoded. By this account, the first level produces edge signals, not lightness levels. And one needs a whole theory of lightness to get from edge signals to lightness levels. What might it mean to say that contrast effects are produced at the first level?

One can say that lateral inhibition produces lightness levels, if one accepts a pointwise conception of lateral inhibition, rather than an edge-encoding conception. Such a pointwise conception is implicit in the familiar scalloped profiles of the simultaneous contrast display after it passes through a lateral inhibitory network (Cornsweet, 1970). But the absence of the scallops predicted to occur in homogeneous surfaces have led most students of lightness to abandon the pointwise conception in favor of the edge-encoding conception.

Lateral inhibition is many things to many people. Cornsweet appears to offer lateral inhibition both in its point-wise form and in its edge-encoding form. But even if the edge-encoding form is accepted, a question arises whether the edges are enhanced or merely encoded. Here it seems that Kingdom equivocates, seeming not to make the distinction. He writes: "By tying lightness to contrast, lightness becomes invariant to changes in overall light level." This implies that edges are encoded. But if edges are encoded and not enhanced, there is no explanation of simultaneous contrast. Kingdom identifies the term contrast with both Paul Whittle on one hand, and Hurvich and Jameson on the other. But these people mean very different things by contrast. For Hurvich and Jameson (1966, p. 85), contrast means an exaggeration of edge differences: "What the contrast mechanism seems to do in all these instances is to magnify the differences in apparent brightness between adjacent areas of different luminances." But for Whittle, contrast merely

---

[3] We recognize that Kingdom does, at one point, attempt to integrate these domains by suggesting that multiscale filtering may be the instantiation of unconscious inference, but for now this remains a promissory note.

[4] Differences on a log scale; that is, ratios.

means that edge differences are encoded. Whittle and Challands (1969, p. 1106) wrote: "On this view the role of 'lateral inhibition' is less obvious than usually assumed. It is involved in determining the size of the edge signal, but simultaneous contrast could in principle be just as great in a system without lateral inhibition." Indeed, Whittle explicitly rejects the use of contrast based on lateral inhibition to explain simultaneous contrast, writing (Whittle, 1994, p. 153): "To explain brightness contrast in terms of lateral inhibition is like explaining the jerky progression of a learner driver in terms of the explosions in the cylinders of the car's engine. The explosions have a place in the causal chain, but regarding them as causes specifically of the jerks is to be mislead by a superficial analogy." Notice that Whittle acknowledges the role of lateral inhibition in the causal chain, but not a role that explains simultaneous contrast.

Adelson (see York transcript on the enclosed CD-ROM) notes that while the lens also plays a role in the causal chain leading to lightness, this does not mean that the lens plays a meaningful role in the explanation of lightness. This is not even to deny that early links in the chain can have an impact on perceptual experience. Malfunction in the lens can make a sharp edge appear blurred. Kingdom cites the fact, reported by Wallach and emphasized by Whittle, that when an increment in one eye overlaps a decrement in the other eye, the result is rivalry. This fact strongly supports the idea that only luminance differences at edges are encoded at the retina. But the enhancement of luminance differences by lateral inhibition has not been established. As Freeman (1967) wrote in a review of the contrast idea: "an experimental analysis of enhanced brightness differences has not, as yet, been performed." Heinemann (1972, p. 147) concurs, outlining the kind of test that would be necessary.

Kingdom supports his claim of multiple levels by citing the different levels of trichromacy and opponency in the widely accepted model of color vision. He notes that the fact that any color can be matched by some combination of three primaries cannot be explained by opponent processes, nor can our inability to see reddish greens or bluish yellows be explained by trichromacy. But this situation does not apply to lightness. It is not too difficult to account for both lightness constancy and simultaneous contrast by a single model and indeed, several such models have been proposed (Gilchrist et al., 1999; Ross and Pessoa, 2000). The explanation of simultaneous contrast does not require a low-level model. Indeed, we have recently shown in a series of experiments, that the anchoring model (which is neither low nor high) successfully predicts the outcome of a series of manipulations of the illusion. In most of these cases, it is either very difficult to derive a prediction from the low-level models, or else those predictions turn out to be wrong.

To describe this work it will first be necessary to briefly describe the anchoring model of lightness and then to consider how the model applies to simultaneous contrast. According to the anchoring model (Gilchrist et al., 1999), complex images can be segmented into frames of reference, or perceptual groups, using the Gestalt grouping principles. Any given target surface will belong to at least two such groups, and a separate lightness value is computed for that target in relation to each group. The perceived lightness of the target is a weighted average of all such values, with the weighting determined by the strength of each group and the degree to which the target belongs to that group. The rules by which lightness values are computed within each group were determined by our studies of anchoring within a simple framework, defined as a single framework that fills the entire visual field (Li and Gilchrist, 1999). In practice this was achieved by placing the observer's head inside an opaque, diffusely illuminated hemisphere, roughly one meter in diameter. Typically a two-part achromatic pattern was painted onto the interior of the hemisphere so as to fill the entire visual field.

We believe that lightness anchoring under these simple conditions can be exhaustively described by three rules. First, the highest luminance within the framework is seen as white, while the lightness of darker shades is determined relative to this. Second, there is a function for area according to which, the larger a surface, the lighter it appears. Third, there is a scale normalization rule. When the luminance range within a framework is less that the canonical range between white and black, the range of perceived grays tends toward the canonical range. More details can be found in Gilchrist et al. (1999).

The application of this model to simultaneous contrast is relatively simple. The contrast display, ignoring any larger context within which it appears, can be said to be composed of three frameworks: two local frameworks, each consisting of a target and its surround, and one global framework, consisting of the entire display. In the global framework, both targets are computed to be middle gray (their true values), relative to the white background, or global maximum. But the local values are different. The target on the black background is assigned the value of white in its local framework because it is the highest luminance within that framework. Thus, its perceived value lies somewhere between middle gray and white, but closer to middle gray because the local framework is very weak.[5]

The target on the white background is computed relative to that background. But due to the scale normalization effect there is a small expansion of the range, and this causes a small darkening of the target. But the darkening of this target is a much smaller effect than the lightening of the other target caused by its local anchoring as the highest luminance. Note that, according to this model, the illusion is caused by local anchoring, not global, and primarily the local anchoring of the target on the black background. We have conducted a series of experiments testing predictions made by the anchoring model against those of a contrast model. Specifically, the anchoring model predicts that:

1. The main error occurs for the target on the black background.

2. No illusion occurs when both targets are increments.

3. The illusion becomes stronger as the target reflectance is lowered.

4. When very light gray targets are used, the main error shifts to the target on

---

[5]It is poorly articulated and poorly segregated.

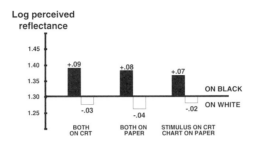

FIGURE 2.2. Results of locus-of-error study.

the white background.

2.2 Methods

Except when otherwise noted, these methods were used. Contrast displays were presented on a CRT screen. Observers sat approximately one meter from the screen and made matches to the targets using a 16-step Munsell chart displayed on the lower part of the screen. Ten naïve observers matched each display.

2.2.1 Locus of Error (Gilchrist et al., 1999)

While the anchoring model makes a clear prediction about the locus of error, it is difficult to derive a prediction from contrast theories. We measured the locus of the error by asking observers to match each target using a Munsell chart containing gray chips mounted on a checkerboard background. Thus each chip bordered equal amounts of white and black. We ran the experiment three times, using three groups of ten observers, once with the contrast display on a CRT screen and the Munsell chart on a paper display, once with both on a CRT screen, and once with both on paper. The results are shown in Figure 2.2. In all three conditions we obtained a much larger error for the target on the black background.

Results consistent with ours have been reported by Logvinenko, Kane, and Ross (2002), who wrote: "the difference in lightness induction between Figs. 2.1 and 2.2 arises from the dark surround." Support also comes from the inspection of several figures in which different workers (Adelson, 2002; Agostini and Bruno, 1996) have created especially strong versions of simultaneous contrast. It is visually obvious in these cases that the enhanced illusion comes primarily from the strong lightening of the target on the black background, not the darkening of the target on the white background.

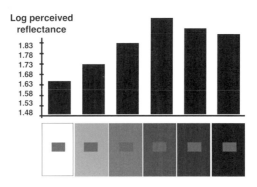

FIGURE 2.3. Results of staircase contrast study.

2.2.2 Staircase Contrast

According to anchoring theory, no contrast illusion should occur for increment targets, that is, when both targets are lighter than their backgrounds. The explanation is quite simple. When both targets are increments, both are assigned the value of white because each is the highest luminance within its local framework. Equal local assignments means no illusion, given that the illusion is held to come from local anchoring. Contrast theories, however, predict no such qualitative change at the increment/decrement boundary. Both Hering (1874) and Cornsweet (1970) present staircase contrast displays like that shown in Figure 2.4b, suggesting that all of the targets appear different. We obtained Munsell matches from ten observers who viewed the display shown in Figure 2.3a. The results, shown in Figure 2.3b, show a clear knee in the curve, at the increment/decrement boundary.

The lack of a contrast effect for double-increment contrast displays has also been reported by Heinemann (1955), Arend and Goldstein (1987), Kozaki (1963, 1965), Gilchrist (1988), and Jacobsen and Gilchrist (1988). A relatively weak contrast effect has been reported by Bressan and Actis-Grosso (2001).

2.2.3 Variation of Target Reflectance

The anchoring model predicts a stronger contrast effect with darker targets because the darker the targets, the larger the difference between the local and global values computed for the target-on-black. That target is always computed to be white in the local framework, but its global value becomes darker as darker targets are used. Thus, with darker targets, the local/global compromise deviates further (upward) from its veridical value. Again, it is not at all clear what predictions would be made by contrast theories. We obtained Munsell matches using targets of three Munsell values (3.0, 6.0, and 7.0), each viewed by ten observers.

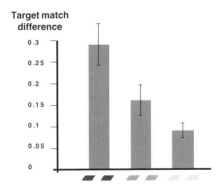

FIGURE 2.4. Size of illusion measured as difference in log reflectance of target matches for dark gray, middle gray, and light gray targets.

The results, shown in Figure 2.4, are consistent with the anchoring predictions.

2.2.4   Shift of Locus

For very light targets, the anchoring model seems to predict that the main error will shift to the target on the white background. We have already seen that the error for the target-on-black approaches zero as that target approaches white. But the smaller error we found for the target-on-white should not decrease in this way. As the targets become lighter, the luminance range of the framework defined by the white background is reduced, producing more expansion according to the scale normalization rule. Depending on the exact coefficient of expansion, the net error for the target-on-white might increase slightly, or at least hold steady, for light gray targets. Thus, the anchoring model seems to predict that a very light gray target on the white background will show more error than the target on the background. We tested target pairs at three more Munsell values (3.5, 5.0, and 8.5), and the results, shown in Figure 2.5, display the predicted shift of locus.

2.3   High-Level Models

Cognitive models such that of Helmholtz, have fared little better than low-level models. For example, Gilchrist (1980) was able to produce strong effects of depth on lightness only by providing the target with a luminance ratio in each plane. He reported that traditional cues to the illumination, such as visibility of the light source, and cast shadows, were not effective in changing perceived lightness. The Helmholtzian idea of taking the illumination into account is too vague to be of much use in modern times. As for the more operationalized intrinsic image mod-

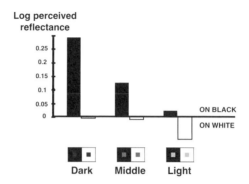

FIGURE 2.5. Size of error for each target in shift-of-focus study, measured as deviation of matched value from actual value. As targets get lighter, the main error shifts to the white background.

els, both Adelson (1993) and Gilchrist (Gilchrist et al., 1999) have retreated from their earlier claims that the retinal image is parsed into reflectance and illuminance layers. Both of these workers have concluded that the empirical evidence shows that the visual system does not construct the full-blown representation of the physical world implied by the intrinsic image models. In one piece of evidence leading to this shift, Todorovic (1997) showed that when the corrugated Mondrian is modified into a staircase configuration, the illusion remains. This result is inconsistent with an explanation in terms of perceived illumination.

2.4   Conclusions

Our inability to explain in concrete terms how lightness is computed by the visual system has left us with an uneconomical and unfalsifiable dual explanation, attributing some aspects of lightness to bodily processes and others to mental processes. The failure of low-level accounts based on contrast mechanisms to cope with either the results of lightness constancy work in 3D displays or the new wave of configuration-based lightness illusions has led to a renewed interest in a dual high-low account. But this retreat is not necessary. The feasibility of a coherent monistic account was shown by Gestalt theory early in this century, and recent work on lightness anchoring has extended this approach and made it more concrete. Indeed, in a recent series of experiments testing variations on simultaneous contrast, the anchoring approach was shown to make specific predictions while the low-level account did not.

References

Adelson, E. (2000). Lightness perception and lightness illusions. In M. Gazzaniga (ed.), *The Cognitive Neurosciences*, 2nd ed., pp. 339–351. MIT Press: Cambridge, MA.

Agostini, T. and Bruno, N. (1996). Lightness contrast in CRT and paper-and-illuminant displays. *Percept. Psychophys.*, 58: 250–258.

Agostini, T. and Galmonte, A. (2002). Perceptual grouping overcomes the effect of local surround in determining simultaneous lightness contrast. *Psycholog. Sci.*, 13: 88–92.

Arend, L. E. and Goldstein, R. (1987). Simultaneous constancy, lightness, and brightness. *J. Opt. Soc. Am. A*, 4: 2281–2285.

Boring, E. (1942). *Sensation and Perception in the History of Experimental Psychology*. New York: Appleton Century Crofts.

Bressan, P. (2001). Explaining lightness illusions. *Percept.*, 30: 1031–1046.

Bressan, P. and Actis-Grosso, R. (2001). Simultaneous lightness contrast with double increments. *Percept.*, 30: 889–897.

Cornsweet, T. N. (1970). *Visual Perception*. New York: Academic Press.

De Valois, R. L., and De Valois, K. K. (1988). *Spatial Vision*, Oxford: Oxford University Press.

Economou, E., Annan, V. and Gilchrist, A. L. (1999). Contrast depends on anchoring within perceptual groups. *Invest. Ophthal. Vis. Sci.* 39: S857.

Freeman, R. B. (1967). Contrast interpretation of brightness constancy. *Psychol. Bul.* 67: 165–187.

Gelb, A. (1929). Die "Farbenkonstanz" der Sehdinge. *Handbuch der normalen und pathologischen Psychologie*. W. A. von Bethe. 12: 594–678.

Gilchrist, A. L. (1979). The perception of surface blacks and whites. *Sci. Am.* 240: 112–123.

Gilchrist, A. L. (1980). When does perceived lightness depend on perceived spatial arrangement? *Percept. Psychophys.*, 28: 527–538.

Gilchrist, A. L. (1988). Lightness contrast and failures of constancy: a common explanation. *Percept. Psychophys.* 43: 415–424.

Gilchrist, A. L., Delman, S. and Jacobsen, A. (1983). The classification and integration of edges as critical to the perception of reflectance and illumination. *Percept. Psychophys.* 33: 425–436.

Gilchrist, A. L., Kossyfidis, C., Bonato, F., Agostini, T., Cataliotti, J., Li, X., Spehar, B., Annan, V., Economou, E. (1999). An anchoring theory of lightness perception. *Psychol. Rev.* 106: 795–834.

Heinemann, E. G. (1955). Simultaneous brightness induction as a function of inducing- and test-field luminances. *J. Exp. Psych.*, 50: 89–96.

Heinemann, E. G. (1972). Simultaneous brightness induction. In D. Jameson and L. Hurvich (eds.), *Handbook of Sensory Physiology VII-4 Visual Psychophysics*, Berlin: Springer-Verlag.

von Helmholtz, H. (1866). *Helmholtz's Treatise on Physiological Optics*. New York: Optical Society of America.

Hering, E. (1874, 1964). *Outlines of a Theory of the Light Sense*. Cambridge, MA: Harvard University Press.

Hurvich, L. and Jameson, D. (1966). *The Perception of Brightness and Darkness*. Boston, MA: Allyn and Bacon.

Jacobsen, A. and Gilchrist, A. (1988). Hess and Pretori revisited: Resolution of some old contradictions. *Percept. Psychophys.*, 43: 7–14.

Jameson, D. and Hurvich, L. M. (1964). Theory of brightness and color contrast in human vision. *Vis. Res.*, 4: 135–154.

Kardos, L. (1934). Ding und Schatten. *Zeitschrift für Psychologie Erg.* Bd 23.

Koffka, K. (1935). *Principles of Gestalt Psychology*. New York: Harcourt, Brace, and World.

Kozaki, A. (1963). A further study in the relationship between brightness constancy and contrast. *Japanese Psych. Res.*, 5: 129–136.

Kozaki, A. (1965). The effect of coexistent stimuli other than the test stimulus on brightness constancy. *Japanese Psych. Res.* 7: 138–147.

Li, X. and Gilchrist, A. L. (1999). Relative area and relative luminance combine to anchor surface lightness values. *Percept. Psychophys.*, 61: 771–785.

Logvinenko, A. D., Kane, J. and Ross, D. A. (2002). Is lightness induction a pictoral illusion? *Perception*, 31: 73–82.

Ross, W. D. and Pessoa, L. (2000). Lightness from contrast: A selective integration model. *Percept. Psychophys.*, 62: 1160–1181.

Todorovic, D. (1997). Lightness and junctions. *Percept.*, 26: 379–394.

White, M. (1981). The effect of the nature of the surround on the perceived lightness of grey bars within square-wave test gratings. *Percept.*, 10: 215–230.

Whittle, P. (1994). Contrast brightness and ordinary seeing. In A. L. Gilchrist (ed.), *Lightness, Brightness, and Transparency*, pp. 111–158. Hillsdale, NY: Erlbaum.

Whittle, P. and Challands, P. D. C. (1969). The effect of background luminance on the brightness of flashes. *Vis. Res.*, 9: 1095–1110.

3

Levels of Brightness Perception

Frederick A. A. Kingdom

3.1   Introduction

Most vision scientists are comfortable with the idea that perception is a multi-level process. For example, we take it for granted that the two best-understood properties of human colour vision, trichromacy and colour-opponency, are under-pinned by physiological mechanisms operating at different stages in the visual pathway. The observation that any colour can be matched by a suitable mixture of three primaries — the definition of trichromacy — is understood to be a consequence of having three cones rather than two or four. On the other hand, our inability to perceive reddish greens or bluish yellows, one of the hallmarks of colour-opponency, is understood to be a consequence of the particular way the three cones are combined postreceptorally. In other words, trichromacy and colour-opponency have independently measurable behavioural consequences that reflect their different physiological origins.

In this chapter, I will argue that a multilevel approach is also the right approach for brightness and lightness perception. No single process mechanism can, in my view, account for the many fascinating brightness/lightness phenomena that presently fill the pages of journals and textbooks alike. This admittedly unglamorous viewpoint is not, as one might expect, shared by all. Notably, Gilchrist and Economou (this volume) argue that all brightness/lightness phenomena can be explained within a single theoretical framework. Their approach, inspired by Gestalt psychology, rejects the very idea of "levels" in perception. Their viewpoint has come to the fore at the same time as a renaissance of interest in contextual effects on surface colour appearance (e.g., see the recent special editions of *Perception*, 1997, Vol. 26, Nos. 4 and 7). Today's emphasis is on configurational relationships, and these are believed to be the major, if not the sole determinant of the perceived pattern of brightness/lightness variations in the image.

I will argue that such contextual effects are best considered within a multi-level framework that includes both low-level contrast and mid-level configurational mechanisms. While the famous Gestalt maxim "the whole is greater than the sum of the parts" undoubtedly holds for brightness/lightness perception, it will be argued that the parts, when considered as levels or stages, nevertheless do exist and can be behaviourally identified.

It should be emphasised that his chapter is not a review of the rapidly expanding literature on lightness and brightness perception. Nor does it present a fully fledged model of lightness/brightness perception. It is essentially a polemic, and only a handful of studies are described that are necessary to make the point. There are many excellent published studies that are highly relevant to the issues dealt with here that are not discussed, and I apologise in advance to anyone who feels their work should have been mentioned but was not.

3.2   Simultaneous Brightness Contrast

Simultaneous brightness contrast, or SBC, is the observation that a grey patch looks brighter on a dark compared to a bright surround. The phenomenon illustrated in Figure 3.1a has intrigued philosophers and scientists for over two millenia (see Wade, 1996, for a historical overview), and it is a sobering thought that even after all this time there still seems no consensus as to why such an apparently simple effect occurs.

When discussing SBC, I will be somewhat cavalier in my use of the terms 'lightness,' or perceived reflectance, and 'brightness,' or perceived luminance. A good discussion of the definitions, uses, and misuses of these terms can be found in the accompanying chapter by Blakeslee and McCourt. Here, I will assume that for figures without an explicit illumination component, such as Figure 3.1a, it is immaterial whether one makes a relative brightness or a relative lightness judgement. Indeed it would be equally valid to refer to Figure 3.1a as an example of simultaneous lightness, rather than brightness contrast. The situation is very different, however, with Figure 3.1b, where there is a pictorial impression of a change in illumination, or specifically a highlight. The distinction between brightness and lightness, as we shall see, becomes critical in any discussion of such stimuli.

Many are familiar with the controversy over SBC that began with Hering and Helmholtz in the nineteenth century. They disagreed as to whether SBC was based on peripheral sensory processes sensitive to contrast (Hering's view), or central influences involving assumptions about the configuration of the display as a whole (Helmholtz's view) (and see Kingdom, 1997). Up to thirty years ago, the dominant view was that contrast underlied SBC, a view sustained by Hurvich and Jameson (e.g., Hurvich and Jameson, 1966), whose own ideas were inspired by Hering (1874/1964). The undergraduate textbook explanation for SBC in Figure 3.2a, which is that SBC results from the operation of filters with centre-surround receptive fields (such as retinal ganglion cells), is the modern version of Hering's explanation of SBC in terms of "lateral inhibition." The idea that low-level filters sensitive to contrast underlies SBC has been the principle theme behind a new generation of brightness models whose other defining characteric is filtering at multiple spatial scales (e.g., Kingdom and Moulden, 1992; Blakeslee and

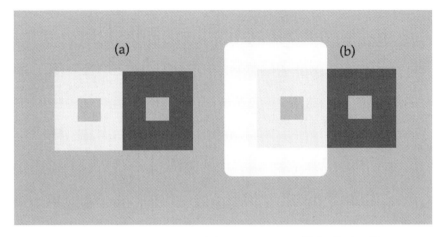

FIGURE 3.1. (a) Standard simultaneous brightness contrast (SBC) display. The two grey patches are equal in luminance, yet appear different in brightness. (b) Simulation of arrangement employed by Gilchrist (1979). One half of the display appears to be lit by a highlight rather than being of different reflectance. The luminances of the grey patches and their immediate surrounds are the same in both displays. Gilchrist used real rather than simulated illumination, and its effect on the magnitude of SBC was reportedly much greater than seen here.

McCourt, 1999; 2001a,b).

During the last decade, the pendulum has swung away from the idea that errors such as SBC are due to contrast, and more toward the idea that they result from mechanisms sensitive to the overall configuration of the display. I have argued previously (Kingdom, 1999) that the studies of Gilchrist and his colleagues in the seventies and eighties (e.g., Gilchrist, 1977, 1979; Gilchrist, Delman and Jacobsen, 1983; see also Gilchrist, 1988) were instrumental in precipitating this change in opinion. One of the stimuli that was central to establishing the new way of thinking is illustrated in Figure 3.1b. It must be made clear at the outset that Figure 3.1b is only an illustration; Gilchrist's original experiments used Munsell papers and real illuminations, and the effects were reportedly much greater than can be seen here. In Figure 3.1b the standard SBC display has been replaced by one consisting of a uniform background with one-half illuminated by a bright light. The luminances of the test patches and their surrounds are, however, identical to the standard display. With Gilchrist's stimulus, subjects reported an enhancement of the lightness difference between the two patches in the part-highlighted display. Since the contrasts of the patches with their surrounds are the same under both configurations (we will return to a critical examination of this assumption later on), the enhancement of SBC cannot, it seems be due to the effects of contrast. It must instead be due to the way subjects interpreted the display as a whole.

Gilchrist et al.'s experiments laid the foundation for many recent demonstrations on a similar theme (e.g., Knill and Kersten, 1991; Adelson, 1993; Anderson,

FIGURE 3.2. (a) Explanation for SBC in terms of centre-surround receptive-field filters, the modern version of Hering's explanation in terms of "lateral inhibition." (b) Explanation of SBC based on Helmholtz's veiling hypothesis. See text for details.

1997; Logvinenko, 1999). With the aid of modern computer graphics, pictorial representations of complex three-dimensional patterns with vivid impressions of shadows, transparency and shading have replaced the conventional SBC display, and with impressive results. I have provided an example of my own in Figure 3.12, a figure inspired by Adelson's (unpublished) checkerboard-shadow illusion. The allure of these new demonstrations is the sheer magnitude of their illusory brightness differences, which far surpass that found with standard SBC displays. Yet there is a negative side. For some protagonists it has meant downplaying the value of not just simple forms of SBC such as the standard display, but more importantly their explanation in terms of contrast. After all, if such stunning illusions are apparently inexplicable in terms of contrast, is contrast really that important? I will argue in the next section that it is. Moreover, demonstrations suggesting that contrast may be insufficient to account for certain brightness/lightness phenomena are best considered in terms of the multilevel framework advocated here, where contrast forms an essential component. Let us therefore now look to the evidence that contrast plays a central role in brightness/lightness perception.

3.3   Contrast Brightness and Low-Level Filtering

3.3.1   A Common Transducer Function for Brightness Discrimination and Brightness Scaling

My first piece of evidence comes from the work of Paul Whittle. Whittle's quantitative measurements of brightness obtained under a variety of task conditions have provided some of the best evidence for the role of contrast in brightness perception. A comprehensive exposition of Whittle's findings and their theoretical implications is provided in two book chapters, Whittle (1994a,b), and here the reader will obtain the full story of "contrast brightness," the term Whittle used to capture the idea of the intimate relationship between contrast and brightness. I consider here a subset of Whittle's findings that for me provides the most succinct evidence for contrast-brightness. Figure 3.3 shows data taken from Whittle (1992) (see also Whittle, 1994a), along with my own illustration of the two types of measurement involved, namely brightness discrimination and brightness scaling. In the brightness discrimination task, subjects were required to detect a difference in the luminance of two patches, where one of the patches served as a baseline, or "pedestal." With this task, the term "brightness discrimination" is synonymous with both "luminance discrimination" and "contrast discrimination," as it is a *threshold* task involving a comparison between two patches against the *same* background.

Results for one background are shown as the crosses in Figure 3.3 (data originally from Whittle, 1986). For increments, the thresholds rise with pedestal luminance, whereas for decrements the function is inverse U-shaped. The different shapes of the increment and decrement functions is of interest in itself (e.g., see Whittle, 1986, 1994a; Kingdom and Whittle, 1996), but for the present purpose one need only assume that both functions reflect the shape of the underlying transducer function for contrast. The second set of measurements in Figure 3.3, the closed squares, are from the brightness scaling experiment (original data from Whittle, 1992). In this task subjects were required to set the luminances of a series of patches so that they appeared to be equally different in brightness. As with the threshold task, one of the patches served as a pedestal. In Figure 3.3 the difference in luminance between adjacent pairs of patches is plotted as a function of pedestal luminance. When the brightness discrimination thresholds were scaled upward by a suitable factor so that they could be compared directly to the brightness scaling data, the two functions superimposed almost perfectly. This strongly suggests that the underlying transducer function for the threshold brightness discrimination task is the same as that for the suprathreshold brightness scaling task. Given that the detection of threshold differences in brightness/luminance/contrast is universally believed to be mediated by bandpass filters in the visual cortex, Whittle's experiment provides powerful evidence that the same filters are also involved in signalling suprathreshold brightnesses.

3.3.2   *Illusory Gratings Facilitate the Detection of Real Gratings*

Whittle's experiment demonstrated that a critical behavioural signature for contrast transduction could be revealed in data from a prototypical brightness task. A similar rationale lay behind an experiment I recently conducted in collaboration with Mark McCourt (McCourt and Kingdom, 1996), my second piece of evidence for the role of contrast in brightness perception. We used a form of SBC known as grating induction that was first demonstrated by McCourt (1982;

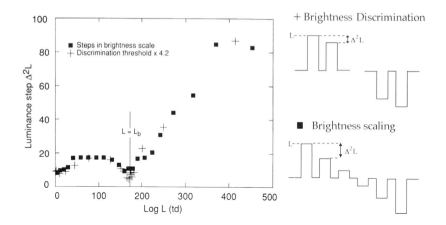

FIGURE 3.3. Left: data from two types of task, contrast discrimination (crosses) and brightness scaling (filled squares), from Whittle (1992). Right: my illustration of the two types of task. Left: reprinted from *Vis. Res.*, 32: 1493–1507, 1992, Whittle, P., Brightness, discriminabililty and the "crispening effect," with permission from Elsevier Science.

see also Blakeslee and McCourt, 1997, for the evidence that grating induction is SBC). Figure 3.4 shows an example grating induction stimulus. An opposite-phase illusory sinewave grating is observed in the uniform mid-grey stripe that runs horizontally through an "inducer" sinewave grating. Grating induction is a useful tool for studying SBC as it lends itself easily to parametric manipulation, and under some circumstances can be quite compelling (e.g., see Figure 3.7). We reasoned that if the induced brightness variations in grating induction were signalled by the same mechanisms that detect real gratings — and here we come to the idea of a critical behavioural signature — an illusory grating should facilitate the detection of a real grating. Facilitation, as used here, means a reduction in the threshold for detecting a stimulus as a result of the presence of another stimulus. The best-known form this takes is the "dipper" observed in the function relating contrast increment thresholds to pedestal contrast (Campbell and Kulikowski, 1966; Foley and Legge, 1981). When the test is added to a different type of stimulus, the pedestal is usually referred to as a mask.

Our experiment is illustrated in Figure 3.5. We first measured increment thresholds for real gratings whose spatial characteristics were the same as the illusory gratings that formed the main part of the study (Figure 3.5a). We then repeated the experiment this time using *illusory* rather than real pedestals (Figure 3.5b). By varying the contrast of the inducer we were able to vary the apparent contrast of the illusory pedestal. The real test grating was added in phase with the illusory pedestal (which at very low inducer contrasts was not visible) in one of the two forced-choice intervals, and subjects had to decide which interval contained the

FIGURE 3.4. Grating induction stimulus, first described by McCourt (1982).

test. Finally, we used a matching technique to find the contrast of a real grating that matched that of the illusory grating at each inducer contrast (Figure 3.5c). This allowed us to recast the contrast of the inducer in terms of "equivalent" real grating contrast.

Figure 3.6 shows results from the grating spatial frequency that produced one cycle of modulation across the display (0.0625 cpd). The data for the real and illusory grating pedestals superimpose almost perfectly when the contrast of the inducer is couched in terms of equivalent contrast. This shows that, at least for one set of spatial characteristics, an illusory grating acts as an almost perfect metamer of a real grating of the same apparent contrast, in terms of its ability to facilitate (and mask) the detection of a superimposed real grating. I see no alternative explanation for these results other than that illusory gratings are signalled by the same mechanisms that detect real gratings (and see Kingdom, McCourt and Blakeslee, 1997, for further evidence in support of this conclusion). Given the abundance of evidence that real gratings are detected by narrowband filters in the visual cortex, one is once again drawn irrevocably to the conclusion that the same filters are involved in signalling brightness variations, in this case illusory ones.

Besides this quantitative evidence, there are some simple demonstrations of grating induction that provide additional evidence for a central role for low-level contrast-sensitive filters. Two of my favourites are shown in Figures 3.7 and 3.8. Figure 3.7 shows two patterns, each appearing to consist of a low-contrast, single-cycle sinewave grating in a narrow horizontal stripe on a uniform surround (based on a similar figure in Moulden and Kingdom, 1991). However, only one of the two patterns accords physically with this description — the one at the top. In the bottom pattern it is the stripe that is uniform, and the surround that contains the sinewave; hence the sinewave in the stripe is illusory. It is hard to tell the two patterns apart. With scrutiny, the digital quantization of the low-amplitude luminance gradients gives it away, but the metamerism of the two patterns is nevertheless striking. Also striking is that in the bottom figure the illusory grating is more visible than the surround grating that induces it.

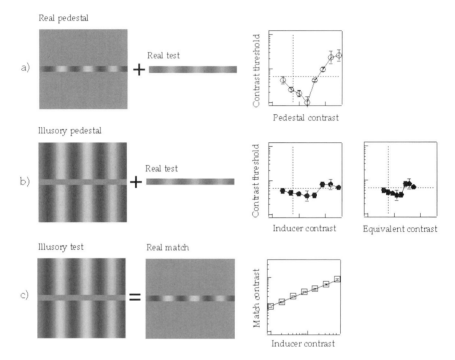

FIGURE 3.5. Method employed by McCourt and Kingdom. In (a) increment thresholds are measured as a function of pedestal contrast for a real grating gated into a narrow stripe. (b) Detection thresholds are measured for a real grating added in phase to an illusory grating, for various contrasts of inducer. (c) The apparent contrast of the illusory gratings was measured by matching them to real gratings. This allowed the contrast of the inducer to be recast in terms of "equivalent" real grating contrast. Based on Figure 1 of *Vis. Res.*, 36: 2563–2573, McCourt, M.E. and Kingdom, F. A. A., Facilitation of luminance grating detection by induced gratings, with permission from Elsevier Science.

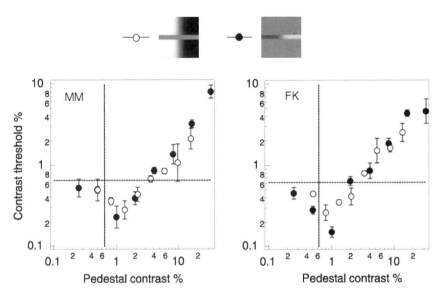

FIGURE 3.6. Results from McCourt and Kingdom using 0.0625 cpd gratings. Filled circles are increment thresholds for a real grating plotted against real grating pedestal contrast. Open circles are thresholds for detecting a real grating on an illusory grating pedestal. The contrast of the illusory pedestal is given as the equivalant contrast of a matched real grating. Note how the functions of the real and illusory gratings superimpose neatly. Data taken from Figure 3 of *Vis. Res.*, 36: 2563–2573, 1996, McCourt, M. E. and Kingdom, F. A. A., Facilitation of luminance grating detection by induced gratings, with permission from Elsevier Science.

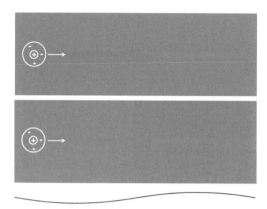

FIGURE 3.7. Top: a single cycle of a real grating runs along a narrow stripe in the middle of a uniform background. Bottom: a uniform stripe lies in the middle of a single-cycle inducer grating. The illusory grating in the bottom figure is more visible than the real grating that induces it. The appearance of both stimuli is most parsimoniously explained by the convolution response (shown below) of a bandpass filter whose receptive field centre is similar in diameter to the height of the stripe.

It is easy to explain the appearance of both patterns in Figure 3.7 with filtering. At the bottom of the figure is shown the horizontal convolution response of a centre-surround filter, obtained when centred on either stripe. Because both the real (top) and illusory (bottom) gratings are gated into narrow stripes, the filter giving the biggest response is one whose centre diameter is approximately the same as the height of the stripe. The surround grating in the bottom pattern, however, will only weakly stimulate the same filter because its dominant spatial frequency lies almost outside the filter's passband.

In Figure 3.8a, stripes containing ramps in luminance alternate with uniform stripes (see also Moulden and Kingdom, 1991). It is hard to distinguish the stripes containing ramps from those that are uniform. In this instance the induced brightness variations are almost as salient as the inducing brightness variations. Once again, the filtering explanation suffices. A filter matched to the height of the stripe produces a response of opposite phase to the ramp and uniform stripes, but of more-or-less identical amplitude, in accord with the percept. Finally, Figure 3.8b shows that at higher contrasts the illusion begins to break down, in that one can easily distinguish the ramp from the uniform stripes. I will discuss the significance of this last demonstration in the following section.

3.3.3   Increment and Decrement Perception is Categorical

My final piece of evidence for a low-level contrast mechanism involves an examination of the differences between "increments" and "decrements." I refer here to the sign, or polarity of contrast of relatively small, usually closed regions in the image. When I began my research into brightness perception in the 1980s I

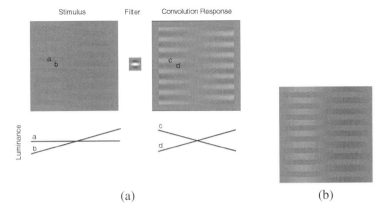

FIGURE 3.8. (a) Ramp-induced brightness. The stimulus on the top left consists of horizontal uniform stripes (e.g., a) alternating with luminance stripes (e.g., b). It is difficult
to tell which stripes are uniform and which are ramps. On the top right is the convolution
response of a centre-surround filter whose centre diameter is approximately the same as
the height of the stripes. The response accords with the percept of the stimulus, as shown
in the luminance profiles of c and d. (b) The ramps and uniform stripes become more dis-
criminable at higher contrasts. Figure reprinted from Kingdom, F. A. A., Guest editorial:
Comments on Lonvinenko "Lightness induction revisited," *Percept.*, 28: 929–934, 1999,
with permission from Pion Ltd.

was often struck, and frequently irritated, by just how difficult it was to find a luminance setting of an increment that matched the brightness of a decrement, and viceversa. Somehow they never quite looked the same. More often than not, increments looked brighter than decrements whatever their luminance (see Whittle, 1994a), and sometimes they even seemed to take on a slightly different hue. These observations complement a substantial psychophysical and neurophysiological literature suggesting that increments and decrements are processed by different mechanisms, specifically the "on" and "off" pathways of the mammalian visual system that begin at the retina (Schiller, 1982; and for a review see Fiorentini et al., 1990).

A simple demonstration of the categorical nature of increment and decrement perception is shown in Figure 3.9, which is based on an early finding by Whittle (1965). Fusion of the two stereo-halves reveals two fusable, and one rivalrous stereo-pair. The difference in luminance between the top two increments, and between the bottom two decrements, is greater than between the increment–decrement pair in the middle, yet only the top and bottom pair fuse to produce patches more-or-less midway in brightness between their monocular half-images. The categorical nature of increment and decrement perception, with its ready physiological substrate in the form of "on" and "off" pathways, shows that our brightness perception is in part a result of low-level physiological processes.

The unique perceptual properties of increments and decrements also pose a special challenge for modellers. The output of brightness models (e.g., Kingdom and Moulden, 1992; Blakeslee and McCourt, 1999, 2001a, b) is a map of (relative) brightness values. For example, with SBC a "successful" prediction is a lower value for the patch on the white background compared to the patch on the black background. Although this accords with our perception that one patch looks darker than the other, it does not capture the categorical nature of the difference.

3.4   Multiscale Filtering and Edge-Based Filling-In

So far I have not discussed details of the filters involved, showing instead how in principle filtering is a valid and simple explanatory tool. In Figures 3.7 and 3.8, a single, linear, circularly symmetric filter captured the qualitative appearance of the stimuli. This is, of course, a gross over-simplification. We know that contrast (and hence brightness) coding is a multiscale process, involving cortical filters tuned to a range of scales and orientations. A full multiscale (such as wavelet) transform of an image produces a veridical output, and if this is what the visual system performed, illusions such as SBC would not occur. One of the main reasons why filtering results in brightness illusions is our relative insensitivity to low spatial frequencies, which is particularly marked at low-contrasts. In Figure 3.8c, unlike its low contrast version in Figure 3.8a, the uniform and ramp stripes have different perceived amplitudes. This is almost certainly due to the increased involvement of filters tuned to relatively low spatial frequencies.

FIGURE 3.9. Increment and decrement perception is categorical. When free-fused, the two decrements (top) and two increments (bottom) easily fuse, but the increment–decrement pair in the middle is rivalrous. The difference in luminance between the increment pair, and also between the decrement pair, is actually bigger than between the increment–decrement pair.

I suggest that all sizes of filters that are active contribute to our percept of brightness/lightness. Precisely how the outputs of filters at different scales (and orientations) are combined for brightness and lightness perception, and in particular what types of nonlinearities are involved, is not fully understood. Yet the most successful attempts at modelling brightness phenomena in terms of filtering have employed filters at multiple spatial scales (Kingdom and Moulden, 1992; and especially Blakeslee and McCourt, 1999, 2001a, b).

Given the abundance of evidence that the early stages of vision involve multiscale filtering, it is somewhat surprising that its importance for brightness/lightness perception has yet to be fully appreciated. I believe one reason for this is a wrong idea that has become entrenched over the years, namely that it is the contrast (or ratio) *in the immediate vicinity of the edge* that is critical to brightness/lightness perception. This idea follows from one of the most enduring themes in the recent history of this topic, namely that the visual system first locates edges, and then "fills-in" the gaps between them by some kind of spreading of neural activity (Ratcliff, 1972; Gilchrist, 1979; Grossberg and Todorovic, 1988; see review by Kingdom and Moulden, 1988; see discussion by Blakeslee and McCourt, this volume). In this view, the luminance relationships between those parts of a stimulus that lie at a distance from the edge exercise little influence on brightness. An almost anecdotal but nevertheless striking demonstration of the importance of distal luminance relationships is illustrated in Figure 3.10. If brightness perception is critically dependent on the luminance relationships at the edge, then it must follow that blurring the edge should *at the very least* reduce the magnitude of any perceived brightness variations. Yet the opposite is found. As can be seen in Figure 3.10b, blurring the edges if anything increases the magnitude of SBC, and this has been confirmed quantitatively by McCourt and Blakeslee (1993) using the grating induction stimulus. On the other hand selective removal of the low spatial frequencies, which define the more distal luminance relationships, substantially reduces SBC, as shown in Figure 3.10c.

This is not to say that the luminance relationships at the edge play no role in brightness perception. The Craik–Cornsweet–O'Brien illusion (e.g., Cornsweet, 1970; Todorovic, 1987; see a weak version of the illusion in Figure 3.10c), in which an illusory brightness difference is observed on either side of a highpass-filtered (or equivalent) edge, suggests that an edge-based filling-in mechanism may contribute to brightness, perhaps even playing a crucial role in the perceived uniformity of physically uniform regions. The point being made here, and Figure 3.10 seems persuasive evidence, is that the role played by an edge-based filling-in mechanism is probably quite minor.

There are important ramifications to the idea that lightness/brightness perception is a multiscale process. An often-heard refrain against contrast theories of brightness/lightness is that two patches with the same luminance and edge contrast can nevertheless appear very different in brightness/lightness (e.g., see the discussion of Figure 3.1; Gilchrist, 1979; Gilchrist et al., 1999). However, once we accept the idea that contrast-sensitive mechanisms operate at multiple spatial scales, we cannot reject an explanation base on contrast merely because of what happens at the edge. We must also consider the distal luminance relationships. Bearing in mind this caveat, let us now turn to a consideration of those brightness/lightness phenomena that appear to defy explanation in terms of contrast.

3.5   Helmholtz and the Illumination-Interpretative Approach

In previous sections I considered the evidence for a contrast-sensitive mechanism based on multiscale filtering. One purpose of such a mechanism is to achieve lightness constancy with respect to the ambient level of illumination (Whittle, 1994a,b). By tying lightness to contrast, lightness becomes invariant to changes in light level. There is a penalty, however: errors such as SBC.

In this section I examine the component of brightness/lightness perception that is thought to be involved in discounting spatial, as opposed to ambient changes in illumination such as shadows, highlights, shading, and transparency. Although the last of these, transparency, is a material property, its luminance relationships are identical to those of shadows. The distinction between lightness and brightness becomes very important when considering spatially varying illumination. Consider, for example, the natural scene in Figure 3.11. Two judgements concerning the shadowed region at **a** can be made. On the one hand we observe that it is darker than its surround — a relative brightness judgement. On the other hand we infer that it is the same shade of grey as its surround — a relative lightness judgement. While it is conceivable that a clever artist might have painted the grass and

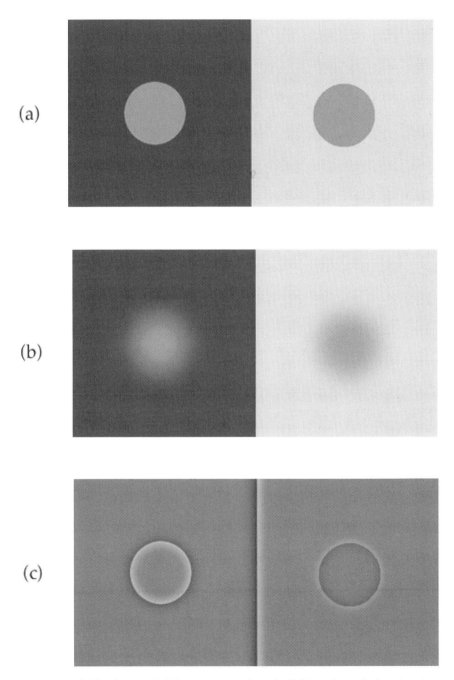

FIGURE 3.10. Simultaneous brightness contrast (top) is slightly enhanced when the stimulus is lowpass filtered (middle), but diminished when highpass filtered (bottom).

FIGURE 3.11. Natural shadow.

road with dark paint to simulate the effect of a shadow, which would make our lightness judgement wrong, this is not our impression.

When discussing the effects of spatial variations in illumination on brightness/lightness, it is instructive to begin with Helmholtz. Helmholtz (1866/1962) mainly considered the chromatic version of SBC, simultaneous colour contrast (SCC), in which a grey patch appears tinted with the complementary colour of its surround. Helmholtz believed that all forms of SCC resulted from "errors of judgement." In some cases SCC occurred because of the mistaken assumption that the grey patch was covered by a transparent veil the colour of the surround, the eye compensating for the veil when estimating the colour of the patch (pp. 282–287). Helmholtz was influenced by an earlier experiment conducted by Heinrich Meyer in 1855. Meyer had shown that the red tinge seen in a grey patch on an intense green background became even redder when both were overlaid with a piece of transparent white paper, which had the effect of desaturating the green background. Helmholtz suggested that the overlay of transparent white paper helped create the illusion that the grey patch was being viewed through a green veil. However, because the eye received from the grey patch a composition of light normally associated with grey, an inference was made that the patch must be pinkish, as the effect of the green veil would be to absorb the long wavelengths associated with the pinkish tint. Thus, according to Helmholtz, we have learned to "correct" for the effects of intervening, transparent media, just as we have learned to "correct" for the prevailing illumination in assessing the intrinsic lightness of objects. A Helmholtzian account of achromatic SBC would be based on an analogous argument, as illustrated in Figure 3.2. We assume that the patch on the bright background is more intensely illuminated than the one on the dark background. However, because the intensity of light reaching the eye is the same for both patches, an inference is made that the patch on the bright background must be of lower reflectance, and that is how it is perceived.

Helmholtz believed that other types of judgement error were also involved in SCC (e.g., see Helmholtz, pp. 274-278; also Turner, 1994, pp. 108-113, for a recent review), but it is interesting that it is his veiling hypothesis that William James seized upon when discussing SCC in his classic work *The Principles of Psychology*. James criticized Helmholtz's veiling hypothesis because SCC occurred under conditions where it was quite implausible to suppose that the test regions were illuminated differently (James, 1890/1981, pp. 662-674). For example, James describes how a pinkish tinge can be seen in grey concentric rings that alternate with green concentric rings, yet one has no impression that any one part of the stimulus is illuminated differently from any other. James's argument is an important one because it suggests not only that there are other explanations for SCC besides the veiling hypothesis, but that one needs to have visible illumination borders before entertaining what I refer to here as an "illumination-interpretative" explanation of SCC. Notwithstanding James's critique, it is the way Helmholtz's veiling hypothesis anticipated the remarkable series of demonstrations alluded to in the Introduction and now considered in more detail that makes his ideas so prescient.

Figure 3.12 is my own figure that was inspired by Adelson's (unpublished) checkerboard-shadow illusion. I multiplied a black-white checkerboard by a low-amplitude, single cycle of near-sinusoidal luminance modulation, such that the luminance of the dark square at **a** in the bright shaded region is identical to the light square at **b** in the dark shaded region. In an important sense this figure is a brightness and not a lightness illusion. The checks **a** and **b** look different in brightness, yet have the same luminance. However, once we attribute the slowly varying luminance component of the figure to shading, we are correct to judge the lightnesses of **a** and **b** as different, even though physically on the page they are the same. The illusion appears to demonstrate our ability to parse the image into its illumination and reflectance components, or its "intrinsic images" (Bergstrom, 1977; Barrow and Tenenbaum, 1978; Adelson and Pentland, 1996).

What is striking about Figure 3.12 is the way our brightness perception appears to be so dominated by our lightness perception. It is as if in discounting the shading we ceased to be aware of its presence altogether, and as a result conclude that **a** and **b** must be different in brightness and not just lightness. Our "intrinsic image" processing seems to work well for lightness, but fails for brightness perception. One can legitimately argue that the goal of the system is lightness constancy, and thus brightness per se is unimportant. Be that as it may, observers often express incredulity when told that **a** and **b** have the same luminance (or told that they are the same shade of grey), which suggests that at the very least they feel they ought to be able to correctly judge their relative brightnesses.

The illusion in Figure 3.12 is strongly suggestive of the involvement of a Helmholtzian, lightness constancy mechanism that discounts spatially varying illumination, i.e., is "illumination-interpretative." But before jumping to this conclusion, we must be careful, lest we miss the fact that **a** and **b** are surrounded by different luminances. Is the brightness illusion really illumination-interpretative, or is it simply a result of contrast? To answer this question we must demonstrate that the pictorial representation of shading enhances the brightness illusion over and above that due to contrast, and for this we need a "control" stimulus with

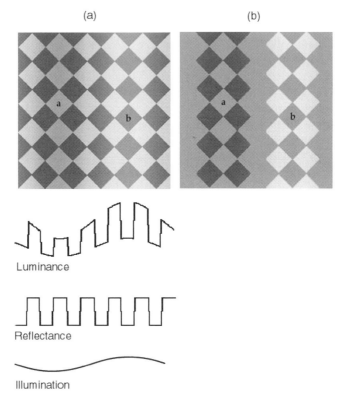

FIGURE 3.12. (a) Checkboard-shading illusion, similar to Adelson's (unpublished) checkerboard-shadow illusion. Areas **a** and **b** are identical in luminance, as shown in the luminance profile. The image can be decomposed into its "intrinsic images": a reflectance and illumination layer, as illustrated below. In (b) the three columns of diamonds centred respectively on **a** and **b** have been placed on a background of the same luminance as **a** and **b**. The brightness difference between **a** and **b** is markedly reduced.

the same pattern of luminance, but without the impression of shading. But herein lies the rub. To remove the impression of shading we must change the arrangement of luminances. Can we be certain when doing this that we have not inadvertently altered contrast, bearing in mind what was said in the previous section about the importance of distal luminance relationships in contrast processing? In Gilchrist's classic experiment illustrated in Figure 3.1b, the highlight increases the area around the test square that is surrounded by a high luminance, and this alone might have caused the patch to appear darker than in the standard display (Figure 3.1a), irrespective of how the surround was interpreted. Consider also my attempt in Figure 3.12b to provide a suitable control. The three columns of diamonds centred on **a** and **b** have been placed on a background of uniform grey the same luminance as **a** and **b**. The impression of shading disappears, and the brightness difference between **a** and **b** is unquestionably reduced. Although it is therefore difficult to see how contrast alone could account for the dramatic reduction in the strength of the illusion, even when taking into account the distal luminance relationships in the figure, one can not be certain. The point being made here is not that illumination-interpretative processes do not influence brightness, on the contrary, but that one must be very careful before rejecting an explanation based on contrast.

Ideally, what one would like are two stimuli whose test regions are surrounded by near-identical patterns of luminance, but whose perceived pattern of illumination is nevertheless very different. Such a stimulus would then *isolate* the putative illumination-interpretative mechanism from the effects of contrast. I think Figure 3.13 goes some way toward achieving this. When free-fused, one sees four figures in stereoscopic depth, each consisting of a simulated transparency in front of a background, with equal-in-luminance test diamonds either on the transparency or on the background. The pattern of luminances surrounding the test diamonds is more-or-less identical, at least in the monocular view of all four figures, and importantly not just at the edges of the test diamonds. Most observers agree that the test diamond on the white background behind the dark transparency looks both brighter and lighter than the others. This is in keeping with the Helmholtzian idea that the lightness attributed to the test diamond is what it would be if the transparency was removed, with the added observation that the brightness of each test diamond is strongly influenced by its lightness.

In a recent experiment, Barbara Blakeslee, Mark McCourt and I measured the brightness of test patches perceived to lie either behind a simulated transparency, as in Figure 3.13, or on a reflectance background with a near-identical pattern of surround luminance (Kingdom, Blakeslee, and McCourt, 1997). We found that the perception of transparency did affect brightness in the expected direction, though in general the effects were quite small (the biggest effect we found was about a factor of two). Thus in spite of the concerns expressed above about the potentially confounding effects of contrast, our experiment confirmed the findings of Gilchrist (1979), Adelson (1993), Logvinenko (1999) and others, and provided additional evidence for an illumination-interpretative component of brightness and lightness perception.

FIGURE 3.13. Effect of transparency on brightness. When free-fused one sees four figures: two consist of a bright transparency in front of a dark background, two of a dark
transparency in front of a bright background. Equal-in-luminance test diamonds lie either
on a transparency or on a background. For most observers the grey patch on the bright
background behind the dark transparency looks brighter than the other test diamonds. Note
that the pattern of luminances surrounding all test diamonds is near-identical in the monocular view.

3.6   Integration and Anchoring

I have now argued that two mechanisms contribute to brightness/lightness perception, a low-level contrast-sensitive, and a mid-level illumination-interpretative
mechanism. Other processes are presumably also involved, and a few remarks
will be made about just two of them.

The first is integration. If contrast-sensitive mechanisms operate locally, some
method of combining their signals across the image may be necessary, one purpose being to compare brightesses/lightnesses across a distance. Since contrast is
a differencing operation, the putative mechanism, if it exists, is arguably analogous
to mathematical integration (see Kingdom and Moulden, 1988; Gilchrist, 1994;
Whittle, 1994b; Arend, 1994 for reviews). Whittle (1994b) has suggested that an
important function of integration is to achieve lightness constancy with respect to
the surround, so that surfaces viewed against different backgrounds do not appear to differ in lightness. Whittle refers to this type of lightness constancy as
Type II, as distinct from Type I, which is constancy with respect to the ambient
level. Whittle includes constancy with respect to spatially varying illumination,
the illumination-interpretative constancy mechanism I described in the previous
section, Type II. The putative integration stage would work in the opposite direction to contrast, serving to mitigate its effects and derive a more veridical representation (Whittle, 1994b). That SBC exists at all is testament to the fact that such
a mechanism is, however, unable to fully override the effects of contrast. Tradi-

tionally, the integration stage has been assumed to operate on edge contrasts, or ratios, perhaps most famously in the Retinex model of Land and McCann (1971). However, Land (1986) and Hurlbert and Poggio (1988) have suggested that lightness constancy with respect to spatial variations in illumination might be achieved directly via the use of filters with small receptive field centres and much larger receptive field surrounds, without need for an explicit integration stage. This raises the tantalising possibility that what appears to be integration might in fact be large-scale filtering.

The second process that deserves to be mentioned is "anchoring," the term coined by Gilchrist et al. (1999) for the mechanism that turns relative lightness judgements into absolute ones. In mathematical terms, anchoring is traditionally associated with the restoration of the d.c. level. Gilchrist et al. have provided evidence that relative lightness values are anchored to the highest luminance in the display, which is ascribed white, an idea suggested by Wallach (1976) and incorporated into models of lightness constancy such as the Retinex (Land and McCann, 1971). Moreover, Gilchrist et al. suggest that anchoring is itself responsible for errors such as SBC, because it operates not only globally but also locally within different perceptual frameworks. This is an interesting idea worth pursuing. In the debate that accompanies this chapter I describe how the anchoring model predicts SBC in the standard display, and offer a critical appraisal of the model's plausibility when applied to other types of SBC, such as grating induction.

3.7   Conclusions

Much can be learnt about how we perceive brightness and lightness from the errors we make when doing so. Brightness and lightness perception involve a number of mechanisms operating at different levels of visual processing. One mechanism is low-level, and processes spatial variations in brightness via multiscale filtering. It serves to achieve lightness constancy with respect to the ambient level of illumination. However, it comes at a cost: errors such as simultaneous brightness contrast. A second, mid-level mechanism aims to achieve lightness constancy with respect to spatially varying illumination such as shading, shadows, highlights and transparency. The cost in this case is an enhancement of errors in brightness judgement.

Acknowledgements

I would like to thank Beverley Mullings for her valuable help with many of the figures, and Alan Gilchrist for helpful comments regarding terminology.

References

Adelson, E. H. (1993). Perceptual organization and the judgement of brightness. *Science*, 262: 2042–2044.

Adelson, E. H. and Pentland, A. P. (1996). Lightness perception and lightness illusions. In M. Gazzaniga (ed.), *The Cognitive Neurosciences*, 2nd ed., Cambridge, MA: MIT Press.

Anderson, B. L. (1997). A theory of illusory lightness and transparency in monocular and binocular images. *Percept.*, 26: 419–453.

Arend, L. E. (1994). Surface colors, illumination, and surface geometry: Intrinsic-image models of human color perception. In A. L. Gilchrist (ed.), *Lightness, Brightness, and Transparency*, pp. 159–213. Hillsdale, IL: Erlbaum.

Barrow, H. G. and Tenenbaum, J. (1978). Recovering intrinsic scene characteristics from images. In A. R. Hanson and E. M. Riseman (eds.), *Computer Vision Systems*, pp. 3–26. Orlando, Fl: Academic Press.

Bergstrom, S. S. (1977). Common and relative components of reflective light as information about the illumination, colour, and three-dimensional form of objects. *Scandinavia J. Psych.*, 18: 180–186.

Blakeslee, B. and McCourt, M. E. (1997). Similar mechanisms underlie simultaneous brightness contrast and grating induction. *Vis. Res.*, 37: 2849–2869.

Blakeslee, B. and McCourt, M. E. (1999). A multiscale spatial filtering account of the White effect, simultaneous brightness contrast and grating induction. *Vis. Res.*, 39: 4361–4377.

Blakeslee, B. and McCourt, M. E. (2001a). A multiscale spatial filtering account of the Wertheimer–Benary effect and the corrugated Mondrian. *Vis. Res.*, 41: 2487–2502.

Blakeslee, B. and McCourt, M. E. (2001b). A multiscale spatial filtering account of brightness perception. In L. R. Harris and M. Jenkin (eds.), *Levels of Perception*, New York: Springer Verlag.

Campbell, F. W. and Kulikowski, J. J. (1966). Orientation selectivity of the human visual system. *J. Physiol. Lond.*, 187: 437–445.

Cornsweet, T. (1970). *Visual Perception*. New York: Academic Press. York, NY.

Fiorentini, A., Baumgartner, G., Magnussen, S., Schiller, P. H. and Thomas, J. P. (1990). The perception of brightness and darkness. In L. Spillman and J. S. Werner (eds.), *Visual Perception: The Neurophysiological Foundations*. San Diego, CA: Academic Press.

Foley, J. M. and Legge, G. E. (1981). Contrast detection and near-threshold discrimination in human vision. *Vis. Res.*, 21: 1041–1053.

Gilchrist, A. L. (1977). Perceived lightness depends on perceived spatial arrangement. *Science*, 195: 185–187.

Gilchrist, A. L. (1979). The perception of surface blacks and whites. *Sci. Am.*, 240: 112–123.

Gilchrist, A. L. (1988). Lightness contrast and failures of lightness constancy: a common explanation. *Percept. Psychophys.*, 43: 415–424.

Gilchrist, A. L. (1994). Absolute versus relative theories of lightness perception. In A. L. Gilchrist (ed.), *Lightness, Brightness, and Transparency*, pp. 1–33. Hillsdale: Erlbaum.

Gilchrist, A. L., Delman, S. and Jacobsen, A. (1983). The classification and integration of edges as critical to the perception of reflectance and illumination. *Percept. Psychophys.*, 33: 425–436.

Gilchrist, A. L. and Economou, E. (2002). Dualistic versus monistic accounts of lightness perception. In L. R. Harris and M. Jenkin (Eds.), *Levels of Perception*, New York: Springer-Verlag.

Gilchrist, A. L., Kossyfidis, C., Bonato, F., Agostini, T., Cataliotti, J., Li, X., Spehar, B., Annan, V. and Economou, E. (1999). An anchoring theory of lightness perception. *Psych. Rev.* 106: 795–834.

Grossberg, S. and Todorovic, D. (1988). Neural dynamics of 1-D and 2-D brightness perception: A unified model of classical and recent phenomena. *Percept. Psychophys.*, 43: 241–277.

von Helmholtz, H. (1962). *Treatise on Physiological Optics*, Vol II, Trans. J. P. L. Southall, New York: Dover Publications, pp. 264–301. (Vol II originally published in 1866.)

Hering, E. (1964). *Outlines of a Theory of the Light Sense*, L. M. Hurvich and D. Jameson, Trans. Cambridge, MA: Harvard University Press. (Original work published in 1874.)

Hurlbert, A. and Poggio, T. (1988). Synthesizing a color algorithm from examples. *Science*, 239: 484–485.

Hurvich, L. M. and Jameson, D. (1966). *The Perception of Brightness and Darkness.* Boston: Allyn and Bacon, Inc.

James, W. (1981) *The Principles of Psychology.* Cambridge, MA: Harvard University Press. (Original work published in 1890.)

Kingdom, F. A. A. (1997). Simultaneous contrast: the legacies of Hering and Helmholtz. *Percept.*, 26: 673–677.

Kingdom, F. A. A. (1999). Commentary on Logvinenko "Lightness induction revisited." *Percept.*, 28: 929–934.

Kingdom, F. A. A., Blakeslee, B. and McCourt, M. E. (1997). Brightness with and without perceived transparency: When does it make a difference? *Percept.*, 26: 493–506.

Kingdom, F. A. A., McCourt, M. E. and Blakeslee, B. (1997). In defence of "lateral inhibition" as the underlying cause of induced brightness phenomena. A reply to Spehar, Gilchrist, and Arend. *Vis. Res.*, 37: 1039–1044.

Kingdom, F. A. A. and Moulden, B. (1988). Border effects on brightness: A review of findings, models and issues. *Spat. Vis.*, 3: 225–262.

Kingdom, F. A. A. and Moulden, B. (1992). A multichannel approach to brightness coding. *Vis. Res.*, 32: 1565–1582.

Kingdom, F. A. A. and Whittle, P. (1996). Contrast discrimination at high contrasts reveals the influence of local light adaptation on contrast processing. *Vis. Res.*, 36: 817–829.

Knill, D. C. and Kersten, D. (1991). Apparent surface curvature affects lightness perception. *Nature*, 351: 228–230.

Land, E. H. (1986). An alternative technique for the computation of the designator in the retinex theory of color vision. *Proc. Nat. Acad. Sci. USA*, 83: 3078–3080.

Land, E. H. and McCann, J. J. (1971). Lightness and retinex theory. *J. Opt. Soc. Am.*, 61: 1–11.

Logvinenko, A. D. (1999). Lightness induction revisited. *Percept.*, 28: 803–816.

McCourt, M. E. (1982). A spatial frequency dependent grating induction effect. *Vis. Res.*, 22: 119–134.

McCourt, M. E. and Blakeslee, B. (1993). The effect of edge blur on grating induction magnitude. *Vis. Res.*, 33: 2499–2508.

McCourt, M. E. and Kingdom, F. A. A. (1996). Facilitation of luminance grating detection by induced gratings. *Vis. Res.*, 36: 2563–2573.

Moulden, B. and Kingdom, F. A. A. (1991). The local border mechanism in brightness induction. *Vis. Res.*, 31: 1999–2008.

Ratcliff, F. (1972). Contour and contrast. *Sci. Am.*, 226: 91–101.

Schiller, P. H. (1982). Central connections of the retinal ON and OFF pathways. *Nature*, 297: 580–583.

Todorovic, D. (1987). The Craik–O'Brien–Cornweet effect: New varieties and their theoretical implications. *Percept. Psychophys.*, 42: 545–560.

Turner, R. S. (1994). *In the Eye's Mind: Vision and the Helmholtz-Hering Controversy.* Princeton, NJ: Princeton University Press.

Wade, N. J. (1996). Descriptions of visual phenomena from Aristotle to Wheatstone. *Percept.*, 25: 1137–1175.

Wallach, H. (1976). *On Perception.* New York: Quadrangle.

Whittle, P. (1965). Binocular rivalry and the contrast at contours. *Q. J. Exp. Psych.*, 17: 217–226.

Whittle, P. (1986). Increments and decrements: Luminance discrimination. *Vis. Res.*, 26: 1677–1691.

Whittle, P. (1992). Brightness, discriminability and the "crispening effect." *Vis. Res.*, 32: 1493–1507.

Whittle, P. (1994a). The psychophysics of contrast brightness. In A. L. Gilchrist (ed.) *Lightness, Brightness, and Transparency*, pp. 35–110. Hillsdale, IL: Erlbaum.

Whittle, P. (1994b). Contrast brightness and ordinary seeing. In A. L. Gilchrist (ed.) *Lightness, Brightness, and Transparency*, pp. 111–157. Hillsdale, IL: Erlbaum.

4

A Multiscale Spatial Filtering Account of Brightness Phenomena

Barbara Blakeslee and Mark E. McCourt

4.1 The Central Problem, and a Consideration of Terminology

Brightness is a fundamental quality of human vision. A central problem in the study of brightness perception is understanding how and when the visual system is able to separate the physically invariant reflectances of surfaces from their potentially changing illumination. Reflectance and illumination are confounded since their product determines luminance — the amount of light reaching the eye from a particular surface. Before proceeding further, however, we need to come to terms with several definitional problems currently plaguing the field. Brightness is defined by the CIE (1970) as the attribute according to which a visual stimulus appears to be more or less intense, or to emit more or less light. Thus, unrelated achromatic colors (i.e., stimuli presented alone in a dark field) vary only in brightness (CIE, 1970). Variations in brightness range from bright to dim. Brightness is highly correlated with the photometric quantity luminance, especially for unrelated stimuli, and therefore another common definition of brightness is perceived luminance (the Trieste group uses this definition: see Arend, 1993). The CIE adds the property of lightness to related achromatic stimuli (i.e., stimuli presented in a display containing multiple stimuli). Lightness, as defined by the CIE, is the attribute according to which a visual stimulus appears to emit more or less light in proportion to that emitted by a similarly illuminated area perceived as "white." Thus, the CIE definition of lightness is actually relative brightness. Variations in lightness range from very light or white, to very dark or black. Although an unrelated color can appear white, only related colors have a grey or black component. Related colors thus possess a perceptual dimension (blackness) that does not exist for unrelated colors; this added dimension arises through spatial interactions, revealed in some instances by induction effects, which can occur only between related stimuli (Wyszecki and Stiles, 1982; Wyszecki, 1986; Lennie and D'Zmura, 1988; Pokorny, Shevell and Smith, 1991).

The Trieste group's definition of the term lightness, viz., perceived reflectance, is rather different. While the CIE term for lightness (relative brightness) is correlated with reflectance under the stimulus conditions specified by their definition, the Trieste group's definition of "lightness" refers directly to a surface property of the stimulus (namely, it's reflectance), without further qualification. A problem with this definition arises, however, because as mentioned above, reflectance is not given directly, but must in all cases be assigned based on either direct knowledge or assumptions (conscious or unconscious) about the illumination of the stimulus. Studies by Arend and Spehar (1993a, b) make it clear that lightness, defined only as perceived reflectance, is underspecified and in fact refers to several very different types of judgments. Arend and Spehar (1993a, b) identified three dimensions along which achromatic judgments could vary: brightness (defined as perceived luminance), brightness contrast (defined as perceived differences in brightness), and lightness (defined as perceived reflectance). Judgments along these three dimensions were separable, however, only in complex visual displays that included an unambiguous illumination component (i.e., where illumination across regions of the display was visibly nonuniform). It is important to note that lightness judgments representing this third dimension of achromatic perception were based on an inferential judgment involving discounting the visible illumination component.

Under all other stimulus conditions, only two dimensions of achromatic perception were available for matching such that judgments of lightness collapsed upon those for either brightness or brightness contrast. Which of these attributes captured lightness judgments depended on the subject's interpretation of the stimulus.

In light of this brief discussion it is clear that close attention to the operational definitions of these various terms — *brightness, brightness contrast* and *lightness* — is essential to correctly interpret their intended (and sometimes unintended) meaning in any particular experiment. Based on the seminal (but underappreciated) experiments of Arend and Spehar (1993a, b), however, we propose the following. First, as suggested by the CIE (1970) the term brightness should be reserved to describe the perceived intensity of a stimulus. Note that, depending on the stimulus conditions, brightness may or may not correlate strongly with either the luminance or the reflectance of a surface. The stimulus quality being judged, however, is simply perceived intensity, which is a primary sensation and can be reliably measured. We further suggest that a distinction be made between the terms lightness (as perceived reflectance) and inferred lightness (as inferred reflectance). Inferred lightness refers to the third dimension, separate from brightness and brightness contrast, along which achromatic judgments can vary. Inferred lightness is always the outcome of a perceptual inference; it is a cognitive interpretation or appraisal of a stimulus property rather than a primary sensory quality. Lightness, as perceived reflectance, is a more general term referring to those stimulus situations that do not contain an explicit illumination component. In these instances lightness judgments will, depending on the conscious or unconscious assumptions made by the observer concerning stimulus illumination, be based on the sensory qualities of brightness or brightness contrast. Note that under these conditions it is only experimenter preference for a description in terms of reflectance that dictates the use of the general term lightness, as opposed to the more specific terms of brightness or brightness contrast, to describe the sensory quality being judged. Therefore, it might be useful to use the term brightness-lightness, or b-lightness, and brightness-contrast lightness, or bc-lightness, to clarify these situations.[1]

4.2 Brightness Illusions: Levels of Explanation

Regardless of whether veridical surface perception (lightness constancy) is actually achieved through the successful separation of reflectance and illumination or is only approximated, we know that all three dimensions of achromatic perception result from the interaction of information derived from multiple surfaces in the field of view. Physiologically this must be accomplished through lateral spatial interactions between receptive fields and/or by temporal interactions within receptive fields. In addition, such interactions may occur at one site or at multiple sites with a parallel and/or hierarchical organization. Perceptual illusions are potentially informative regarding the mechanisms underlying normal visual perception, including that of brightness and lightness, and their study has historically been and continues to be a productive topic of research.

While a large and growing number of intriguing brightness illusions have been introduced, a survey of the literature reveals that the number of proposed explanations for these illusions is itself cumbersome (Kingdom and Moulden, 1988; Fiorentini et al., 1990; Gilchrist et al., 1999; Adelson, 2000). In addition, although phenomenal brightness demonstrations are often exploited to support various theories or proposed mechanisms of brightness coding, far too few quantitative data are actually offered in support of these claims. The goal of this chapter is to summarize our recent research efforts (Blakeslee and McCourt, 1997, 1999, 2001), which have aimed to remedy these deficiencies by investigating and modeling the spatial interactions between different areas of the visual field through the quantitative study of brightness illusions. The collection of quantitative psychophysical data on brightness effects enlarges the quantitative database and critically tests various theories of brightness perception. In addition, these data inform the continued development of a mechanistic model of brightness perception, the oriented difference of Gaussians (ODOG) model of Blakeslee and McCourt (1999). This model has, in our view, been extremely successful in simplifying our understanding of the mechanisms underlying brightness perception by simultaneously encompassing a large number of seemingly diverse brightness phenomena with a history of different explanations. As we will discuss in greater detail below, these various explanations include (i) low-level filtering mechanisms which are the

---

[1]Readers interested in an amplified treatment of these definitial issues are referred to the accompanying commentary that appears on the compact disc supplement to this volume.

modern equivalent of lateral inhibition originally proposed by Mach (1838–1916) and developed by Hering (1834–1918), (ii) explanations in terms of T– and X– junctions (Todorovic, 1997; Zaidi, Spehar and Shy, 1997), (iii) higher-level mechanisms involving perceptual inferences about depth and/or transparency, such as those first proposed for illumination by Helmholtz (1821–1894), and (iv) explanations in which the key factor is visual grouping (Gilchrist et al., 1999; Ross and Pessoa, 2000), based on such concepts as the Gestalt principle of "belongingness."

The defining features of the ODOG model (i.e., multiscale spatial frequency filtering, orientation selectivity, and response normalization) are all characteristics of early visual processing in both cat, and monkey (Geisler and Albrecht, 1995; Gilbert et al., 1996; Rossi, Rittenhouse and Paradiso, 1996; Rossi and Paradiso, 1999). We contend that explanations couched in terms of "higher-level" mechanisms are not required to explain the majority of the wide variety of brightness effects that we have examined (Blakeslee and McCourt, 1997, 1999, 2001), and that these effects are more parsimoniously accounted for by the ODOG model. The success of the model suggests that many brightness illusions primarily reflect the operations of early-stage cortical filtering. We note, in addition, that the mechanistic explanation offered by the ODOG model does not necessarily conflict with junction or grouping analyses, and may actually represent a mechanistic basis for both. Finally, there are a number of effects that remain unexplained by the ODOG model. We look forward to a careful analysis of these effects, because they may help us to refine the model or to determine the circumstances under which higher-level factors do, in fact, exert unique influences on brightness perception.

4.2.1   Simultaneous Brightness Contrast and Grating Induction

It has long been known that the brightness of a region of visual space is not related solely to that region's luminance, but depends also upon the luminances of adjacent regions. This phenomenon, known as brightness induction, includes both brightness assimilation and contrast effects. Assimilation refers to the situation in which the brightness of a test region changes such that it appears more similar in brightness to the adjacent regions. In general, assimilation effects are restricted to complex displays with small (high-frequency) patterns (Helson, 1963; Smith, Jin and Pokorny, 2001). Contrast effects refer to the opposite situation, in which the brightness of a test region changes such that it appears more different in brightness to the adjacent regions. A well-known demonstration of this is simultaneous brightness contrast (SBC). SBC produces a (nearly) homogeneous brightness change within an enclosed test field such that a grey patch on a white background looks darker than an equiluminant grey patch on a black background (Figure 4.1a). This effect has been well quantified with respect to inducing background and test field luminance (Heinemann, 1955). Although SBC decreases with increasing test field size, brightness induction occurs for test fields as large as 10 deg (Yund and Armington, 1975). Since this distance far exceeds the dimen-

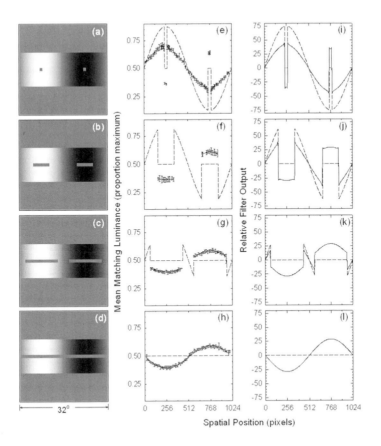

FIGURE 4.1. (a-d) Four of the stimuli used to measure the effect of test field width on induction magnitude. Display width is 32 deg; test field widths of 1 deg, 6 deg, 12 deg and 32 deg are illustrated. Test field height is 1 deg. Sinewave-inducing contrast was constant at 0.75. Test field luminance was set to the mean of the display (50 cd/m^2). Note that panel (a) is a "classical" simultaneous brightness contrast (SBC) stimulus, and that panel (d) is a standard grating induction (GI) stimulus. (e-h) Point-by-point brightness matches (at 0.25 deg intervals) across the test fields of displays illustrated in panels (a-d). Open symbols plot mean brightness matches made to the test fields (proportion mean luminance ±1 s.e.m.); filled symbols in (e) are brightness matches to the inducing grating. The dashed line depicts the veridical luminance profile of the stimulus display along a horizontal line through the vertical center of the test field and display. (i-l) Solid lines represent slices taken through the ODOG model filter output for each of the stimulus displays in panels (a-d). Note the excellent qualitative and quantitative agreement between the ODOG model output and the corresponding point-by-point brightness matching data. The model captures the magnitude and structure of brightness induction within the homogeneous test fields (i-l), as well as the nonveridical perception of the inducing grating itself (i). SBC and GI are thus demonstrated to be congruent phenomena, which are both accounted for by the ODOG model. After *Vis. Res.*, 37: 2849–2869,1997, Blakeslee, B. and McCourt, M. E., Similar mechanisms underlie simultaneous brightness contrast and grating induction, with permission from Elsevier Science.

sions of retinal or LGN receptive fields in monkey (DeValois and Pease, 1971; Yund, Snodderly, Hepler and DeValois, 1977; DeValois and DeValois, 1988), a common explanation for SBC has been that the brightness of the test field is determined by the information at the edges of the bounded region (e.g., by average perimeter contrast), and is subsequently filled in or assigned to the entire enclosed area (Cornsweet and Teller, 1965; Shapley and Enroth-Cugell, 1984; Grossberg and Todorovic, 1988; Paradiso and Nakayama, 1991; Paradiso and Hahn, 1996; Rossi and Paradiso, 1996; for review, see Kingdom and Moulden, 1988). Evidence has slowly accumulated, however, that this explanation is too simple, and that distal factors must also play a role in SBC (Arend, Buehler and Lockhead, 1971; Land and McCann, 1971; Heinemann, 1972; Shapley and Reid, 1985; Grossberg and Todorovic, 1988; Reid and Shapley, 1988).

Grating induction (GI), unlike SBC, is a brightness contrast effect that produces a spatial brightness variation (a grating) in an extended test field (Figure 4.1d). The perceived contrast of the induced grating decreases with increasing inducing grating frequency and with increasing test field height (McCourt, 1982), such that GI magnitude is constant for a constant product of inducing frequency and test field height (McCourt, 1982; Foley and McCourt, 1985). GI, like SBC, extends over large distances, since it is still observed in test fields at least as large as 6 deg (Blakeslee and McCourt, 1997). Unlike SBC, however, homogeneous brightness fill-in cannot account for GI. For example, a fill-in mechanism dependent on average perimeter contrast does not predict the appearance of a pattern in a GI test field because only a single value (average perimeter contrast) determines the assignment of brightness.

A homogeneous fill-in mechanism that computed brightness based on local contrast rather than on average perimeter contrast (and which could therefore in principle produce both positive and negative brightness signals originating from the opposite-polarity test field edges) still cannot produce a patterned test field (i.e., an induced grating). This is so because, without boundaries within the test field to arrest the propagation of these putative brightness signals, induced brightness and darkness will diffuse and average to produce the percept of a homogeneous test field. Several more complex brightness models have been proposed that incorporate nonhomogeneous fill-in mechanisms (Grossberg and Mingolla, 1987; Pessoa, Mingolla and Neumann, 1995), however, these models have not been applied to grating induction. Another suggestion (pursued by Blakeslee and McCourt, 1997) is that GI might be understood in terms of the output of parallel spatial filtering across multiple spatial scales (Moulden and Kingdom, 1991). An attractive feature of this approach is that both the low-pass spatial frequency response of GI, and the invariance of induction magnitude with viewing distance (i.e., the direct tradeoff between the effects of inducing grating spatial frequency and test field height), are both explained parsimoniously by multiscale spatial filtering.

Despite the fact that SBC is typically considered a homogeneous brightness effect dependent on a fill-in mechanism, whereas the defining characteristic of grating induction is that it possesses spatial structure and cannot be produced by a homogeneous fill-in mechanism, it has nevertheless been suggested both that SBC is a special low-frequency instance of grating induction (McCourt, 1982), and that GI is a particular case of SBC (Zaidi, 1989; Moulden and Kingdom, 1991). Blakeslee and McCourt (1997) explored this issue, asking whether the mechanism(s) underlying GI could account for SBC as well, or if fundamentally different brightness mechanisms were required to explain these two effects. The structure and magnitude of induction in both GI and SBC stimuli were measured where the inducing conditions for the two effects were rendered as similar as possible by employing one cycle of a low-frequency sinewave grating as the inducer. Test field dimensions spanned a range that incorporated both classic SBC and GI configurations (see Figures 4.1a-d). At each of three test field heights (1 deg, 3 deg, and 6 deg), point-by-point brightness matches (Heinemann, 1972; McCourt, 1994) were obtained at intervals of 0.25 deg, for test field widths of 32 deg (the GI condition), 14 deg, 12 deg, 8 deg, 6 deg, 3 deg and 1 deg (Figure 4.1e-h). Point-by-point brightness matches were analyzed to assess systematic changes in induction structure (i.e., departures from the sinusoidal brightness variation seen in the 32 deg wide test field GI condition) and in the average magnitude of brightness and darkness induction within the test fields, as a function of test field height and width. In the widest test fields (14 deg and 12 deg) induction structure was well accounted for by the sinewave pattern observed in the GI condition.

As test field width decreased further, the sinewave amplitude of the induced structure in the test field decreased (i.e., the pattern flattened), and eventually became negative (i.e., showed a reverse cusping) at the narrower test field widths. Both the structure and magnitude of brightness induction as a function of changing test field height and width were accounted parsimoniously for by the output of a differentially weighted, octave-interval array of seven difference-of-Gaussian (DOG) filters (Figure 4.1 i-l). This array of spatial filters differed from those previously employed to model various aspects of spatial vision in that it included filters tuned to much lower spatial frequencies. While cells with receptive field sizes corresponding to the largest DOG filters used in this study do not exist at the level of the retina, such filters are postulated to exist at those levels of the nervous system where brightness percepts are determined. Recent evidence suggests that a significant number of cells in cat primary visual cortex respond in a manner correlated with perceived brightness, and that they do so over distances far exceeding the size of their "classical" receptive fields mapped using conventional techniques (Rossi et al., 1996). Recent evidence from primate anatomy and physiology also indicates that at the earliest cortical levels (V1) the substrate exists for providing cells with input from relatively large regions of the visual field, and that the response properties of cells are modulated by stimuli lying far outside the "classical" receptive field (Gilbert et al., 1996). Thus, it appears that heretofore unappreciated lateral interactions at early levels of visual processing, or feedback from hierarchically higher processing areas (Lamme and Roelfsema, 2000) may provide for an area of visual integration an order-of-magnitude larger than that revealed by the "classical" receptive field, making the inclusion of large filters in a multiscale array less implausible than previously believed.

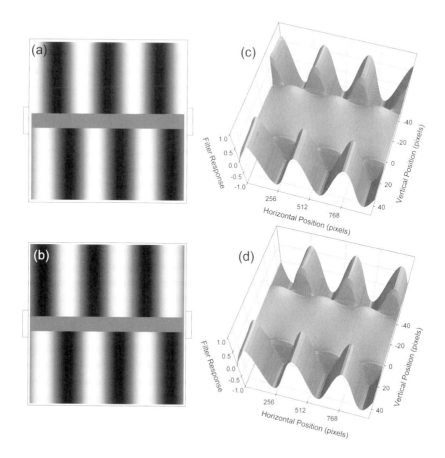

FIGURE 4.2. (a, b) Two GI displays like those used by Zaidi (1989) to argue that the orientation and spatial frequency of grating induction was governed by proximal cues. The relative spatial phase of the upper and lower inducing gratings in panel (a) is 90 deg, whereas it is 180 deg in panel (b). Despite the vertical orientation of the inducing gratings, note that the induced grating in panel (a) tilts across the test field. There is no induced grating in panel (b), but bright and dark meniscuses are apparent in the test field at the margins of the inducing grating. (c, d) 3D mesh plots illustrating the ODOG model output following convolution with the stimuli shown in the corresponding left-hand panels. Mesh plots are close-up views of the test field regions indicated by the lateral tabs on the displays of panels (a) and (b). Model output closely approximates the appearance of the induced brightness in the test fields. The model explains both the tilted orientation of the induced grating in panel (a) as well as the meniscuses within the test field of panel (b). From *Vis. Res.*, 37: 2849–2869, 1997, Blakeslee, B. and McCourt, M. E., Similar mechanisms underlie simultaneous brightness contrast and grating induction, with permission from Elsevier Science.

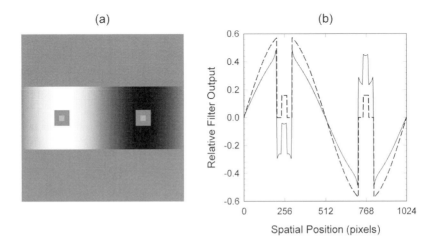

FIGURE 4.3. (a) A version of the stimulus used by Shapley and Reid (1985) to demonstrate brightness contrast and "assimilation." (b) The dashed line plots the luminance profile across this stimulus at the vertical center of the display. The solid line represents a slice taken through the ODOG model output at a corresponding spatial location. The ODOG model output clearly predicts induction within the equiluminant surrounds, as well as "assimilation" within the equiluminant central test fields. From *Vis. Res.*, 37: 2849–2869, 1997, Blakeslee, B. and McCourt, M. E., Similar mechanisms underlie simultaneous brightness contrast and grating induction, with permission from Elsevier Science.

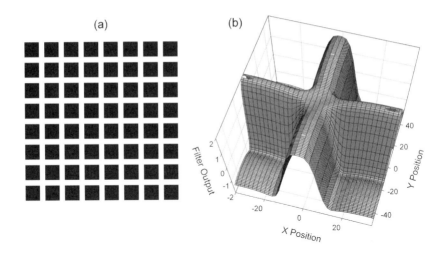

FIGURE 4.4. (a) An example of the Hermann grid stimulus. Dark spots are perceived in the intersections. (b) 3D mesh plot (close-up of one street intersection) illustrating the ODOG model output following convolution with the Hermann grid stimulus. The model produces localized output minima at locations where dark spots are observed. From *Vis. Res.*, 37: 2849–2869, 1997, Blakeslee, B. and McCourt, M. E., Similar mechanisms underlie simultaneous brightness contrast and grating induction, with permission from Elsevier Science.

It is significant that this relatively simple filtering explanation can be generalized to account for several other important brightness phenomena including: Zaidi's (1989) GI demonstrations showing both local and distal effects (Figure 4.2a-b); Shapley and Reid's (1985) contrast and assimilation demonstration (Figure 4.3a-b) modeled as due to the integration of local contrasts across space; and the induced spots seen at the street intersections of the Hermann grid (Figure 4.4a-b) classically explained in terms of on- and off-center receptive fields (Fiorentini et al., 1990). Thus, the model of Blakeslee and McCourt (1997) brings together with a common explanation a variety of seemingly diverse brightness phenomena with a history of different explanations that include local filtering, filling in, and edge integration.

4.2.2   White's Effect and Todorovic's SBC Demonstration

In a subsequent paper Blakeslee and McCourt (1999) specifically addressed a group of effects, including the White effect (White, 1979) (Figure 4.5a), and a SBC demonstration of Todorovic (1997) (Figure 4.5b), which cannot be accounted for by isotropic contrast models such as the DOG model and the edge-dependent models discussed earlier. In the White effect, grey test patches of identical luminance placed on the black and white bars of a square-wave grating appear different in brightness. What makes the effect so interesting, however, is that

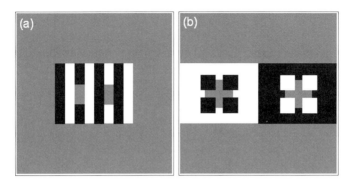

FIGURE 4.5. (a) An example of the White stimulus. (b) Todorovic's variation of a simultaneous brightness contrast stimulus. In both stimuli the grey patches are equiluminant, but appear different in brightness. These effects are not explicable by contrast models utilizing isotropic spatial filters. After *Vis. Res.*, 39: 4361–4377, 1999, Blakeslee, B. and McCourt, M. E., A multiscale spatial filtering account of the White effect, simultaneous brightness contrast, and grating induction, with permission from Elsevier Science.

the direction of the brightness change is independent of the aspect ratio of the test patch. This means that, unlike SBC and GI, the White effect does not correlate with the amount of black or white border in immediate contact with the test patch, or in its general vicinity. For example, when the grey patch is a vertically oriented rectangle sitting atop a white stripe of a vertical grating, it has two short sides that are in contact (above and below) with the coaxial white bar upon which it sits, and two long sides (left and right) that are in contact with the flanking black bars (see Figure 4.5a). This configuration describes a test patch having more extensive contact with the dark flanking bars, yet the grey patch appears darker than a similar grey patch flanked by white bars. This is not simply an assimilation effect, however, since if the height of the test patch is reduced until it has more extensive border contact with the bar on which it is sitting (i.e., the coaxial white bar), the direction of the effect is unchanged (White, 1979; 1981). From observations of this type White (1979) concluded that explanations (whether contrast or assimilation) that depended simply on the relative amounts of black and white surrounding the grey elements could not explain the effect, and that directional (orientation) properties of the inducing grating must be important.

A number of qualitative filtering explanations have been offered for the White effect. White himself proposed a mechanism called "pattern-specific inhibition" (White, 1981), based on the notion that elongated cortical filters having similar preferred orientation and spatial frequency selectivity, and which received their input from adjacent retinal locations, might tend to inhibit one another and thus produce the effect. In a similar vein Foley and McCourt (1985) suggested that hypercomplex-like cortical filters with small centers and elongated surrounds might be responsible for the effect. Moulden and Kingdom (1989) proposed a dual mechanism model to explain the results of an investigation in which they varied the height of both the flanking and coaxial inducing bars. They concluded that a local mechanism, mediated by circularly symmetric center-surround receptive fields, operated along the borders of the test patch and produced a particularly strong signal at the corner intersections of the test patch with the coaxial bar. According to their model it is this corner signal that in some (unspecified) manner disproportionately weights the coaxial bar relative to the flank and induces brightness into the test patch. Additionally, they proposed that a more spatially extensive mechanism was required to allow the coaxial bar to exert an influence on the brightness of the test patch throughout its length. This mechanism was seen as possibly implicating the operation of neurons with small centers and elongated surrounds similar to those proposed by Foley and McCourt (1985).

Numerous other attempts have also been made to explain the White effect on the basis of higher-order perceptual inferences involving depth and/or transparency, and the Gestalt notion of "belongingness" (Agostini and Profitt, 1993; Taya, Ehrenstein and Cavonius, 1995; Spehar, Gilchrist and Arend, 1995; Anderson, 1997; Ross and Pessoa, 2000). According to the Gestalt approach perceptual organizations (such as relative depth relations in the White stimulus) influence brightness contrast such that surfaces predominantly interact (i.e., contrast) with other surfaces with which they are grouped. Using this reasoning, Agostini and Profitt (1993) and Gilchrist et al. (1999) argued that in the White effect the test patch appears lighter (or darker) when it is on the black (or white) bar because of the phenomenal impression that it "belongs to" or has been "grouped with" that bar. According to Gilchrist et al. (1999) the principal grouping factor at work here is the T-junction, which is thought to signal depth through occlusion. Of note, however, is that both Zaidi et al. (1997) and Todorovic (1997) argue that while an explanation based on an analysis of local junctions in the stimulus, specifically T-junctions, can account for White's effect, it does so without any requirement that T-junctions contribute to perceptual organization. In other words, both studies show that it is not the depth-inducing aspect of T-junctions that is responsible for the effect. The T-junction rule states that the brightness of regions that share edges with several other regions, and whose corners involve T-junctions, is predominantly dependent on the luminance of collinear regions and is in the direction of a SBC effect. In the White stimulus the flanking bars form the tops of the four T-junctions that define the corners of the test patch. The stems of the T-junctions are formed by the test patch and the coaxial bar on which it is superimposed.

Although both the T-junction and grouping analyses offer useful rules for qualitatively predicting the appearance of various brightness effects, they fall short of identifying an underlying mechanism. Blakeslee and McCourt (1999) were able to provide such a mechanistic explanation in the form of an oriented-difference-of Gaussians (ODOG) model. The oriented filters of the ODOG model were produced by setting the ratio of DOG center/surround space constants to 1:2 in one orientation and to 1:1 in the orthogonal orientation (Table 4.1). A grey-level representation of such an ODOG filter appears in Figure 4.6a. Note that although the center is circular, the surround extends beyond the center for a distance of twice

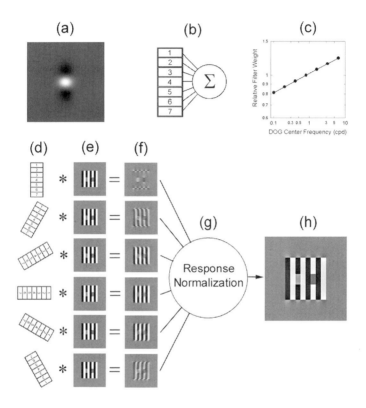

FIGURE 4.6. A diagrammatic representation of the oriented difference-of-Gaussians (ODOG) model of brightness perception. (a) Illustration of a two-dimensional oriented difference-of-Gaussian (ODOG) filter. (b) Seven filters, with center frequencies spaced at octave intervals, are summed within orientation after being weighted across frequency (c) using a power function with a slope of 0.1. (d) The resulting six multiscale spatial filters, one for each orientation, are convolved with stimuli of interest (e), in this instance a White stimulus. The convolution outputs (f) are pooled across orientation according to their space-averaged root-mean-square (RMS) activity level (g) to produce a resultant output (h). From *Vis. Res.*, 39: 4361–4377, 1999, Blakeslee, B. and McCourt, M. E., A multiscale spatial filtering account of the White effect, simultaneous brightness contrast, and grating induction, with permission from Elsevier Science.

Mechanism	Space Constant	
	Center	Surround
1	.047 deg	.093 deg
2	.094 deg	.188 deg
3	.188 deg	.375 deg
4	.375 deg	.75 deg
5	.75 deg	1.5 deg
6	1.5 deg	3 deg
7	3 deg	6 deg

TABLE 4.1. Difference of Gaussian space constants

the center size in one orientation, but is the same size as the center in the orthogonal orientation. These filters can be described as Gaussian blobs with inhibitory flanks or as simple-like cells (such as those found in the cortex of monkey or cat) that are orientation and spatial frequency selective. The ODOG model is implemented in six orientations (0, 30, 60, 90, -30, and -60 degrees). Each orientation is represented by seven volume-balanced filters that possess center frequencies arranged at octave intervals (from 0.1 to 6.5 c/d). The seven spatial frequency filters (Figure 4.6b) within each orientation are summed after weighting across frequency using a power function with a slope of 0.1 (Figure 4.6c). This slope is consistent with the shallow low-frequency fall-off of the suprathreshold contrast sensitivity function that is expected to be associated with the high-contrast stimuli under investigation (Georgeson and Sullivan, 1975). The resulting six multiscale spatial filters, one per orientation, are convolved with the stimulus of interest (Figure 4.6d-e). The filter outputs (Figure 4.6f) are pooled across orientation according to their space-averaged root-mean-square (RMS) activity level, as computed across the entire image. The pooling is in accord with a simple response normalization in which the filter outputs are weighted such that the RMS contrast in the "neural images" across orientation channels are equated (Figure 4.6g). Response nonlinearities found in neurons in cat and monkey visual cortex, such as contrast gain control and the rapidly accelerating increase in response at low contrast and saturation at high contrast, may represent the physiological substrate for this type of response normalization (for an overview, see Geisler and Albrecht, 1995). Note that when the filters of the ODOG model are summed linearly across the full range of orientations within each spatial frequency these filters combine to produce a DOG filter. Thus, the DOG model of Blakeslee and McCourt (1997) is simply a subset of the ODOG model in which the filter outputs are pooled linearly.

As mentioned previously, the defining features of the ODOG model, e.g., multiscale spatial frequency filtering, orientation selectivity, and response normalization, are response characteristics that are observed routinely at early cortical stages of visual processing in both cat, and monkey (Rossi and Paradiso, 1999; Rossi et al., 1996; Gilbert et al., 1996; Geisler and Albrecht, 1995). It is specifically the addition of orientation selectivity and response normalization, however, that allows the model to account for anisotropic effects such as the White effect. An intuitive sense for the model can be obtained from examining Figure 4.6d-f. When the long axis of the multiscale ODOG filter is vertical, as it is in the orientation represented by the top row of Figure 4.6d-f, the convolution output of this filter with the White stimulus shows the greatest activity in the region of the test patches and produces the White effect. Although the top and bottom edges of the inducing grating are also a good stimulus for this filter, the inducing grating is not. This situation is largely reversed in the convolution output of the multiscale filter with a horizontal orientation (represented in the fourth row of Figure 4.6d-f). Here the activity generated by the inducing grating is high compared to that for the test patches. Added together, however, these two filter orientations represent both the test patches and the inducing grating. Response normalization prior to summation simply weights the features extracted by these two filters equally. This prevents high contrast features (such as the inducing grating) captured at one orientation from swamping lower contrast features (such as the test patches) captured at another orientation.

Blakeslee and McCourt (1999) showed that the ODOG model qualitatively predicts the relative brightness of the test patches in the White effect, the Todorovic SBC demonstration, GI and SBC, and quantitatively predicts the relative magnitudes of these brightness effects as measured psychophysically using brightness matching (Figure 4.7). This mechanistic explanation does not necessarily conflict with T-junction or grouping analyses, but may, at least to some extent, serve as a mechanism for both. Indeed, to the extent that junctions influence "higher-level" grouping, and to the extent that filters of the ODOG model capture the operations of junctions and grouping, one might expect all these approaches to yield similar results (Todorovic, 1997; Blakeslee and McCourt, 1999). The ODOG model has the advantage, however, in that it makes quantitative predictions about the relative size of various brightness effects and provides an explanation for a larger variety of brightness effects. For example, SBC and GI do not contain T-junctions or X-junctions and, therefore, cannot be addressed by a junction analysis (Blakeslee and McCourt, 1999). There is also no explanation for GI based on either Gestalt grouping or Gilchrist's anchoring hypothesis (Gilchrist et al., 1999). In addition, Blakeslee and McCourt (1999) showed that the ODOG model accounts for the smooth transition in mean brightness seen in the White effect (Figure 4.8) when the relative phase of the test patch is varied relative to the inducing grating (White and White, 1985) (Figure 4.9 a-f). Significantly, this smooth transition is not readily explained by a T-junction or grouping analysis. Finally, point-by-point brightness matching revealed brightness variations across the test patches of White stimuli (Figure 4.9 g-l) (Blakeslee and McCourt, 1999), as well as GI and SBC stimuli (Blakeslee and McCourt, 1997) that accord with ODOG model predictions. Only spatial filtering can easily account for these types of brightness gradients.

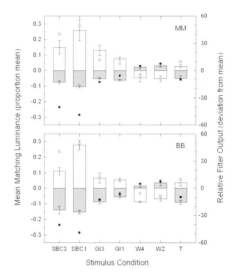

FIGURE 4.7. The bar graph (read against left ordinate) plots the deviation of mean matching luminance from the mean luminance (as a proportion of mean luminance) for various brightness stimulus conditions. The error bars are 95% confidence limits. Data from two subjects appear in the upper and lower panels. Condition SBC3 refers to brightness matches obtained in simultaneous brightness contrast conditions where test patch height and width were 3 deg; condition SBC1 refers to 1 deg test patches. The bars extending above the mean represent brightness matches for test patches on the dark background (which appear brighter than the mean), while the bars extending below the mean represent the test patch matches on the bright background (which appear darker than the mean). Next are matches for two GI displays: a 0.03125 cyc/deg sine wave inducing grating with a test field height of 3 deg (GI3), and a 0.125 cyc/deg sine wave inducing grating with a test field height of 1 deg (GI1). The conditions labeled W4 and W8 plot the magnitude of the White effect for a 0.25 cyc/deg and a 0.5 cyc/deg square wave inducing grating, respectively. For the 0.25 cyc/deg inducing grating, test patch width was 2 deg and test patch height was 4 deg. For the 0.5 cyc/deg inducing grating, test patch width was 1 deg and test patch height was 2 deg. Note that for these two conditions the bars extending above the line represent matches to test patches located on the dark bars of the inducing grating while those below the line are matches to the test patches located on the bright bars of the inducing grating. The final condition, (T), plots the magnitude of brightness induction in the Todorovic stimulus (see Figure 4.5b). The bar extending above the mean luminance represents the match to the test patch on the dark inducing background with the overlapping white squares. The bar extending below the mean is the match to the test patch on the white background with the overlapping black squares. Inducing patterns of 100% contrast were used in all brightness displays. The symbols are read against the right-hand ordinate and represent the ODOG model outputs to the test fields in each stimulus condition. The filled symbols are the predictions for the matches that appear as dark bars and the open symbols are the predictions for the matches that appear as white bars. The ODOG model output and the empirical brightness matching data are clearly similar across a wide variety of brightness phenomena. From *Vis. Res.*, 39: 4361–4377, 1999, Blakeslee, B. and McCourt, M. E., A multiscale spatial filtering account of the White effect, simultaneous brightness contrast, and grating induction, with permission from Elsevier Science.

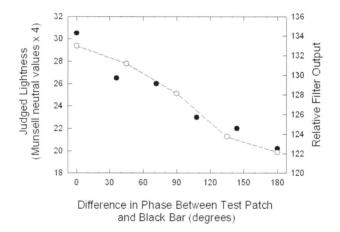

FIGURE 4.8. Judged lightness (filled symbols, read against left ordinate), replotted from White and White (1985), as a function of test patch spatial phase. Open symbols plot predicted test patch brightness from ODOG model output averaged across the width of the test patch (right-hand ordinate). ODOG model output accurately predicts the linear phase-brightness relationship reported by White and White (1985). After *Vis. Res.*, 39: 4361–4377, 1999, Blakeslee, B. and McCourt, M. E., A multiscale spatial filtering account of the White effect, simultaneous brightness contrast, and grating induction, with permission from Elsevier Science.

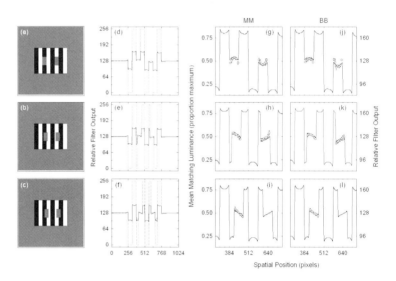

FIGURE 4.9. (a-c) The White stimulus illustrating the effect of shifting the phase of the test patch relative to the inducing grating. (a) In the standard configuration the grey test patch is in a 0 deg phase relationship with the black bar, and a 180 deg phase relationship with the white bar. In panel (b) both test patches have been shifted to the right by 45 deg phase angle; this reduces the magnitude of the effect by half (see Figure 4.8). The test patches in panel (c) have been shifted by 90 deg phase angle, completely eliminating the effect. (d-f) The dashed lines depict the veridical luminance profiles of the stimulus displays taken along a horizontal line through the vertical center of the test field and display. Solid lines represent corresponding slices taken through the ODOG model filter output. (g-l) Magnified views of the ODOG model output (solid lines) illustrated in panels (d-f), and point-by-point brightness matches (with 95% confidence intervals) obtained at seven locations across each 2 deg test patch (open symbols, as read against left ordinate). Data from two subjects (MM and BB) are shown. ODOG model output closely parallels the observed brightness variations across the test patches in these stimuli. After *Vis. Res.*, 39: 4361–4377, 1999, Blakeslee, B. and McCourt, M. E., A multiscale spatial filtering account of the White effect, simultaneous brightness contrast, and grating induction, with permission from Elsevier Science.

4.2.3   The Wertheimer–Benary Effect and the Corrugated Mondrian

A recent paper (Blakeslee and McCourt, 2001) extended the ODOG model to include two additional brightness illusions: the Wertheimer–Benary effect (Benary, 1924; Todorovic, 1997) and the corrugated Mondrian effect (Adelson, 1993; Todorovic, 1997). These effects, like the White effect and Todorovic's SBC demonstration, cannot be accounted for on the basis of isotropic contrast models, and offered another opportunity to test the generality of the ODOG model. Figures 4.10a, b illustrate the Wertheimer–Benary stimuli used by Blakeslee and McCourt (2001). Note that in the stimulus in Figure 4.10a, known as the Benary cross, and for the left half of Todorovic's (1997) version of this effect (Figure 4.10b), the two grey triangles are identical in luminance but appear different in brightness despite having identical border contrast. A frequently referenced qualitative explanation for this effect, based on the Gestalt concept of "good whole" or "belonging" (Wertheimer, 1923; 1958), states that the triangle embedded in the arm of the black cross appears to belong to the cross and therefore contrasts with it and appears lighter. The triangle on the white background likewise appears to belong to the white background and thus contrasts with it and appears darker (Benary, 1924; Mikesell and Bentley, 1930; Jenkins, 1930; Gilchrist, 1988). Note both that the right half of Todorovic's Wertheimer–Benary figure (Figure 4.10b) is simply a reverse contrast version of the same effect, and that a similar explanation can be applied.

Todorovic (1997) and Zaidi et al. (1997) argued, however, that like the White effect, the Wertheimer–Benary effect can be explained on the basis of structural factors or T-junctions alone and is not dependent on the "higher-level" Gestalt grouping factors mentioned above. In the example of the Wertheimer–Benary effect seen in Figure 4.10(a), the triangle situated within the black cross has one T-junction associated with it. In the original version of the Benary cross this triangle is shifted away from the center of the cross and is associated with two T-junctions, however, the same analysis applies. The grey triangle and the black background are collinear regions and the white background is the flanking region. Therefore, the triangle contrasts with the black collinear region and appears lighter. The triangle on the white background is associated with two T-junctions. In both instances the white background forms the collinear edge and the black cross forms the flanking edge. Therefore, this triangle contrasts with the white background and appears darker. A similar analysis can be applied to the Todorovic (1997) version of the effect in Figure 4.10b.

Figures 4.10(c, d) illustrate Adelson's (1993) original corrugated Mondrian stimuli and Figure 4.10(e) is a novel configuration created by Todorovic (1997). The grey test patches appearing in the third column of the second and fourth rows in each panel are of identical luminance. The luminances of the other patches also remain fixed. Thus, the only difference between the panels is in the geometrical shape and arrangement of the patches. There are, nevertheless, obvious differences in test patch brightness both within and between the various configu-

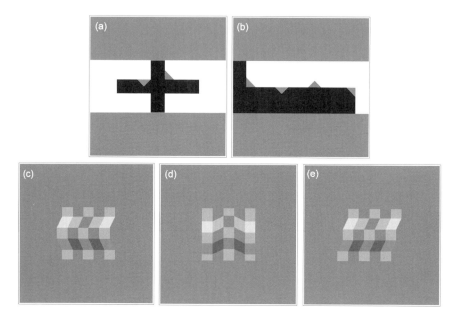

FIGURE 4.10. Illustrations of the typical Wertheimer–Benary stimulus (a), and Todor-ovic's (1997) version of this stimulus (b). The grey triangles in each stimulus are equilumi-nant, yet appear different in brightness. The ODOG model successfully accounts for the di-rection and magnitude of this brightness effect. (c-d) Illustrations of Adelson's (1993) orig-inal corrugated Mondrian stimuli. (e) A novel configuration created by Todorovic (1997). The grey test patches appearing in the third column of the second and fourth rows in each panel are of identical luminance. The luminances of the other patches also remain fixed. The only difference between the panels is the geometrical shape and arrangement of the patches. There are, nevertheless, obvious differences in test patch brightness, both within and between the various configurations. The ODOG model successfully predicts the dif-ferences in brightness in the various stimuli, calling into question the need for "mid-level" explanations. After *Vis. Res.*, 41: 2487–2502, 2001, Blakeslee, B. and McCourt, M. E., A multiscale spatial filtering account of the Wertheimer-Benary effect and the corrugated Mondrian, with permission from Elsevier Science.

rations. Adelson (1993) offered what he called a "mid-level" explanation (based on Figures 4.10c, d) in which the Mondrians are seen as 3-D objects with different amounts of illumination falling on the different planes, and in which the perceived illumination influences the brightness of the test patches. According to this account the upper test patch in Figure 4.10c is seen as a dark grey patch that is brightly lit, while the lower test patch is seen as a light grey patch that is dimly lit. In Figure 4.10d the two patches are perceived in the same plane, thus sharing the same illumination, and should therefore appear similar in brightness.

Adelson (1993) attributed the small residual brightness difference he measured in this condition to a low-level (lateral inhibitory) process. Todorovic (1997) challenged this explanation, favoring instead an explanation in terms of local junctions rather than perceived illumination and 3-D structure. To account for this effect, Todorovic (1997) extended the T-junction analysis to include X-junctions where four regions come together. For X-junctions the brightness rule similarly states that the lightness of the grey patch is predominantly affected by the luminance of its collinear neighbors and that the direction of the effect is as in SBC. Using this rule Todorovic was able to predict the direction of the brightness effect in Adelson's (1993) corrugated Mondrians. In Figure 4.10c the upper test patch and the lower test patch are collinear with their horizontal neighbors and therefore contrast with them. Since the collinear neighbors for the upper patch are lighter than those for the lower patch, the upper patch appears darker than the lower patch. In Figure 4.10d, however, the upper and lower test patches are collinear with their vertical neighbors and these neighbors all have the same luminance. Therefore, an analysis of X-junctions predicts no brightness difference between the upper and lower test patches. The small difference that does persist is consistent with the hypothesis that collinear regions predominantly, but not exclusively, influence brightness, and that lateral regions may induce a residual effect (Todorovic, 1997).

Gilchrist et al. (1999) offered another interpretation of the corrugated Mondrian based on Gestalt grouping and the anchoring hypothesis. In their formulation the anchor in a given framework is the luminance that appears white. The appearance of each darker region in the framework depends on its relationship to the anchor. According to this analysis grouping by rows (Figure 4.10c) produces the brightness effect because the highest luminance in the row to which the lower test patch belongs is lower than the highest luminance in the row to which the upper test patch belongs. Therefore, the lower test patch has a higher local lightness assignment than does the upper test patch. Grouping by columns (Figure 4.10d) produces no effect since both test patches share the same group and are anchored to the same highest luminance. Grouping by local retinal adjacency produces a weak effect in the same direction as grouping by rows and is held responsible for the small residual brightness effect seen in Figure 4.10d.

Figure 4.10e is a staircase version of the corrugated Mondrian produced by Todorovic (1997) which refutes Adelson's (1993) illumination hypothesis. Since rows two and four in this configuration are seen as lying in parallel planes, they should, according to Adelson (1993), be perceived to receive the same illumination and the upper and lower test patches should appear equally bright. This is clearly not the case since the test patches appear dissimilar (Todorovic, 1997). This brightness difference is predicted by the 4-junction analysis (Todorovic, 1997), however, because the test patches are again collinear with their horizontal neighbors as in Figure 4.10c. This brightness difference would also be predicted by the anchoring hypothesis (Gilchrist et al., 1999) since this configuration has not interrupted the grouping by horizontal rows.

As discussed previously for the White effect, although both the T-junction and grouping analyses offer useful rules for qualitatively predicting the appearance of various brightness effects, they fall short of identifying an underlying mechanism. Blakeslee and McCourt (2001) quantitatively measured the size of the brightness effect for the two Wertheimer–Benary stimuli (Figure 4.10a, b) and for a low- and high-contrast set of the corrugated Mondrian stimuli (Figure 4.10c-e). The ODOG model accounted for the relative brightness of the test patches in the Benary cross (Benary, 1924), the corrugated Mondrian (Adelson, 1993) and in Todorovic's (1997) versions of these effects. In addition, the model also predicted the effect of manipulating contrast in the corrugated Mondrian.

4.2.4   Conclusions and Discussion of Possible Higher-Level Influences: Transparency

It is clear from the above review that the ODOG model can successfully account for a large constellation of diverse brightness effects that have been explained previously by appealing to a wide variety of different proposed brightness mechanisms. These explanations include low-level filtering, filling in, edge integration, and junction analysis, as well as higher-level mechanisms involving perceptual inferences about depth and/or transparency, and explanations in which the key factor is visual grouping based on such concepts as the Gestalt principle of "belongingness." The fact that all of the induced brightness effects reviewed here can be accounted for parsimoniously by the ODOG model suggests that these particular effects primarily reflect the operations of early-stage cortical filtering, and that explanations in terms of "higher-level" grouping mechanisms are not required. Indeed, in our view "grouping" is instantiated by the nonlinear summation of the oriented DOG filters within the ODOG model. The orientation selectivity and filter output normalization prior to summation perform a "low-level" or "filtering" version of grouping by rows, columns, and diagonals. Thus, one could argue that, in some instances, the ODOG model actually provides a mechanistic, mathematically precise account of grouping.

There may, however, be other situations in which higher-order effects on brightness do occur. For example, several claims have been made (including our own) for an effect of transparency on perceived brightness (Adelson, 1993; Anderson, 1997; Kingdom, Blakeslee and McCourt, 1997). In the interests of parsimony, however, careful study is required to determine the circumstances under which higher-order factors, such as transparency, exert a unique influence on brightness, and to determine the magnitudes of these higher-order effects. For example, in a carefully controlled study, Kingdom et al. (1997) demonstrated a small effect of perceived transparency on the brightness of the test patch in a SBC stimulus. Multiplicative transparency affected brightness in such a way that subjects perceived the test patch to be brighter than in other configuration conditions. Somewhat surprisingly, this is consistent with an explanation whereby the transparency was partially discounted from the brightness of the test patch. Carefully sorting out those brightness effects that are and are not accounted for by low-level mechanisms, as well as measuring their relative magnitudes, will provide needed direction, precision, and insight into the investigation of brightness perception and the role of higher-order mechanisms.

Acknowledgement

The research described in this chapter was supported by NSF Grants IBN-9514201 and IBN-9816690.

References

Adelson, E. H. (1993). Perceptual organization and the judgment of brightness. *Science*, 262: 2042–2044.

Adelson, E. H. (2000). Lightness perception and lightness illusions. In M. Gazzaniga (Ed.), *The New Cognitive Neurosciences*, 2nd ed., Cambridge, MA: MIT Press.

Agostini, T. and Proffitt, D. R. (1993). Perceptual organization evokes simultaneous lightness contrast. *Percept.*, 22: 263–272.

Anderson, B. (1997). A theory of illusory lightness and transparency in monocular and binocular images: The role of contour junctions. *Percept.*, 26: 419–453.

Arend, L. E. (1993). Mesopic lightness, brightness, and brightness contrast. *Percept. Psychophys.*, 54: 469–476.

Arend, L. E., Buehler, J. N. and Lockhead, G. R. (1971). Difference information in brightness perception. *Percept. Psychophys.*, 9: 367–370.

Arend, L. E. and Spehar, B. (1993a). Lightness, brightness, and brightness contrast: 1. Illuminance variation. *Percept. Psychophys.*, 54: 446–456.

Arend, L. E. and Spehar, B. (1993b). Lightness, brightness, and brightness contrast: 2. Reflectance variation. *Percept. Psychophys.*, 54: 457–468.

Benary, W. (1924). Beobachtungen zu einem Experiment über Helligkeitskontrast. *Psychologische Forschung*, 5: 131–142.

Blakeslee, B. and McCourt, M. E. (1997). Similar mechanisms underlie simultaneous brightness contrast and grating induction. *Vis. Res.*, 37: 2849–2869.

Blakeslee, B. and McCourt, M. E. (1999). A multiscale spatial filtering account of the White effect, simultaneous brightness contrast, and grating induction. *Vis. Res.*, 39: 4361–4377.

Blakeslee, B. and McCourt, M. E. (2001). A multiscale spatial filtering account of the Wertheimer–Benary effect and the corrugated Mondrian. *Vis. Res.*, 41: 2487–2502.

CIE (1970). *International Lighting Vocabulary*. 3rd Ed. Publ. No. 17 (E1.1), Paris: Bureau Central de la CIE.

Cornsweet, T. N. and Teller, D. (1965). Relation of increment thresholds to brightness and luminance. *J. Opt. Soc. Am.*, 55: 1303–1308.

DeValois, R. L. and DeValois, K. K. (1988). *Spatial Vision*. New York: Oxford University Press.

DeValois, R. L. and Pease, P. L. (1971). Contours and contrast: Responses of monkey lateral geniculate cells to luminance and color figures. *Science*, 171: 694–696.

Fiorentini, A., Baumgartner, G., Magnussen, S., Schiller, P. H. and Thomas, J. P. (1990). The perception of brightness and darkness: Relations to neuronal receptive fields. In L. Spillman and J. S. Werner (eds), *Visual Perception: The Neurophysiological Foundations*. San Diego, CA: Academic Press.

Foley, J. M. and McCourt, M. E. (1985). Visual grating induction. *J. Opt. Soc. Am. A*, 2: 1220–1230.

Geisler, W. S. and Albrecht, D. G. (1995). Bayesian analysis of identification performance in monkey visual cortex: Nonlinear mechanisms and stimulus certainty. *Vis. Res.*, 35: 2723–2730.

Georgeson, M. A. and Sullivan, G. D. (1975). Contrast constancy: Deblurring in human vision by spatial frequency channels. *J. Physiol. (Lond.)*, 252: 627–656.

Gilbert, C. D., Das, A., Ito, M., Kapadia, M. and Westheimer, G. (1996). Spatial integration and cortical dynamics. *Proc. Nat. Acad. Sci. USA*, 93: 615–622.

Gilchrist, A. L. (1988). Lightness contrast and failures of constancy: A common explanation. *Percept. Psychophys.*, 43: 415–424.

Gilchrist, A. L. , Kossyfidis, C., Bonato, F., Agostini, T., Cataliotti, J., Li, X., Spehar, B., Annan, V. and Economou, E. (1999). An anchoring theory of lightness perception. *Psych. Rev.*, 106: 795–834.

Grossberg, S. and Mingolla, E. (1987). Neural dynamics of surface perception: Boundary webs, illuminants, and shape-from-shading. *Comp. Vis. Graph. and Image Proc.*, 37: 116–165.

Grossberg, S. and Todorovic, D. (1988). Neural dynamics of 1-D and 2-D brightness perception: A unified model of classical and recent phenomena. *Percept. Psychophys.*, 43: 241–277.

Heinemann, E. G. (1955). Simultaneous brightness induction as a function of inducing and test-field luminances. *J. Exp. Psych.*, 50: 89–96.

Heinemann, E. G. (1972). Simultaneous brightness induction. In D. Jameson and L. M. Hurvich (eds), *Handbook of Sensory Physiology, VII-4 Visual Psychophysics*. Springer-Verlag: Berlin.

Helson, H. (1963). Studies of anomalous contrast and assimilation. *J. Opt. Soc. Am.*, 53: 179–184.

Jameson, D. and Hurvich, L. M. (1989). Essay concerning color constancy. *Ann. Rev. Psych.*, 40: 1–22.

Jenkins, J. G. (1930). Perceptual determinants in plain designs. *J. Exp. Psych.*, 13: 24–46.

Kingdom, F. A. A. (1997). Simultaneous contrast: The legacies of Hering and Helmholtz. *Percept.*, 26: 673–677.

Kingdom, F. A. A., Blakeslee, B. and McCourt, M. E. (1997). Brightness with and without perceived transparency: When does it make a difference? *Percept.*, 26: 493–506.

Kingdom, F. A. A. and Moulden, B. (1988). Border effects on brightness: A review of findings, models and issues. *Spat. Vis.*, 3: 225–262.

Lamme, V. A. F. and Roelfsema, P. R. (2000). The distinct modes of vision offered by feedforward and recurrent processing. *Trends in Neurosci.*, 23: 571–579.

Land, E. H. and McCann, J. J. (1971). The retinex theory of vision. *J. Opt. Soc. Am.*, 61: 1–11.

Lennie, P. and D'Zmura, M. (1988). Mechanisms of color vision. *CRC Crit. Rev. Neurobiol.*, 3: 333–400.

McCourt, M. E. (1982). A spatial frequency dependent grating-induction effect. *Vis. Res.*, 22: 119–134.

McCourt, M. E. (1994). Grating induction: A new explanation for stationary visual phantoms. *Vis. Res.*, 34: 1609–1618.

Mikesell, W. H. and Bentley, M. (1930). Configuration and brightness. *J. Exp. Psych.*, 13: 1–23.

Moulden, B. and Kingdom, F. A. A. (1989). White's effect: A dual mechanism. *Vis. Res.*, 29: 1245–1259.

Moulden, B. and Kingdom, F. A. A. (1991). The local border mechanism in grating induction. *Vis. Res.*, 31: 1999–2008.

Paradiso, M. A. and Hahn, S. (1996). Filling-in percepts produced by luminance modulation. *Vis. Res.*, 36: 2657–2663.

Paradiso, M. A. and Nakayama, K. (1991). Brightness perception and filling-in. *Vis. Res.*, 31: 1221–1236.

Pessoa, L., Mingolla, E. and Neumann, H. (1995). A contrast- and luminance-driven multiscale network model of brightness perception. *Vis. Res.*, 35: 2201–2223.

Pokorny, J., Shevell, S. K. and Smith, V. C. (1991). Colour appearance and colour constancy. In P. Gouras (ed), *Vision and Visual Dysfunction, Vol. 6, The Perception of Color*, pp. 43–61, Boca Raton, FL: CRC Press.

Reid, C. R. and Shapley, R. (1988). Brightness induction by local contrast and the spatial dependence of assimilation. *Vis. Res.*, 28: 115–132.

Rossi, A. F. and Paradiso, M. A. (1996). Temporal limits of brightness induction and mechanisms of brightness perception. *Vis. Res.*, 36: 1391–1398.

Rossi, A. F. and Paradiso, M. A. (1999). Neural correlates of perceived brightness in the retina, lateral geniculate nucleus, and striate cortex. *J. Neurosci.*, 19: 6145–6156.

Ross, W. D. and Pessoa, L. (2000). Lightness from contrast: A selective integration model. *Percept. Psychophys.*, 62: 1160–1181.

Rossi, A. F., Rittenhouse, C. D. and Paradiso, M. A. (1996). The representation of brightness in primary visual cortex. *Science*, 273: 1104–1107.

Shapley, R. and Enroth-Cugell, C. (1984). Visual adaptation and retinal gain-controls. *Prog. Retinal Res.*, 3: 263–346.

Shapley, R. and Reid, C. R. (1985). Contrast and assimilation in the perception of brightness. *Proc. Nat. Acad. Sci. USA*, 82: 5983–5986.

Smith, V. C., Jin, P. Q. and Pokorny, J. (2001). The role of spatial frequency in color induction. *Vis. Res.*, 41: 1007–1021.

Spehar, B., Gilchrist, A. L. and Arend, L. E. (1995). The critical role of relative luminance relations in White's effect and grating induction. *Vis. Res.*, 35: 2603–2614.

Taya, R., Ehrenstein, W. and Cavonius, C. (1995). Varying the strength of the Munker–White effect by stereoscopic viewing. *Percept.* 24: 685–694.

Todorovic, D. (1997). Lightness and junctions, *Percept.* 26: 379–395.

Wertheimer, M. (1923). Untersuchungen zur Lehre von der Gestalt, II. *Psychologische Forschung*, 4: 301–350.

Wertheimer, M. (1958). Principles of perceptual organization. In D. C. Beardslee and M. Wertheimer (eds.), *Readings in Perception*, Princeton, NJ: D. Van Nostrand Company, Inc., 115–135.

White, M. (1979). A new effect of pattern on perceived lightness. *Percept.*, 8: 413–416.

White, M. (1981). The effect of the nature of the surround on the perceived lightness of grey bars within square-wave test gratings. *Percept.*, 10: 215–230.

White, M. and White, T. (1985) Counterphase lightness induction. *Vis. Res.*, 25: 1331–1335.

Wyszecki, G. (1986). Color appearance. In K. R. Boff, L. Kaufman and J. B. Thomas (Eds), *Handbook of Perception and Human Performance, Vol 1, Sensory Processes and Perception*, New York: John Wiley and Sons.

Wyszecki, G. and Stiles, W. S. (1982). *Color Science: Concepts and Methods, Quantitative Data and Formulae*, pp. 487. New York: John Wiley and Sons.

Yund, E. W. and Armington, J. C. (1975). Color and brightness contrast effects as a function of spatial variables. *Vis. Res.*, 15: 917–929.

Yund, E. W., Snodderly, D. M., Hepler, N. K. and DeValois, R. L. (1977). Brightness contrast effects in monkey lateral geniculate nucleus. *Sensory Processes*, 1: 260–271.

Zaidi, Q. (1989). Local and distal factors in visual grating induction. *Vis. Res.*, 29: 691–697.

Zaidi, Q., Spehar, B. and Shy, M. (1997). Induced effects of background and foregrounds in three- dimensional configurations: The role of T-junctions. *Percept.*, 26: 395–408.

Part II

Levels of Perception

5

Levels of Motion Perception

Stuart Anstis

I shall present some new, or newish, illusions to show that motion signals in the early parts of the visual system are profoundly altered by stimulus luminance and contrast. I shall show that contrast affects (1) motion strength in time till breakdown; (2) motion strength in crossover motion; (3) speed in the footsteps illusion; (4) direction in the plaid-motion illusion; an (5) direction: split dots.

I shall then consider how it is that higher perceptual processes massage these neural motion signals into the perception of moving objects. For instance, moving line terminators help to solve the aperture problem. But these solutions are modified by stimulus contrast in the plaid-motion illusion and in the peripheral-oblique illusion. In the chopstick and sliding rings illusion, the motion of terminators propagates along straight lines and is blindly (and incorrectly) assigned to the motion of the central intersection.

Finally, a new display of moving dots alternates perceptually between two radically different perceptual interpretations. Usually the local percept (trees) is seen first, but the global interpretation (forest) gradually takes over in the course of time.

5.1 Introduction

In this chapter I shall review my own work, with my various co-workers, on levels of motion perception. I shall start with motion signals in the early parts of the visual system and show (with the aid of some newish illusions) how these are greatly affected by stimulus luminance and contrast. I shall then consider how it is that higher perceptual or cognitive processes manage to interpret these neural motion signals in order to achieve the real purpose, namely the perception of moving objects.

5.2 Illusory Rotation of a Spoked Wheel

Brian Rogers and I have studied apparent movement that occurs in a disk that is divided into sixteen stationary sectors of different grey levels. All the greys are stepped synchronously clockwise around the sectors, driven by color cycling

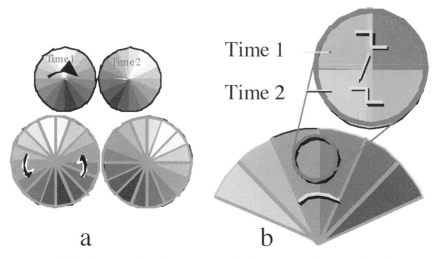

FIGURE 5.1. (a) A sectored disk jumps around clockwise (upper diagram). Superimposed stationary, unchanging grey spokes (lower diagram) appear to rotate counterclockwise. (b) Magnified view of a vertical spoke. A light/dark border at the right-hand edge of a spoke at time 1 jumps to the left-hand edge of the spoke at time 2. See text.

in the computer palette. The edges of the sectors never move; only the colors change. As a result, the sectors show continuous apparent clockwise motion. Figure 5.1 shows that the sectors rotate 45 deg clockwise between successive time frames. The interesting feature is the behaviour of the thin, stationary grey radial lines that lie along the edges of the sectors, looking rather like bicycle spokes. As the sectors appeared to rotate clockwise during the colour cycling, these radial lines appeared to rotate vigorously counterclockwise. When the color cycling was stopped after 20 seconds, all observers reported a strong clockwise motion aftereffect. This clockwise aftereffect cannot come from the sectors that had been jumping clockwise, but must come from the bicycle spokes that had been apparently moving counterclockwise. The presence of a motion aftereffect makes it likely that the "illusory" rotation of the spokes actually contains motion energy that stimulates neural motion detectors (Braddick, 1974; Adelson and Bergen, 1985). The spokes appeared to move only when they were thin (~5 min arc in foveal vision), and when their luminance matched the sectors they abutted.

Various control experiments ruled out induced brightness and induced motion as explanations. Instead, Figure 5.1b shows our explanation. When the radial spoke straddles the luminance of two adjacent sectors, then as the sectors jump clockwise the right-hand edge of the spoke at time 1 jumps to the left at time 2, through the width of the spoke (only 5 min arc). Thus, as 16 sectors jump clockwise through a whole sector width, a spoke edge makes a tiny jump to the left. But this is sufficient to give the strong impression of motion, followed after inspection by a motion aftereffect from the spokes. We conclude that in motion, less is more. There must be many neural motion detectors with tiny receptive fields (or subfields: see Barlow and Levick, 1965), in the order of 5 min arc in width. Motion detectors wide enough to respond to the huge jumps made by the sectors must be far less numerous or less sensitive. In conclusion, the apparent motion that we observed in our spoked wheel illusion results from actual small displacements of luminance contours in the reversed direction. However, our phenomenon does reveal that (i) the small displacements in the reversed direction are a more powerful stimulus than the sector motion in the forward direction, (ii) the effect can be seen only if the spoke width is small (< 5 min of arc) when the stimulus is viewed foveally, and (iii) the asynchronous movements of the spokes are seen as being distributed uniformly and synchronously over the whole display. These simple observations tell us something about the spatial and temporal characteristics of the human motion system.

5.3   Contrast Affects Motion Strength

It is well known that perceived speed can depend on contrast (Thompson, 1976, 1982; Campbell and Maffei, 1979, 1981; Kooi et al., 1992; Stone and Thompson, 1992; Hawken et al., 1994; Ledgeway and Smith, 1995; Gegenfurtner and Hawken, 1996; Smith and Derrington, 1996; Thompson et al., 1996; Thompson and Stone, 1997; Blakemore and Snowden, 1999). We have recently investigated five fresh examples of this contrast dependence, namely:

1. Motion strength: Time till breakdown

2. Motion strength: Crossover motion

3. Speed: The footsteps illusion

4. Direction: The plaid-motion illusion

5. Direction: Split dots

1. In *time till breakdown*, the strength of apparent motion depends on its contrast (Figure 5.2) and can be measured by timing its durability. A spot or bar that jumps back and forth between two positions gives an impression of apparent motion (AM), but prolonged inspection of the stimulus causes the initial impression of AM to degrade to flicker (Kolers, 1964; Anstis, Giaschi and Cogan, 1985). This adaptation effect was measured for a single jumping bar (Smith and Anstis, in press) and was found to depend on bar contrast; the time till breakdown TTB (the time at which the impression of AM first degraded into flicker) was greater for a higher-contrast bar and fell off as contrast was reduced, but it was independent of the bar's luminance polarity (Figure 5.3).

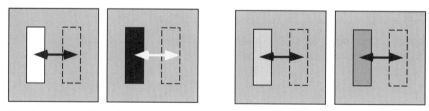

High contrast bars have longer TTBs than.....Low contrast bars.

FIGURE 5.2. Contrast bars

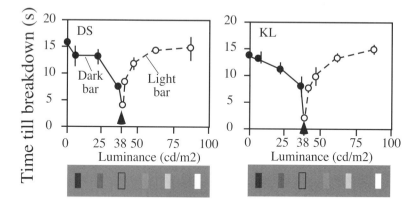

FIGURE 5.3. Graph shows that time till breakdown (TTB) increases with contrast, for bars either lighter or darker than the surround. Polarity does not matter.

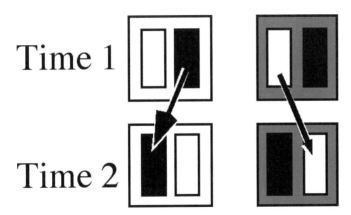

FIGURE 5.4. Crossover motion. A black and a white bar exchange luminances. On a light surround, the dark bar appears to jump. On a dark surround the white bar appears to jump. Thus, the bar with higher contrast has the stronger motion signal.

2. Another motion phenomenon that depends upon contrast is "crossover motion" (Anstis and Mather, 1985; Mather and Anstis, 1995; Anstis, Smith and Mather, 2000; Smith and Anstis, in press). Two parallel bars side by side, one dark and one light, switch luminances repetitively over time. This generates a stimulus that is consistent with two potential competing bar motions; one dark and the other light.

Whether observers see the light bar or the dark bar as moving depends critically on the luminances of the bars and their surround. On a dark surround, the light bar is seen as moving. On a light surround, the dark bar is seen as moving. The bar differing more from the surround luminance dominates the motion percept (Anstis and Mather, 1985).

1+2. *Time till breakdown for crossover motion.* After measuring the time till breakdown (TTB) for a single jumping bar, we measured the TTB for a crossover motion stimulus in which two bars, of different contrasts, jumped in opposite directions (Smith and Antis, in press). A single jumping bar had a short TTB at low contrasts, and TTB grew longer as contrast increased. A single black bar had a long TTB (about 15 s at an alternation rate of 3.75 Hz). When a grey bar was added in opposite motion, creating a crossover stimulus, the TTB of the two combined bars was slightly shortened by a low-contrast opposing bar and was considerably shortened by a high-contrast opposing bar (Figure 5.5). Thus we measured the penalty (in terms of "strength" of apparent motion percept) that the winning motion signal paid in overcoming the simultaneous presence of the other motion signal. In short, TTB was a monotonic function of the *difference in contrast* between the two bars (Smith and Anstis, in press). For one bar's motion to dominate the percept, the visual system must "discount" the motion of the competing bar. This discounting was not like a winner-take-all mecha-

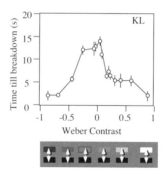

FIGURE 5.5. Competing opposite motions: Time till breakdown (TTB) for a black bar decreases when high-contrast crossover motion is added in the opposite direction. (Courtesy of David Smith.)

nism, in which the losing signal has no effect upon the dominating percept, but instead showed some inhibition from the losing signal. Winner-take-all would be like a horse race, in which the losing horses do not slow down the winner in any way, whereas the mutual inhibition that we discovered for TTB is like a tug of war in which the losers certainly impede the progress of the winners. Note that this motion inhibition pooled across both spatial decrements and increments.

3. In the "footsteps" illusion (Anstis, 2001) two grey squares, one light and one dark, move horizontally against a background of black and white vertical stripes (Figure 5.6). When the dark grey square moves across a white vertical stripe its edges have high contrast and it appears to speed up. When it moves across a black vertical stripe its edges have low contrast and it appears to slow down. The opposite is true for the light grey square. As a result the two squares appear to speed up and slow down in alternation, like the footsteps of a walking person. These changes in apparent speed are maximal for almost-white and almost-black squares, and they fall to zero for a mid-grey square.

This mid-grey lay at the arithmetic mean, not the geometric mean, of the black and white stripes. Thus if the black and white stripes had relative luminances of 1% and 100%, the footsteps illusion went away for a grey bar of luminance 50.5%, not 10%. At first this seems inconsistent with the well-known fact that the visual system applies a log transform to luminance stimuli. But we noted that a 50.5% mid-grey square had the same Weber contrast on the black surround as on a white surround, and we concluded that the apparent speed depended upon Weber contrast. (The mid-grey square had a Weber contrast of $(50.5 - 1)/50.5 = +0.98$ on a black surround, and $(50.5 - 100)/50.5 = -0.98$ on a white surround).

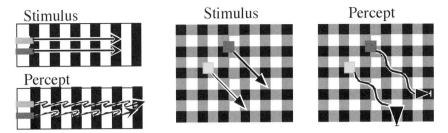

FIGURE 5.6. Artist's impression of the footsteps and plaid illusions.(2-D speed changes on plaid can change perceived directions.) See text.

4. *The plaid-motion illusion.* This was simply a 2-D version of the footsteps illusion. Two gratings were crossed to form a plaid that tiled the background with black, white, and grey squares. The two moving squares, one light grey and the other dark grey, were the same size as the plaid tiles. They moved in synchrony along parallel oblique paths at 45 deg to the orientation of the plaid (Figure 5.6b), but subjectively they appeared to wiggle in and out toward each other, changing their directions repetitively as they pursued their common oblique path (Figure 5.6c). To understand why, consider a light square moving down to the right, at the instant when it crossed over a "corner" of the plaid. When its leading right-hand edge moved on to black, the rightward motion of the square appeared to speed, because of the high contrast of the square's leading edge. At the same instant its leading edge at the bottom moved on to white, so the downward motion of the square appeared to slow down because of the low contrast of the square's bottom leading edge. Consequently the square seemed to veer toward the horizontal. When the light grey square moved across the next "corner" of the plaid, which had opposite luminance polarities, it seemed to veer toward the vertical. Corresponding arguments apply to the light square. As a result the two squares appeared to move along counterphasing wiggly paths.

5. *Split moving dots* In crossover motion, bars of different contrast moved in opposite directions. Our split-dot effects will now show that two combined orthogonal motions of different contrasts appear to move in a direction that is a function of the relative contrasts. Our time till breakdown (TTB) experiments established that two opposed motions (at $180°$ to each other) mutually inhibited each other in a kind of tug of war. Now imagine a peculiar tug of war in which two teams of unequal strength are pulling at right angles, with the stronger team pulling north while the weaker team pulls west. The rope would move at some intermediate angle such as north-northwest. Hiro Ito and I have found that when two dots of different contrasts cross each other at right angles, they can combine into a single vector, or perceived direction of motion. This vector gives a sensitive measure of the relative motion strengths of the two dots.

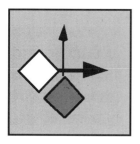

FIGURE 5.7. (left panel) Basic stimulus was two dots that crossed. (center panel) Vector summation of two dots of same polarity gave oblique motion (grey arrow), favoring dot of higher contrast. (right panel) Vector summation of two dots of opposite polarity gave perceived direction (dark grey arrow) outside the range of the two stimulus motions.

The dots in each pair were adjacent and touching, because Qian, Andersen and Adelson (1994a, b, c) showed that when fields of random dots drift over each other they separate out into sheets moving in different directions, so-called "transparent" motion. It is only if dots are arranged in touching pairs that they fuse together to give the "coherent" motion that we want. Braddick (1997) and Curran and Braddick (2000) have found that the visual system extracts the vector sum of these fused coherent dots in assigning a perceived direction, and we entirely agree. Unlike these authors, we were primarily interested in the effects of contrast upon the direction of perceived movement.

The basic stimulus was a pair of touching dots that moved along crossing, orthogonal paths (Figure 5.7). One dot jumped back and forth horizontally, whilst at the same instant the other dot jumped back and forth vertically. The dots had luminance values ranging from 4% (black) to 100% (white) on a 50% (mid-grey) surround. This gave to the dots Weber contrasts ranging from -0.6 (spatial decrements) to +0.6 (spatial increments). Luminances of the horizontally moving dots are shown on the abscissa, and of the vertically moving dots on the ordinate.

Ito and I set up a 7 * 7 array of dot pairs, and observers were invited to set a line to match the perceived direction of each of the 49 dot pairs. Results are shown in Figure 5.8b. Broadly speaking, the perceived directions radiate out from the center of the graph. We treated the orientation of each motion arrow drawn by the observer as the vector sum of a horizontal and a vertical motion, and the length of each of these vectors is taken as an index of its relative motion strength. In summary, we found that perceived motion strength varied linearly with contrast, for dots of the same polarity. The same linear law was true if the dots had opposite polarity, with one dot being lighter than the grey surround and the other darker; but now the contrast of the decremental dot had to be taken as a negative number.

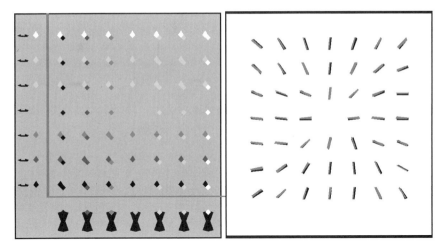

FIGURE 5.8. (left panel) Split-dot display comprises pairs of dots that jump back and forth, as in Figure 5.7. Luminance of horizontally jumping dots is shown on x-axis, and of vertically jumping dots on the y-axis. (right panel) Perceived directions of motion show vector summation and differencing of dots with different contrasts.

To summarize, our experiments show that contrast affects motion strength and apparent speed. Contrast increases motion strength, or resistance to breakdown into flicker, as measured by time till breakdown (TTB). In crossover motion there is a contest between two potential motions in opposite directions (at 180 deg to each other), and the higher contrast wins. However, the net strength of the winning motion is reduced because the losing motion of the other bar is subtracted from it — so the situation resembles a tug of war more than a horse race. We demonstrated this reduction in net strength in our time till breakdown experiments. In our split-dot motion experiments we found that the visual system takes the vector sum of two orthogonal motions, as weighted by their relative contrasts. In the footsteps illusions, contrast affects perceived speed of a drifting square. When a background plaid generates two orthogonal footsteps illusions, alternating over time, the relative motion strengths are expressed as changes in direction of the drifting square. Finally in the peripheral-oblique illusion, the center and terminators of a line are in competition, and the relative visibility of the center and the ends, as determined by the line's contrast, determines the perceived net direction.

5.3.1   Contrast and Motion: Conclusions

- High-contrast motion looks faster, and is more durable, than low-contrast motion.

- Speed changes are local in space and time (in the footsteps illusion)

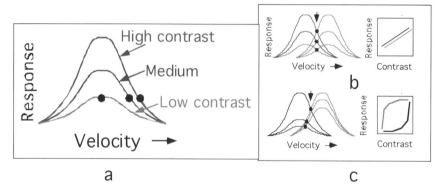

FIGURE 5.9. (a) A hypothetical cell in MT, tuned to velocity but sensitive to contrast, would confound three stimuli (spots) of different velocities and contrast. (b) Adding a second unit tuned to higher velocities would disambiguate velocity from contrast. Three stimuli (spots) of the same velocity but of different contrasts, would evoke the same firing ratio (here 1:1) from the two units. (c) However, nonlinear responses to contrast in the two units could allow contrast to distort the velocity signals again.

- 2-D speed changes (across a plaid) can change perceived directions

- Split dots, whether of the same or opposite polarities, give motion signals proportional to their contrast.

- These dot motions are combined by vector summation

What kind of visual codes for motion will be susceptible to distortion by stimulus contrast? One simple candidate code would *totally* confound contrast and velocity. Consider a visual neuron in MT that is tuned to a preferred range of velocities, as described by Maunsell and van Essen (1983). The response of such cells also depends upon stimulus contrast (Sclar and Freeman, 1982), so the *hypothetical* responses of such a cell to velocity and contrast are diagrammed in Figure 5.9. The cell would give a maximum response to a high-contrast stimulus moving at its preferred velocity, and would be less responsive to higher or lower velocities. Its response would also fall when the stimulus contrast was reduced. Figure 5.9a shows three hypothetical stimuli of different velocities and contrasts. The left-hand spot shows a low-contrast stimulus moving at the unit's rather slow preferred velocity. The right-hand spot shows a stimulus of high contrast but moving faster than the unit prefers. The middle spot is in between. All three spots have the same vertical height; in other words, they all elicit the same firing rate from the unit. By the principle of univariance (Estevez and Spekreijse, 1982) the firing rate of a neuron cannot distinguish between a low-contrast stimulus at the preferred velocity from a high-contrast stimulus at a less-preferred velocity.

This problem of confounding velocity with contrast can be solved by adding additional tuned units (Figure 5.9b, c). In this figure the three spots show three stimuli of the same velocity but of different contrasts. A single tuned unit responds differently to all three. However, the addition of a second unit tuned to a higher preferred velocity disambiguates the stimulus. All three stimuli provoke the same ratio of firing between the two units, and this is the velocity code. A higher contrast makes both units fire more rapidly, but their ratio of firing is preserved (Figure 5.9b). This is similar to the coding of hue by three broadly tuned retinal cones, and is one of the class of models known as banks of tuned filters (Regan, 2000).

All is well, provided that everything is linear. But suppose that the gamma, or contrast response, of the two hypothetical units is different. The result will be that velocity and contrast are again confounded (Figure 5.9c), although not on the wholesale scale committed by a single tuned unit. I conjecture that this type of nonlinearity is responsible for the illusory changes in apparent speed produced by changes in contrast. Note that this dependence of motion on contrast as a motion analog of the Bezold-Brucke hue shift. When spectral light increases in luminance, the hues change. Normally, long-wavelength light becomes increasingly yellow, and short-wavelength light turns blue or blue-green. This is caused mainly by nonlinear responses of colour-opponent P cells in the retina (Ejima and Takahashi, 1984) and in the lateral geniculate nucleus (Valberg, Lange-Malecki and Seim, 1991). Note that in the Bezold-Brucke phenomenon, $x =$ luminance, y = hue, whilst in our effects x = contrast, y = strength of motion signal.

So far we have shown that low-level motion signals increase as the stimulus contrast increases. Of course, this gives rise to perceptual errors. Its advantage may be that increases in contrast increase the salience, and hence the perceived reliability, of motion signals at the expense of an accurate representation of absolute speed (Clifford and Wenderoth, 1999).

5.4   From Low-Level to High-Level

So far we have shown that low-level neural motion signals are considerably distorted (increased) as the stimulus contrast increases. We turn now to high-level processes whose job is integrate these signals into the perception of moving objects. For instance, each side of a moving polygon generates local motion signals that are usually different from the direction in which the whole polygon moves — the so-called "aperture problem." We are perversely interested in partial and complete failures in high-level solutions of the aperture problem. Moving line terminators help to solve the aperture problem, but these solutions are modified by stimulus contrast in the plaid-motion illusion and in the peripheral-oblique illusion. Thus, perceptual combinations of contrast-distorted motion signals lead to distorted trajectories for moving objects — a minor failing. We also recapitulate our "chopstick" illusion (Anstis, 1990) and "sliding rings" illusion, in which the motion of terminators propagates along straight lines and is blindly (and incorrectly) assigned to the motion of the central intersection. Both illusions show grossly erroneous integration of line motions and terminators — an almost complete failure.

5.5   Terminators and the Aperture Problem

The motion of a long straight line is ambiguous if its ends are hidden — for example when an observer views it through a round aperture — or when it passes across the round receptive field of a motion-sensitive neuron. The line is invariant under translation along its own length, so its motion is ambiguous, and oblique motion cannot be discriminated from orthogonal (Adelson and Movshon, 1982; Movshon et al., 1983; Hildreth, 1983; Wilson, 1994; Shiffrar and Pavel, 1991; Lorenceau and Shiffrar, 1992; Duncan, Albright and Stoner, 2000). This is the aperture problem. A vertical line that moves to the right gives the same retinal stimulus whether it be moving to the right, or up-to-the-right, or down-to-the-right. (The difficulty arises not from the aperture itself but from the fact that the ends of the lines are not seen, so it should really be called the "endless problem"). This raises the problem of how we can correctly see a moving square. Adelson and Movshon (1982) suggested that a vertical line is constrained to move along any of the set of arrows shown in Figure 5.10a in order to reach its right-hand position. If this is the right-hand side of a square that is moving obliquely down to the right, the bottom edge of the same square undergoes a similar set of constraints that make it move vertically down, or obliquely down (Figure 5.10b). The point at which these two constraint lines intersects (Figure 5.10b) defines the true direction of motion, namely at 45 deg down and to the right.

5.6   Contrast Affects the Aperture Problem: The Plaid-Motion Illusion and Intersections of Constraints

A plaid surround can induce 2-D illusions that change the apparent direction, not just the speed, of moving squares (Anstis, 2001). A plaid was made by superimposing two orthogonal square wave gratings, and a light grey square and a dark grey square drifted obliquely across the plaid. Result: Although the squares followed parallel paths they appeared to vary in direction, seemingly moving in and out toward and away from each other. Consider a light grey square lying on a plaid and jumping back and forth obliquely at 45 deg to the vertical (Figure 5.10). The square has black on either side of it, which enhances its horizontal motion, and it has white above and below it, which de-emphasise its vertical motion. Consequently the light grey square appears to veer toward the horizontal. For corresponding reasons a dark grey square (not shown) appears to veer toward the horizontal.

Contrast acts here in a dynamic fashion, rotating the perceived trajectory of a moving object without perceptually displacing the contours of a stationary object. Contrast subjectively enhances the amplitude of the horizontal motion component and reduces that of the vertical motion (Figure 5.11c), shifting the intersection-of-constraints solution toward the horizontal. It follows that contrast modifies the

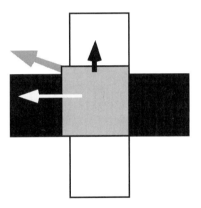 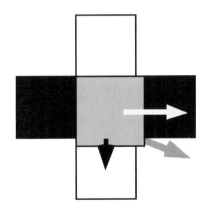

FIGURE 5.10. Explanation of the plaid illusions illustrated in Figure 5.6. When a light grey square makes small back and forth jumps on a black/white plaid, its left and right edges have high contrast against the black surround, which enhances the horizontal component of motion. Its top and bottom edges have low contrast against the white surround, which de-emphasises the vertical component of motion. Result: the oblique motion looks somewhat horizontal.

amplitude of perceived motion before the intersection of constraints is computed (Adelson and Movshon, 1982).

5.7   Contrast Affects the Aperture Problem: The Peripheral-Oblique Illusion

I stumbled by chance on another illusory phenomenon in which stimulus contrast determines the solution to the aperture problem. Faced with the ambiguity of a straight line moving behind an aperture, the default percept is that the line moves at right angles to its own orientation. The line's motion is completely disambiguated if the terminators, or ends of the line, are visible. Then the default solution to the aperture problem is rejected and the motion of the terminators propagates along the whole line, which is seen correctly as moving in the same direction as its terminators. I have found that an aperture problem can arise even without an aperture! Figure 5.12 (top icon on left) depicts a white or grey line, tilted at 45 deg from vertical and moving vertically up and down through a distance of 6 deg at a rate of 1 Hz. The line is 6 deg in length and is viewed with both eyes against a black background at an eccentricity of 15 deg, with strict fixation. Usually the line is correctly seen as moving vertically. However, if the line is made really dim its trajectory appears to veer round toward the oblique, and by the time it is just above threshold it appears to move at 45 deg, at right angles to its own orientation.

At first I thought this was some kind of dark-adaptation effect, perhaps related to Pulfrich's pendulum. But this idea was quickly proved wrong when the line was

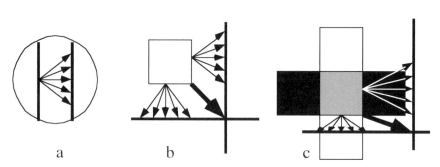

FIGURE 5.11. (a) In the aperture problem the motion of a long straight line seen through a round aperture is ambiguous and could be in the direction of any of the arrows. (b) So how is the moving square seen unambiguously? Adelson and Movshon's (1982) intersection of constraints solution. The thick vertical and horizontal lines form the "envelopes" of possible motions of the right-hand and bottom edges of the square. Their intersection point (bottom right) yields the perceived direction of the square. (c) When the motions of the sides of a square (taken from Figure 5.10) are distorted by contrast, the square's trajectory is distorted. Conclusion: Local neural signals from moving edges undergo contrast distortion before being integrated by intersecting constraints.

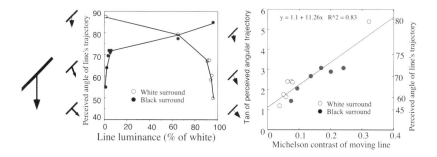

FIGURE 5.12. Perceived direction of tilted line moving in the periphery is veridical (vertical) at high contrast, but driven by motion of line center (obliquely) at low contrast.

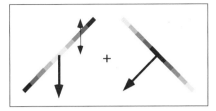 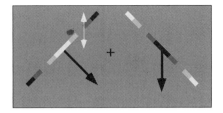

FIGURE 5.13. Spatially graded lines appear to move veridically (vertically) if tips have high contrast, but driven by center (obliquely) if tips have low contrast.

put on a white background. Now a black or dark grey line was seen veridically, and it was an almost-white line that appeared to veer toward 45 deg (Figure 5.12). It was the contrast of the line, not its luminance, that determined its perceived direction of motion. I replotted the lines as a function of their Michelson contrast,

$$abs[(L_{line} - L_{surround})/(L_{line} + L_{surround})]$$

an expression whose value lies between 0 and 1. This made it clear that regardless of polarity a high-contrast line was seen veridically, whilst a low- contrast line was seen moving at right angles to its own length — almost as though it were being viewed through a nonexistent aperture.

I believe that at low contrast and in peripheral vision the terminators start to lose visibility, and with it their ability to influence the perceived direction of motion. Thus, the visual system's ability to solve the aperture problem depends upon the terminators reaching some criterion level of contrast; otherwise they are ignored. To verify this hypothesis I emphasised or de-emphasised the terminators in two moving, spatially graded lines (Figure 5.13). The left-hand line was black at both ends, shading to white at its center. The right-hand line was white at both ends, shading to black at its center. The lines were tilted at +45 deg and −45 deg, and both lines moved vertically up and down in step, on either side of a fixation point. Result: On a white surround, the trajectory of the black-tipped line was seen veridically as vertical, but the trajectory of the white-tipped line showed an illusory inclination (thick arrows in Figure 5.12). On a black surround the opposite was the case. Thus, on both surrounds and regardless of polarity, high-contrast terminators successfully disambiguated the motion, but low-contrast terminators did not.

What difference does eccentricity make? It seems to reduce the visibility of the whole line, in such a way that a low-contrast terminator is not seen clearly in peripheral vision, so it loses its influence on the perceived motion of its line. Why should the terminator be less visible than the rest of the line? Perhaps it is under-sampled, stimulating only one receptive field whilst the central portion of the line has a chance to stimulate a whole row of receptive fields.

Conversely, there is also something special about foveal viewing. In Figure 5.14 an oblique line moves vertically up and down past three stationary dots j, k, l. If one fixates j, positioned 1 deg to the left of the line, then the line is seen veridically as moving up and down. But when k is fixated, the central part of the line takes on

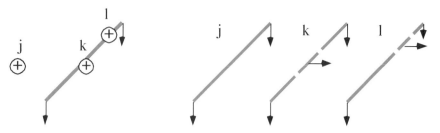

FIGURE 5.14. Oblique line moving downward. The portion near the fixation point appears to move locally to the right.

a life of its own and seems to move horizontally as it passed through *k*. This effect is tied to the foveal location and is not merely a landmark effect, because when point *l* is fixated the portion of the line close to the new fixation point *l* appears to move horizontally.

5.8   Sliding Rods and Rings: The Chopstick Illusion

A long line that moves behind a circular aperture is invariant under translation along its own length, so its motion is ambiguous, and oblique motion cannot be discriminated from orthogonal. This is the aperture problem. Similarly, a circle is invariant under rotation. These invariances can produce strong illusions in the *sliding movements of intersections*. Steve Shimozaki and Dana Ballard from the University of Rochester and I have studied two motion illusions of this kind that reveal links between human perceptual representations and the motor system. In the "chopsticks illusion" (Anstis, 1990) a vertical and a horizontal line overlapped to form a cross, and each line moved along a separate counterclockwise circular path in antiphase, without changing orientation. The intersection of the lines moved clockwise, but it was wrongly perceived as rotating counterclockwise. In the "sliding rings illusion" two rings overlapped in a figure-8 and rotated about the centre of the figure-8. When two dots were added that rotated with the rings, observers reported seeing the two rings as welded together into a rigid 8. Observers could readily track the intersections of the rings. When each dot "floated" so that it lay at 12 o'clock on its ring, observers saw the figure as breaking into two separate rings that slid over each other, and the eyes were unable to track the moving intersections. We conclude that pursuit eye movements are under top-down control and are compelled to rely upon perceptual interpretation of objects.

Not all motions are visible; in particular, the sliding movements of intersections. The oblique motion of an arm sweeping past a horizontal table edge is clearly seen, but the horizontal motion of the intersection of the arm and table edge is never noticed. Yet the same retinal signal of two intersecting edges in some other context could easily give a sensation of motion. We suggest that observers parse intersections as being non-objects and therefore cannot see them

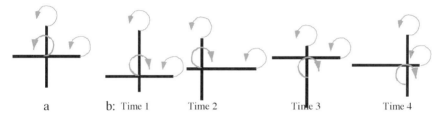

a   b: Time 1   Time 2   Time 3   Time 4

FIGURE 5.15. (a) A rigid cross that follows a counterclockwise circular path is seen veridically. (b) In the chopstick illusion, both rods follow similar counterclockwise paths but with a phase lag between them. Result: The central sliding intersection actually moves clockwise but appears to move counterclockwise. Conclusion: The motion of the line terminators is blindly assigned to the intersection.

move.

We studied intersections as follows. In Figure 5.15a, a cross moves clockwise along a circular "polishing" path, remaining upright like a sponge in the hand of a window cleaner. This control stimulus is always seen veridically. In Figure 5.15b two intersecting rods, one vertical and one horizontal, move clockwise along circular paths, forming a cross with the ends of the rods always visible (Anstis, 1990). The rods move in antiphase so that when the vertical rod is at 12 o'clock on its path the horizontal rod is at 6 o'clock. The central intersection of the two rods actually moves along a counterclockwise path (a Lissajou circle). The motions of these intersections were grossly misperceived. 230 undergraduates viewed videotapes of Figures 5.15a and 5.15b, rotating for 5 s at 2.2 rev/s. In both cases the center moved along exactly the same circular path; only the lengths of the arms changed over time. The control cross in Figure 5.14a was correctly seen as rotating clockwise by 99.6% of the students. Yet 86.8% of students incorrectly reported the center of Figure 5.15b as rotating clockwise, even though it actually moved counterclockwise. Thus they appeared virtually blind to the true motion of the center, and instead wrongly perceived it as moving along the same path as the tips of the rods. The visual system did not parse the sliding intersection as an object, and so refused to perceive its motion directly. Instead, it inferred the intersection's path through space by monitoring the unambiguous clockwise rotation of the terminators (tips) of the rods. This tip motion propagated along the entire length of the rods and was blindly assigned to their intersection. We conclude that intersections were not parsed as objects, and therefore their motion path was not extracted, but instead the motion of the terminators (tips) propagated along the lines and was blindly assigned to the intersection.

Surprisingly, the chopstick illusion still persists if the two rotating rods have their ends clipped by a stationary square window or hole cut in a large, invisible screen. This means that the ends of the vertical rod move back and forth horizontally, and the ends of the horizontal rod move back and forth vertically. Because of the phase lag between the two lines, the center is still rotating counterclockwise — yet it still appears to be rotating clockwise! I am not yet sure why. However, when the aperture is made visible in Figure 5.16b, like a square hole cut in a tex-

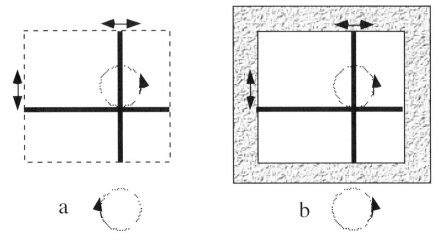

FIGURE 5.16. **a**, Chopstick illusion is still seen even when the tips of the lines travel along straight lines. **b**, Chopstick illusion vanishes. Extrinsic terminators behind screen permit veridical perception.

tured card, the two rods no longer appear to slide over each other, but immediately look like the rigid cross of Figure 5.16a moving coherently counterclockwise. The ends of the rods were are perceived as extrinsic, that is, as occluded by the aperture and extending behind it (Shimojo, Silverman and Nakayama, 1989; Duncan, Albright and Stoner, 2000), and do not influence the perceived motion of the central intersection.

We recorded the eye movements of a naive observer when he attempted track the intersection of the two rods in Figure 5.15a and 5.15b with his eyes. The tracking errors during 20 stimulus revolutions, expressed as the mean deviation or offset between eye and target, were 1.06 deg of visual angle for the rigid rotating cross of Figure 5.15a and 5.6 deg for the chopstick illusion of Figure 5.15b, so the eye tracking errors were four to five times higher for the sliding than the rigid rods.

We then removed the terminators by bending each rod around into a smooth, featureless circular ring. The two rings overlapped to form a figure-8 and rotated at 1.25 rev/s about the center of the 8 (Figure 5.16c). Since each ring was invariant under rotation the figure-8 display was potentially ambiguous, being equally consistent with the two rings sliding over each other or else being welded into a single figure-8.

Four small marks were now placed on each ring, at 3, 6, 9, and 12 o'clock. This provided small cues that radically altered the perceived rotation.

1. Rings marked as in Figure 5.17a were perceived as a rigid welded figure-8, rotating coherently. This satisfies the rigidity constraint 10.

2. Rings marked as in Figure 5.17b were perceived as two separate rings, each remaining upright and sliding over its companion. Thus, perceptual rigidity

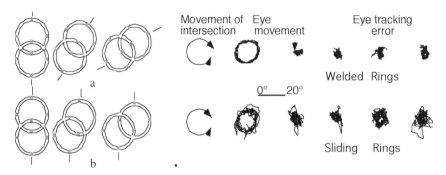

FIGURE 5.17. Two rotating rings overlapped in a figure-8. Short lines indicate the perceived rotation. (a) Rings appeared to rotate rigidly like a welded figure-8. Moving intersections could be tracked easily and accurately. (b) Rings appeared to remain upright and slide over each other. Moving intersections were tracked poorly and inaccurately. (c) Without the four marks on each ring, the rings appeared to slide as in part b, showing that motion perception aimed at minimising local motions, not at conserving rigidity.

was sacrificed in favour of minimising the motion seen within each ring.

Unmarked, featureless rings (not shown) often appear to slide over each other, especially if the rings lay in different planes of stereo depth. This shows that the visual system preferred to minimise local motions within rings rather than to maximise global rigidity of the whole 8 (Ullman, 1979; Braunstein and Andersen, 1984). It cannot be predicted from the vague idea that the visual system prefers "simplicity" or a "good Gestalt" We found that perceptual organisation of the stimuli strongly affected pursuit eye movements. The observer was asked to track the intersection of the two rings in Figure 5.17a or 17b with his eyes. The tracking errors for four observers during 20 stimulus revolutions are shown to the right of the stimuli. The rings of Figure 5.17a, which were seen as a rigid figure-8, were tracked accurately with a mean deviation error of only 1.04 deg of visual angle. However, the rings in Figure 5.17b, which were seen as two rings sliding over each other, gave a mean deviation of 9.93 deg, so the eye tracking error was 9.5 times higher for sliding than for welded rings. This breakdown in tracking performance as a result of a small change in stimulus markings has the surprising implication that smooth pursuit movements, which are normally thought of as a bottom-up servo system based upon retinal feedback (Lisberger, Morris and Tychsen, 1987; Krauzlis, 1994) may be strongly influenced by top-down cognitive processes such as object interpretation (Kowler, 1990).

Thus, although a moving stimulus is usually necessary to initiate smooth voluntary pursuit movements (Ullman,1979; Braunstein and Andersen, 1984; Lisberger et al., 1987) it is not always sufficient. A welded intersection could readily be pursued but a sliding intersection could not, even though the foveal stimulus was identical and the peripheral retinal stimulus nearly so in the two cases. The essential difference lies not in the details of the retinal stimulation but in the higher level cognitive parsing of the objects represented by this retinal image; top-down cognitive processes played a role in enabling or disabling pursuit eye movements.

5.9   Aperture Problem: Conclusions

- Terminators rule!

- They disambiguate motion of line centre (intersection of constraints) — but after contrast has altered perceived motion.

- Chopstick illusion: Motion of terminators blindly assigned to centre . . .

- . . . but not if ends of lines are hidden (extrinsic terminators).

- Intersections are not parsed as objects, and eyes can't track them.

- Dim peripheral terminators do not affect seen motion.

5.10   One Low-Level Stimulus, Two High-Level Interpretations: Local Versus Global Perception of Ambiguous Motion Displays

We have seen how the ambiguous motion of a straight line is disambiguated by combining it with unambiguous motion of the line's terminators. This leads us on to a much more general question — how are motion signals in different parts of the visual field combined? An ideal visual system would successfully combine all the motion signals that arise from a single moving object, while segregating them from signals that arise from other moving objects (Curran and Braddick, 2000). The combination could be done by some kind of global organizing principles, but a local propagation process might achieve the same result move economically by combining adjacent moving regions, providing that the motion paths are similar enough to satisfy some criteria. In 1983, Ramachandran and I examined ambiguous dot quartets, in which the dots at top and bottom corners of an imaginary diamond are flashed up, then replaced by dots at the left and right corners. This is an ambiguous stimulus, in which the top dot is equally often perceived as jumping down to the left, or down to the right. We wondered what happened when a whole field of a dozen or more of these dot quartets was visible at once. Do all the dots move in step, or does each dot quartet follow its own whim, so that about half the top does jump down to the left while the other half jump down to the right? We found a very strong tendency for all the dots to move in the same directions. This display is a dynamic analog of a set of reversible Necker cubes, where one can ask whether all the cubes reverse in step, or independently.

Alecia Dager and I have recently been studying a new multimotion display. A pair of dots rotates clockwise around a common center at 1 rev/sec. Four such

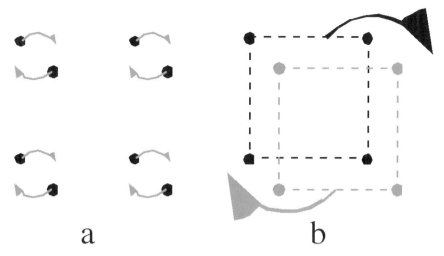

FIGURE 5.18. 8-dot display showing ambiguous binding. (a) At first it looks like four local pairs of dots, each rotating about a common center. (b) Same display looks like two large moving squares with a dot at each corner. Only the dots, not the arrows or dotted lines, were visible in the actual display.

pairs, well separated and rotating in synchrony, are arranged in a square array (Figure 5.18). At first each pair is seen rotating clockwise, but no interactions are seen between different pairs. In other words only "local" motions are perceived. But soon the display undergoes a radical perceptual reorganisation; the dots suddenly coalesce into two large, overlapping squares that slide over each other along circular paths without rotating, somewhat like a glass of water that one is rinsing out, or like the sponge in the hand of a window cleaner. We call this "global" motion. Thus, during local motion the observer sees four small pairs of rotating dots, whilst during global motion the same display looks like two large quartets of dots following a circular path. The display tended to flip back and forth over time between local and global motion, although the physical display never alters. In other words the ambiguity in this display lay not in the motions themselves, but in the perceptual groupings, or solutions that the visual system adopted to the "binding problem."

We find that motion always looks local upon first viewing, but global motion tends to increase over time, both within a single trial and across a sequence of separate trials. Observers view the display for a period of 30 s, striking keys to indicate when they see local or global motion. Ten trials are run and then averaged together, second by second. The resulting average curve shows that the probability of seeing global motion increases steadily during the 30 s observation period. We also noted that the percentage of each successive 30 s trial for which seeing global motion was seen, increased from trial to trial. This suggests two separate perceptual processes, both favouring an increase in global motion, but with different time constants.

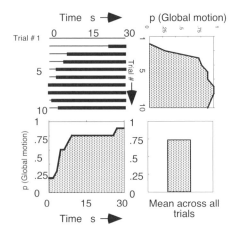

FIGURE 5.19. (top left panel) On each 30 second trial the 8-dot display appears to move locally (thin lines) or globally (thick lines). Ten trials averaged together (bottom left panel) show that probability of seeing global motion increases throughout the 30 second trial. Averaging each trial (top right panel) shows that global motion also increases from one trial to the next. Averaging across all times and trials (bottom right panel) shows that mean probability of seeing global motion was 0.75.

Shifting the display on to a fresh patch of retina restores local motion. Adding more pairs of dots increases the amount of global motion, but increasing the number of dots within a group from 2 to 3 to 4 has the opposite effect, making the display look more local. We also find that we can increase the amount of local or global motion by adding visible cues such as color that provide independent cues to grouping. Making both dots in one circling pair red and coloring another pair green, another pair blue, and so on, greatly increase the chances of seeing local motion. Conversely, making one dot in each circling pair red and the other dot green gives a large square of red dots and a large square of green dots. This greatly increases the chances of seeing global motion.

Local and global motion are two different and incompatible solutions to the problem of binding dots into groups. They are incompatible because it is impossible to see the same dots as partaking in local and global groups simultaneously. We suspect that local motion is preattentive, whereas global motion is attentive.

Acknowledgements

Supported by NIH Grant EY10241 and by a grant from the UCSD Academic Senate. Thanks to my collaborators Dana Ballard, Patrick Cavanagh, Richard Gregory, Hiro Ito, George Mather, Brian Rogers, Steve Shimozaki, David Smith; and to my students Alecia Dager, Shawn Ewbanks, Laura Johnston, Efrat Stark and Megan Tatreau for assistance in data collection.

References

Adelson, E. and Bergen, J. (1985). Spatiotemporal energy models for the perception of motion. *J. Opt. Soc. Am. A.*, 2: 284–299.

Adelson, E. and Movshon, J. A. (1982). Phenomenal coherence of moving visual patterns. *Nature*, 300: 523–525

Anstis, S. M. (1990). Imperceptible intersections: The chopstick illusion. In A. Blake and T. Troscianko (eds), *AI and the Eye*, pp. 105–117. London, UK: Wiley.

Anstis, S. M. (2001). Footsteps and inchworms: Illusions show that contrast modulates apparent speed. *Percep.*, 30: 785–794.

Anstis, S. M., Giaschi, D. and Cogan, A. I. (1985). Adaptation to apparent motion. *Vis. Res.*, 25: 1051–1062.

Anstis, S. M. and Mather, G. (1985). Effects of luminance and contrast on direction of ambiguous apparent motion. *Percep.*, 14: 167–179.

Anstis, S. M., Smith, D. R. R. and Mather, G. (2000). Luminance processing in apparent motion, Vernier offset and stereoscopic depth. *Vis. Res.*, 40: 657–675.

Barlow, H. B. and Levick, W. R. (1965). The mechanism of directionally selective units in rabbit's retina. *J. Physiol. (Lond.)*, 178: 477–504.

Blakemore, M. R. and Snowden, R. J. (1999). The effect of contrast upon perceived speed: a general phenomenon? *Percep.*, 28: 33–48.

Braddick, O. J. (1974). A short-range process in apparent motion. *Vis. Res.*, 14: 519–527.

Braddick, O. J. (1997). Local and global representations of velocity: transparency, opponency, and global direction perception. *Percep.*, 26: 995–1010

Braunstein, M. L. and Andersen, G. J. (1984). *Percep.*, 13: 213–217.

Campbell, F. W. and Maffei, L. (1979). Stopped visual motion. *Nature*, 278: 192.

Campbell, F. W. and Maffei, L. (1981). The influence of spatial frequency and contrast on the perception of moving patterns. *Vis. Res.*, 21: 713–721.

Clifford, C. W. and Wenderoth, P. (1999). Adaptation to temporal modulation can enhance differential speed sensitivity. *Vis. Res.*, 39: 4324–4332.

Curran, W. and Braddick, O. J. (2000). Speed and direction of locally paired dot patterns. *Vis. Res.*, 40: 2115–2124.

Duncan, R. O., Albright, T. D. and Stoner, G. R. (2000). Occlusion and the interpretation of visual motion: perceptual and neuronal effects of context. *J. Neurosci.*, 20: 5885–5897.

Ejima, Y. and Takahashi, S. (1984). Bezold–Brucke hue shift and nonlinearity in opponent-color process. *Vis. Res.*, 24: 1897–1904.

Estevez, O. and Spekreijse, H. (1982). The "silent substitution" method in visual research. *Vis. Res.*, 22: 681–691.

Gegenfurtner, K. R. and Hawken, M. J. (1996). Perceived velocity of luminance, chromatic and non-Fourier stimuli: Influence of contrast and temporal frequency. *Vis. Res.*, 36: 1281–1290.

Hawken, M. J., Gegenfurtner, K. R. and Tang, C. (1994). Contrast dependence of colour and luminance motion mechanisms in human vision. *Nature*, 367: 268–270.

Hildreth, E. C. (1983). *The Measurement of Visual Motion*. Cambridge, MA: MIT Press.

Kolers, P. A. (1964). The illusion of movement. *Sci. Am.*, 211: 98–106.

Kooi, F. K, De Valois, K. K., Grosof, D. H., De Valois, R. L. (1992). Properties of recombination of one-dimensional motion signals into a pattern-motion signal. *Percept. Psychophys.*, 52: 415–424.

Kowler, E. (ed): (1990). *Eye Movements and Their Role in Visual and Cognitive Processes*. New York: Elsevier.

Krauzlis, R. J. (1994), The visual drive for smooth eye movements. In A. T. Smith and R. J. Snowden (eds). *Visual Detection of Motion*. pp. 437–473. New York: Academic Press.

Ledgeway, T. and Smith, A. T. (1995). The perceived speed of second-order motion and its dependence on stimulus contrast. *Vis. Res.*, 35: 1421–1434

Lisberger, S. G., Morris, E. J. and Tychsen, L. (1987). Visual motion processing and sensory-motor integration for smooth pursuit eye movements. *Ann. Rev. Neurosci.*, 10: 97–129.

Lorenceau, J. and Shiffrar, M. (1992). The influence of terminators on motion integration across space. *Vis. Res.*, 32: 263–273

Mather, G. and Anstis, S. (1995). Second-order texture contrast resolves ambiguous apparent motion. *Percept.*, 24: 1373–1382.

Maunsell, J. H. and Van Essen, D. C. (1983). Functional properties of neurons in middle temporal visual area of the macaque monkey. I. Selectivity for stimulus direction, speed, and orientation. *J. Neurophysiol.*, 49: 1127–1147.

Movshon, A., Adelson, E. H., Gizzi, M. S. and Newsome, W. T. (1983). In C. Chagas, R. Gattas and C. G. Gross (eds), *Pattern Recognition Mechanisms*. Rome: Vatican Press.

Qian, N. and Andersen, R. A. (1994). Transparent motion perception as detection of unbalanced motion signals. II. Physiology. *J. Neurosci.*, 14: 7367–7380.

Qian, N., Andersen, R. A. and Adelson, E. H. (1994). Transparent motion perception as detection of unbalanced motion signals. I. Psychophysics. *J. Neurosci.*, 14: 7357–7366.

Qian, N., Andersen, R. A. and Adelson, E. H. (1994). Transparent motion perception as detection of unbalanced motion signals. III. Modeling. *J. Neurosci.*, 14: 7381–7392.

Ramachandran, V. S. and Anstis, S. M. (1983). Perceptual organization in moving displays. *Nature*, 304: 829–831.

Regan, D. M. (2000). *Human Perception of Objects*. Sunderland, MA: Sinauer.

Schiller, P. H. (1992). The ON and OFF channels of the visual system. *Trends in Neurosci.*, 115: 86–92.

Sclar, G. and Freeman, R. D. (1982). Orientation selectivity in the cat's striate cortex is invariant with stimulus contrast. *Exp. Brain Res.*, 46: 457–461.

Shiffrar, M. and Pavel, M. (1991). Percepts of rigid motion within and across apertures. *J. Exp. Psychol. Hum. Percept. and Perf.*, 17: 749–761.

Shimojo, S., Silverman, G. H. and Nakayama, K. (1989). Occlusion and the solution to the aperture problem for motion. *Vis. Res.*, 29: 619–626.

Smith, D. R. R. and Anstis, S. M. (in press). Strength of cross-over motion measured by time till breakdown.

Smith, D. R. R. and Derrington, A. M. (1996). What is the denominator for contrast normalization? *Vis. Res.*, 36: 3759–3766.

Stone, L. S. and Thompson, P. (1992). Human speed perception is contrast dependent. *Vis. Res*, 32: 1535–1549.

Thompson, P. (1976). *Velocity Aftereffects and the Perception of Movement*. PhD Thesis, University of Cambridge, Cambridge, UK,

Thompson, P. (1982). Perceived rate of movement depends on contrast. *Vis. Res.*, 22: 377–380.

Thompson, P. and Stone, L. S. (1997). Contrast affects flicker and speed perception differently. *Vis. Res.*, 37: 1255–1260.

Thompson, P., Stone, L. S. and Swash, S. (1996). Speed estimates from grating patches are not contrast-normalized. *Vis. Res.*, 36: 667–674.

Ullman, S. (1979). *The Interpretation of Visual Motion*. Cambridge, MA: MIT Press.

Valberg, A., Lange-Malecki, B. and Seim, T. (1991). Colour changes as a function of luminance contrast. *Percept.*, 20: 655–668.

Wilson, H. R. (1994). In A. T. Smith and R. J. Snowden (eds), *Visual Detection of Motion*. London: Academic Press.

6

Reconciling Rival Interpretations of Binocular Rivalry

Randolph Blake

6.1 Introduction

The theme of this volume (and the associated conference) — levels of perception — is grounded in the widely held view that vision comprises hierarchically arranged stages of neural information processing. The architecture of those stages remains controversial (e.g., compare Lennie, 1998, to Grossberg, 2000), and there is debate concerning the nature of the representations that pass back and forth among levels of analysis (e.g., Mather, 2001). There is little argument, however, that perception results from bidirectional communication among multiple levels of analysis, and other chapters in this volume document details of some of those levels. Of course, to successfully elucidate the mechanisms of perception we need to develop and utilize strategies for isolating levels of analysis and for measuring their unique contributions to perception.

In his landmark book, Julesz (1971) coined the term "psychoanatomy" to refer to inferential, psychophysical strategies for delineating stages of information processing within human vision. The hallmark of these strategies can be summarized as follows: one process A is placed "prior" to another process B if it can be shown that the second utilizes the output of the first. In essence, stages are identified by changes in the way information is represented following transformation at a given stage; the emphasis is on evolving neural "descriptions" of features, objects and events, not on serially ordered anatomical loci.

Over the years clever psychoanatomical strategies have been developed using revealing stimuli such as random-dot stereograms and using innovative techniques such as transcranial magnetic stimulation. To complement this arsenal of strategies, I have championed the phenomenon of binocular rivalry as a potentially powerful instrument for studying levels of perception (Blake, 1995). The kernel idea is to exploit the dissociation between physical stimulation and perceptual experience that characterizes rivalry. By so doing, it may be possible to learn what aspects of visual information processing remain effective during suppression phases of rivalry and what aspects do not (Blake, 1997). To give an example of this strategy at work, my colleagues and I have shown that a "priming" picture that facilitates object recognition is rendered ineffective when that picture is not consciously perceived, owing to binocular rivalry suppression (Cave et al., 1998). This finding implies that suppression transpires at a site prior to those neural events underlying object identity.

The utility of an "instrument" — whether it's a sextant or a psychophysical procedure — is greatly enhanced when we understand the principles of operation of that instrument. For this reason, I want to use this chapter to examine the "instrument" of binocular rivalry and, therefore, its utility as an effective psychoanatomical strategy. This examination is motivated, in part, by the ongoing debate about whether rivalry involves competition between alternative perceptual interpretations or competition between conflicting monocular image features. Resolution of this debate concerning "what" rivals during rivalry bears importantly on the kinds of conclusions reached using rivalry as a means for partitioning stages of visual information processing.

This issue — what rivals during rivalry — is tantamount to asking at what level of visual representation are the conflicting stimuli competing (Walker, 1978). Surveying the literature on rivalry one can identify two broadly contrasting views on this issue.[1] One view holds that rivalry transpires at an "early" stage of processing concerned with the analysis of primitive image features such as oriented contours (Wade, 1974; Abadi, 1976). Some versions of these "early" models posit that rivalry involves competition between the eyes, in that the neural machinery promoting dominance and suppression comprises populations of neurons differentially responsive to left- and right-eye stimulation (Matsuoko, 1984; Wolfe, 1986; Lehky, 1988; Blake, 1989; Mueller, 1990). It is worth noting, however, that "early" rivalry does not necessarily imply "eye" competition; there are versions of "early" rivalry in which the underlying neural events are embodied in binocular mechanisms (e.g., Grossberg, 1987). One influential stimulus for "early" models of rivalry was Levelt's (1965) seminal monograph documenting that rivalry predominance is governed by "low-level" stimulus attributes such as luminance, contrast, and contour density.

The alternative view holds that rivalry is a "late" or "high-level" process involving competition between more refined descriptions of meaningful objects. "Late" rivalry was, in fact, the prevailing view during the nineteenth and early twentieth century, and it was embraced by leading figures including William James, Hermann von Helmholtz, and Charles Sherrington. For example, in his landmark monograph *On Inhibition*, Sherrington (1906) wrote:

> Only after the sensations initiated from right and left corresponding points have been elaborated, and have reached a dignity and definiteness well amenable to introspection, does interference between the reactions of the two eye-systems occur. The binocular sensation

---

[1]These competing views on the nature of binocular rivalry strongly resemble alternative conceptualizations of attention (Kanwisher and Wojciulik, 2000).

attained seems combined from right and left uniocular sensations elaborated independently...In retinal rivalry we have an involuntarily performed analysis of this sensual bicompound. The binocular perception in that case breaks down, leaving phasic periods of one or other of the simpler component sensations bare to inspection.[2] (p. 379)

This "late" view endured throughout much of the twentieth century. Twenty-three years ago Peter Walker (1978), wrapping up his historical review of selective processes in rivalry, concluded that

the available evidence supports the thesis that the suppressed stimulus in binocular rivalry is being analyzed and that the selective processes involved in the phenomenon are centrally located. (p. 386).

Following an interlude during which models endorsing "early" rivalry dominated, the pendulum quite recently has swung back in the direction of "late" rivalry, thanks in part to results from influential psychophysical experiments (Logothetis et al., 1996; Kovacs et al., 1997) and physiological experiments (Logothetis, 1998). The essence of this contemporary version of "late" rivalry has been well expressed by Logothetis (1998)

[T]he dominance and suppression of a pattern during rivalry reflects the excitation and inhibition of cell populations in the higher visual areas, which are directly involved in the representation of visual patterns. (p. 1815)

So today, where do things now stand on the "early" versus "late" rivalry debate? During the last decade, refined techniques have produced a substantial body of evidence that bears further on this issue. As the result, the consensus seems to be evolving toward a more refined view in which rivalry entails a cascade of neural events that are amplified throughout the visual pathways. The concomitants of rivalry, in other words, are distributed throughout the visual hierarchy, culminating in competing perceptual interpretations.

In the following sections I wish to explore some of this recent evidence bearing on the nature of rivalry. From the outset it should be stressed that this survey is not exhaustive; the goal rather is to provide some flavor for the kinds of evidence marshalled for and against the issue of "early" versus "late" rivalry. Following this selective review, I will conclude by asking whether the distinction between "early" and "late" rivalry is sufficiently clear and meaningful to justify maintaining it.

---

[2]It is interesting to note that Sherrington's "late" view also implies that rivalry occurs between separate monocular components and, in this respect, the idea has somethign in common with versions of "early" rivalry.

6.2   Reasons for Believing That Rivalry Is "Early"

6.2.1   Eye Swapping

While observing binocular rivalry, we are never aware of which eye carries the currently dominant stimulus. Instead, we see one pattern or another: competition appears to transpire between alternative perceptual interpretations. However, impressions can be misleading. Suppose during rivalry we replace the currently dominant stimulus with the currently suppressed one, and vice versa. This can be simply accomplished by interchanging the two rival targets between the eyes, being careful to perform the swap only when the observer reports exclusive dominance of one target and to execute the swap in a way that avoids abrupt transients (Figure 6.1). When this interchange is performed, one clearly sees that the dominant pattern abruptly disappears and the previously suppressed pattern springs into dominance. Evidently, then, it was not a particular stimulus that was dominant but, instead, a region of one eye (Blake et al., 1979), a result consonant with the "early" account of rivalry.

6.2.2   Rivalry Suppression Is Nonselective

During suppression phases of rivalry, a given stimulus is rendered invisible for several seconds at a time — indeed, from a phenomenological standpoint, the invisibility of the stimulus is just as compelling as the experience associated with physical removal of the stimulus. Evidently the neural process underlying suppression involves some potent form of inhibition. However, it is not just that particular stimulus that is rendered temporarily invisible. We know, for example, that probe targets superimposed on a stimulus engaged in rivalry are more difficult to detect when that rival figure is suppressed, compared to detection performance when the same probe appears on the same rival figure when it is dominant (e.g., Wales and Fox, 1970; Fox and Check, 1972; Nguyen et al., 2001). This is true even if the probe itself is highly dissimilar in form or color from the rival target; even one's own name can be impossible to detect if presented during suppression (Blake, 1988). Suppression, then, seems to generalize beyond the configural properties of the initially suppressed stimulus, a characteristic referred to as "nonselectivity" (Fox, 1991).

For that matter, normally conspicuous changes in a rival figure itself can go undetected for several seconds if those changes are introduced during suppression. These unnoticed changes can consist of large angular rotations in contour orientation, large changes in spatial frequency or transitions from incoherent motion to coherent motion (Blake et al., 1998).[3] In recent work in our laboratory, we have

---

[3]Changes in a suppressed target that are accompanied by sharp transients are usually detectable, for abrupt transients prematurely terminate suppression. Thus, for example, an abrupt increase in the contrast of a suppressed grating almost immediately brings that grating into dominance. To avoid this artifact, experiments involving changes to a suppressed target have introduced those changes

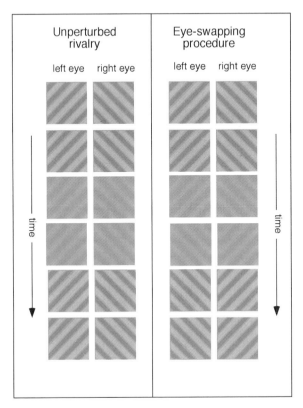

FIGURE 6.1. Schematic of "eye-swap" procedure for contrasting "stimulus" versus "eye" rivalry. Observers dichoptically view two dissimilar targets that engage in binocular rivalry. When one target is dominant in its entirety, the other one being completely suppressed, the observer depresses a key. On some trials key press triggers an exchange of the left- and right-eye targets, such that the currently dominant stimulus is placed in the eye previously viewing the currently suppressed target, and vice versa ("Eye-swap procedure"). The swap is performed by ramping the contrasts of the two patterns off and then back on, to eliminate abrupt transients; the time constant of the contrast ramp is brief (shown schematically here). On other trials key press causes the two targets to be ramped in contrast just as in the "eye-swap" condition, but the targets are not exchanged between the eyes: unperturbed rivalry. The observers' task is to report which target is dominant immediately following the key press; results show that on trials observers continue seeing the currently dominant target, but on "swap" trials the previously suppressed target suddenly assumes dominance.

utilized this "replacement" paradigm with more complex rival targets that, at "later" stages of processing, activate distinctively different brain areas.. We find that when a "face" rival target is replaced by a "house" rival target during suppression, that replacement goes unnoticed for several seconds; the same replacement during dominance is immediately perceived. Of course, the "face" and "house" images both activate neurons in visual area V1, but it is commonly believed that the two images activate distinctly different higher brain areas (the fusiform "face" area versus the parahippocampal "place" area). For more on the difficulty of detecting changes during suppression, see Box 6.1.

The nonselectivity of suppression, which generalizes to complex, familiar figures, implies that the inhibitory events underlying suppression are operating not just on the initially suppressed target but instead more generally on all stimulus features imaged within a given, local region of the eye (Fox, 1991). What would be the most efficient means for implementing widespread suppression? Efficiency argues for an early site of suppression, where neurons registering a wide range of composite image features are anatomically localized (e.g., in a hypercolumn). Suppression at this "early" stage would effectively encompass a very broad range of figures. To implement nonselective suppression "late" in processing would necessitate the coordination of inhibitory events over widely distributed visual areas. (By way of analogy, contrast the difficulty of identifying aliens at the point of entry into a country vs within individual cities and towns scattered throughout the interior of that country.)

6.2.3   Dominance Is Uncontrollable

We tend to think of "high-level" mental processes as amenable to attentional control — for example, we can chose what to pay attention to, such as the words on this page and not the background noises in the room. "Early" processes, on the other hand, are construed as obligatory, or, to borrow a term from Fodor, "cognitively impenetrable." Rivalry is autonomous in that observers cannot willfully maintain dominance of a given figure, except by uninteresting means such as closing one eye or moving the eyes in specific ways that accentuate one figure (e.g., Breese, 1899). Even when required to monitor an interesting, potentially personal rival stimulus, observers are unable to maintain dominance of that stimulus for any extended period of time (Blake, 1988). Thus, binocular rivalry stands in stark contrast to dichotic listening wherein a listener can "shadow" continuously one of two competing messages broadcast to the two ears. The attentional mechanism supporting dichotic listening does not seem to be available to observers experiencing rivalry. Now, it is true that with prolonged practice observers can learn to influence the temporal dynamics of rivalry (Lack, 1978), but this is not the same thing as arresting the alternations in dominance. On the assumption that willful

---

gradually, typically using Gaussian contrast ramps with time constants on the order of 100 msec. For more on the role of transients, see Box 6.1.

Box 6.1: Detecting Change During Binocular Suppression

The following experiment, performed in collaboration with Timothy Vickery and Sang-Hun Lee, illustrates the sensitivity of rivalry suppression to transients. Observers viewed the rival targets shown in Figure 6.2a (a man's face and a checkerboard) through a mirror stereoscope whose two optical channels could be carefully aligned for each observer. In the actual experiment, the checkerboard was contrast reversed at 2 Hz to enhance its predominance. The observer's task was to depres a key on the computer keyboard when a given target was dominant in its entirety, with no hint of the other target. After a brief, 150 msec period, the left eye, the right eye, or both eyes of the "face" target "blinked" and then, 500 msec later, both displays were removed. The observer reported which eye(s) blinked, guessing if necessary; all three "blink" possibilities were equally likely, and error feedback was provided. During some blocks of trials, blinks were initiated while the face was dominant and during other trial blocks blinks were initiated while the checkerboard was dominant.

The salience of the blink was manipulated in the following way (Figure 6.2b). During each 150 msec "blink" presentation, we presented a series of "blended" image frames, with each frame consisting of a weighted average of the face with eyes open and the face with eyes closed. By varying the range of these weights, we could produce "blinks" with amplitudes ranging from subtle ("open" eye image more heavily weighted) to obvious ("closed" eye image more heavily weighted). This weighting manoeuvre, while maintaining a nearly constant space average luminance, produced "blinks" that varied in amplitude and, hence, in perceived abruptness.

Results for four observers (two naïve) are shown in Figure 6.2c, where percent-correct on this 3AFC task is plotted against blink amplitude. Performance during suppression was consistently poorer than during dominance, replicating many earlier studies showing decreased probe sensitivity during suppression phases (e.g., Fox, 1991). These results also document the ability of transients (large-amplitude blinks) to break suppression (see also Walker and Powell, 1979), at least in the immediate vicinity of the transient event.

On a practical note, it is crucial to avoid large, sharp transients when changing a suppressed target or when introducing new information to that target, unless the purpose of the change is explicitly to break suppression (e.g., Wilson et al., 2001). When transients are avoided, large changes in a suppressed target go undetected. In this respect, suppression behaves as if it were "broad" (meaning nonselective) yet "shallow" (meaning easily perturbed).

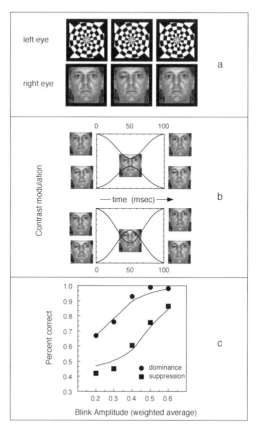

FIGURE 6.2. An otherwise conspicuous change in a suppressed rival target is difficult to detect unless that change is accompanied by a transient of sufficient magnitude. (a) The observer depressed a key when the face pattern was exclusively visible (face dominant condition) or when the checkerboard pattern was exclusively visible (face suppressed condition). This key press caused the left eye, the right eye (shown in this example) or both eyes to "blink" briefly (close then open), with the "amplitude" of the blink varying randomly across trials. (b) Blink amplitude was manipulated by varying the relative proportions of the "open" eye image and the "closed" eye image presented during the change period, which lasted 100 msec; complete interchange of the two images produced a large amplitude transient, and more incomplete interchanges (two examples are shown here) produced smaller amplitude "blinks." Presentation of the mixture was windowed with a temporal gaussian. Following each trial, the observer reported which eye(s) blinked, guessing if necessary. (c) Percent-correct on this three alternative, forced-choice task for blinks occurring during dominance and during suppression; performance is plotted as the function of blink "amplitude." Results are averaged over four observers (two naïve), all of whom showed the same pattern of results. Standard errors are less than the symbol size.

control over what one attends to is a hallmark of higher cognitive function, the inability to maintain dominance raises some doubt about rivalry as a "late" process.

In fairness, though, it should be noted that people experience difficulty controlling perception when viewing bistable figures such as Rubin's vase/face figure or the Necker cube. Indeed, these well-known examples of bistable perception exhibit temporal dynamics resembling binocular rivalry, which could be construed to imply a common underlying mechanism (e.g., Andrew and Purves, 1997). Whether this mechanism is "early" or "late" remains debatable, although the lack of control over what one sees does not comport intuitively with high-level mental processes.

6.2.4   Rivalry Dominance Follows Cortical Magnification

When viewing rival figures, one can perceive what is termed "piecemeal" rivalry: dominance consists of bits and pieces of the two dissimilar figures intermingled in a patchworklike spatial pattern (Meenes, 1930; Hollins, 1980). Larger rival figures are more prone to piecemeal rivalry than are small figures, and it is possible to find a stimulus size where rivalry is unitary, not piecemeal, essentially all the time. Blake et al. (1992) found that foveally viewed rival figures had to be about 0.1 deg visual angle or smaller to preclude piecemeal rivalry. However, this critical angular subtense increased with retinal eccentricity in a manner predictable from cortical magnification estimated for human primary visual cortex. And more recently, Wilson et al. (2001) have found that the spread of dominance over time throughout a given rival figure also corresponds with known properties of human primary visual cortex (Figure 6.3).

It is commonly recognized that the cortical magnification factor varies among visual areas of the brain, being greatest in visual area V1. The close correspondence between the extent of spatial spread of dominance and the magnification factor for area V1 provides another piece of evidence favoring "early" rivalry ("early" in this case being primary visual cortex). (See also the naso-temporal asymmetry in rivalry dominance that mirrors asymmetries in the retinocortical projections, as reported by Fahle, 1987). It should be acknowledged, however, that this line of reasoning by analogy — wherein a "perceptual" function resembles a "physiological" function — is considered to be a rather weak kind of linking proposition (Teller, 1984).

Having summarized circumstantial evidence pointing to an "early" site for binocular rivalry, let's next consider several pieces of evidence favoring a "late" site for rivalry.

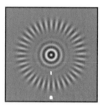

FIGURE 6.3. By crossing or by diverging the eyes, fuse these two rival targets and fixate the central spot. Notice how dominance spreads rapidly and regularly throughout a target as it emerges from suppression. In particular, notice how the annular grating emerges from suppression. Targets like these were used by Wilson et al. (2001) to study the rate at which dominance spreads throughout a previously suppressed rival target. Observers waited until the radial grating was completely suppressed and then depressed and held a key, which produced a brief, spatially localized contrast increment somewhere on the radial grating. This contrast increment triggered the immediate dominance of that portion of the radial grating. The observer monitored the spread of dominance from that trigger point and indicated when dominance reached a given region of the radial grating. By measuring the time between the contrast increment and the arrival of dominance at the monitored location, it was possible to compute the speed at which dominance traveled around the annulus. The speed of this dominance "wave" was found to be constant when expressed in angular units scaled according to V1 cortical magnification.

6.3   Reasons to Believe That Rivalry Is "Late"

6.3.1   Interocular Grouping

Imagine cutting a pair of rival targets into "fragments" and then arranging those fragments so that a given eye received bits and pieces of both targets and the two eyes together received complementary parts of the figures. Examples of this kind of "hybrid" rival target are shown in Figure 6.4. Obviously with rival targets of this sort, if a single, coherent figure achieves dominance during rivalry, that figure must have been assembled from left- and right-eye components; strict "eye" rivalry cannot produce interocular grouping in these cases. But if rivalry transpires at a "late" stage where coherent figures are represented, interocular presentation presents no barrier to rivalry.

Several published reports document the occurrence of global rivalry dominance using hybrid, interocular grouping displays (Diaz-Caneja, 1928; Dörrenhaus, 1975; Kovacs et al., 1997; Alais and Blake, 1999; Ngo et al., 2000). For that matter, even conventional rival targets can produce patterns of dominance that defy "eye" rivalry. As mentioned above, when viewing relatively large rival targets — one complete target viewed by each eye — observers typically experience periods during which bits and pieces of both targets are dominant, the result being a "patchwork" pattern comprising intermingled portions of both eyes' views (Meenes, 1930). So here, too, dominance is distributed between the eyes, although the resulting, dominant stimulus is not coherent. Can these observations be reconciled with the "early/eye" account of rivalry?

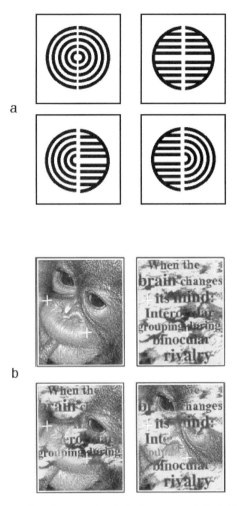

FIGURE 6.4. Two examples of conventional rival targets and "hybrid" versions of those targets produced by interchanging portions of each target within each eye's view. Readers capable of free fusion should compare the incidence of complete dominance of a given figure for the two categories of rival figures. Part (a) is styled after the rival target created by Diaz-Caneja (1928). Interested readers may see the original figure by navigating to: http://www.perceptionweb.com/perabs/p29/p3017.html. Part (b) is a greyscale version of portions of Figure 1 of Kovacs et al. and reproduced with permission from Kovacs, I., Papathomas, T.V., Yang, M. and Fehér, A. (1997). When the brain changes its mind, Interocular grouping during binocular rivalry. *Proc. Nat. Acad. Sci. USA*, 93:15508–15511, with permission. Copyright 1997 National Academy of Science, USA. The original color version may be seen by navigating to: http://www.pnas.org/cgi/content/full/93/26/15508.

It is conceivable that rivalry transpires within local spatial "zones" distributed throughout the binocular visual field, with "global" dominance corresponding to the assemblage of dominant features within the aggregate of zones. On this account, global dominance of an entire rival target would be associated with the conjoint predominance of the components of that target distributed over the array of local zones. And there is no reason those locally dominant zones could not be distributed between the two eyes. Indeed, piecemeal rivalry discloses just such behavior. But what is the mechanism by which conjoint dominance is achieved? If these putative local zones are spatially independent, the probability of coherent, global dominance would correspond to the probability of all requisite zones being dominant simultaneously. Using cross-correlation techniques, Alais and Blake (1999) showed that the incidence of global dominance is greater than that predicted on the basis of probability alone, leading them to posit the existence of interactions among neighboring local zones.

Even with these compelling interocular grouping displays, several pieces of evidence imply that rivalry maintains some neural "signature" concerning the eye in which a given rival feature is imaged. Compare the pattern of rivalry experienced with the two types of rival targets in Figure 6.5. Note in particular the incidence of global dominance of the "face" in these two configurations (rivalry is confined to the lower left-hand quadrant of the figures). Blake et al. (1997) found that the "face" in the top configuration was dominant almost 40% more often than was the "face" in the bottom configuration. The visual system, in other words, remains sensitive to the eye in which the rival components are imaged. Exactly the same result is observed when comparing global dominance in the displays created by Kovacs et al. (1997): global predominance is more frequent when all image components are imaged in the same eye versus distributed between the eyes, as readers can confirm using the rival targets in Figure 6.4b.

Even more convincing evidence for an "eye" signature in these kinds of displays comes from a set of experiments performed by Sang-Hun Lee (in preparation). Using the Kovacs et al. (1997) "monkey vs. jungle" scenes as rival targets (Figure 6.4b), Lee implemented a version of the eye-swap technique (Blake et al., 1979) wherein left and right eye rival targets are switched at the onset of dominance of a given figure. So, for example, suppose the left eye views the monkey's face and the right eye views the jungle scene (upper pair of rival targets in Figure 6.4b). In Lee's experiment, observers depressed a button when the monkey's face was dominant in its entirety, with no hint of the jungle scene. On some trials this button press triggered an exchange of the two eye's rival patterns (in the example here, the monkey face would be switched to the right eye and the jungle scene to the left eye); this swap was achieved using a gaussian ramp with a time constant of 100 msec, to avoid abrupt transients that can disrupt rivalry (Walker and Powell, 1979). On other trials, the two rival patterns were ramped off and on, just as they were in the eye swap trials, but the patterns were **not** exchanged between the eyes. Observers were simply asked to report the dominant feature at a given location of the display immediately following button depression (in most of Lee's experiments, that location corresponded to the monkey's left eye, although

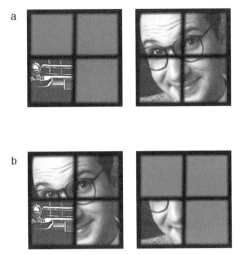

FIGURE 6.5. Readers capable of free fusion should compare the rivalry experienced in the upper pair of targets to that experienced with the lower pair of targets. In particular, note how often the complete face is perceived in pair a and how often in pair b.

the particular location monitored had no influence on the pattern of results). The results were unequivocal: on nonswap trials, observers continued seeing the stimulus that was dominant at the time of button press, but on swap trials observers now saw the stimulus that was suppressed at the time of button press. Dominance, in other words, always corresponded to the stimulus currently imaged on a given region of the eye, not to a given stimulus feature. This pattern of results was found both with conventional monkey/jungle rival targets and with monkey/jungle targets reciprocally distributed between the two eyes (i.e., the hybrid interocular variety illustrated by the bottom pair of rival figures in Figure 6.4b).[4]

So even when rivalry dominance is global and distributed between the two eyes, the mechanism underlying this pattern of dominance seems to retain information about the eye of origin of the components of the dominant pattern. On the assumption that such information is present only at relatively early stages of visual processing, these results favor an "early" interpretation of rivalry. This is not to say, however, that global organizing factors are not involved in the promotion and maintenance of dominance. As pointed out above, global dominance is sensitive to configural properties of the rival targets (e.g., Alais and Blake, 1999). And there is no reason why even more "cognitive" factors such as affect and meaning could not influence dominance. After all, information about the dominant stimulus in

---

[4]Lee's observers found that it took longer to complete trials with displays involving interocular grouping, because the incidence of complete dominance of an entire figure (e.g., entire monkey face) was less frequent with these displays compared to the conventional versions where an entire rival target was presented to a single eye. This trend is also evident in the published data of Kovacs et al. (1997).

rivalry is flowing throughout the same brain areas activated when that stimulus is viewed under nonrivalry conditions. And since the dominant stimulus undergoes the same, full-blown analysis it does ordinarily, any feedback ("top-down") signals normally activated by that stimulus would be engaged during dominance phases as well. According to this view, then, the dominance of a pattern during rivalry does indeed reflect the activity of neurons in higher visual areas, just as Logothetis (1998) concluded. However, according to the view advanced here, the suppressed stimulus is temporarily denied access to those "higher" stages of processing and, hence, to the organizing forces that normally promote perceptual interpretation. Consistent with this supposition, Sobel and Blake (2001) found that global context lengthened the average duration of dominance of a rival target that "fit" within that context, whereas context had no influence on the duration of suppression of that target. Thus, it would be incorrect to conclude that suppression of a pattern during rivalry reflects inhibition of neural activity in higher visual areas; according to the view advanced here, information about a pattern never reaches those areas higher during periods of suppression, being instead inhibited at an early stage where context has no effect but where information about eye of origin is retained.

6.3.2   Dissociation of Color, Motion, and Form During Rivalry

It is commonly assumed that at an "early" stage of visual processing (e.g., V1) neurons are broadly tuned for multiple stimulus features (e.g., orientation, size, and color), the result being that a specific visual quality ("red") is not explicitly represented in the activity of individual neurons. This form of coding is sometimes termed "multiplexing." At later stages, however, explicit representations of stimulus qualities emerge, with different aspects of the visual scene being represented in different visual areas (see review by Grossberg, 2000). To the extent this conceptualization is correct, we would expect a nonselective, "early" rivalry process to operate at once on all characteristics of a given rival target. To illustrate, consider a rival target consisting of obliquely oriented red contours drifting up and to the right. If rivalry were occurring "early" where these stimulus qualities are being "multiplexed" within an array of neurons, we would expect all three of the grating's qualities — color, form, motion — to appear (dominance) and disappear (suppression) in unison during rivalry. If, on the other hand, rivalry is occurring at later stages where these qualities are being represented separately, it is possible that one quality could be suppressed phenomenally while, at the same time, another is dominant in vision.

So, what actually happens when one views multidimensional rival targets? A handful of rivalry studies report survival of one stimulus quality while, at the same time, a companion quality suffers suppression. Take, for example, the report by Carney et al. (1987). They had observers dichoptically view a pair of differently colored gratings both of which underwent counterphase flicker, with the interocular phases of flicker arranged to be unambiguously consistent with a given direction of motion. And consistent motion was what observers experienced even though the two gratings underwent color rivalry. Carney et al. concluded that motion and color signals were being registered in different pathways that could be in different states of rivalry. Along similar lines, Andrews and Blakemore (1999) presented orthogonally oriented, drifting gratings separately to the two eyes and asked observers to monitor the resulting form rivalry and the direction of drift of the dominant grating. In about half of the trials, the dominant grating seemed to drift in a direction specified by the motion vectors for both dominant and suppressed gratings, that is the motion of the grating whose orientation was suppressed remained effective at least some of the time. This finding is reminiscent of the earlier observation that a suppressed half-image can nonetheless contribute to a sensation of apparent motion (Wiesenfelder and Blake, 1991). And to give one more example, Carlson and He (2000) had observers dichoptically view two objects differing in shape and in color, which produced clear binocular rivalry. Each object was also rapidly flickering in luminance, with the flicker rates differing slightly between the two eyes. Remarkably, observers perceived relatively slow variations in luminance corresponding to the difference frequency (or "beats") between the two flickering patterns. Thus, observers were seeing flicker resulting from the combination of the two patterns while, at the same time, being unable to see one of the two patterns. Carlson and He discussed their findings in terms of parvocellular and magnocellular pathways, with rivalry being confined to the latter.

While striking, none of these contemporary findings is as dramatic as the one described many years ago by Creed (1935). He dichoptically viewed pairs of small postage stamps using a stereoscope and noticed that on occasion the color portrayed in one stamp would appear simultaneously with the outline forms depicted in the other stamp. To quote Creed (p. 391), "A contour which prevails in rivalry does not necessarily bring with it the colour of its own object." This observation is particularly noteworthy for it implies that rivalry may proceed independently along different dimensions, an observation difficult to reconcile with "early" rivalry (see, also, Treisman, 1962, for observations that bear on this point). For that matter, it is not obvious how "late" rivalry based on the neural representation of a fully elaborated object would account for this finding, which constitutes a form of illusory conjunction.[5]

6.3.3 Visual Adaptation Survives Suppression

Many well-known visual aftereffects of adaptation can be generated even when the adapting stimulus itself is suppressed phenomenally during the adaptation period. This is true for the grating threshold elevation aftereffect (Blake and Fox, 1974), the motion aftereffect (Lehmkuhle and Fox, 1976; O'Shea and Crassini,

---

[5]Color/form rivaly is a fascinating subject that deserves more careful study. Interested readers are invited to study some of the rival pairs involving color shown on the author's rivalry webpage: http://www.psy.vanderbilt.edu/faculty/blake/rivalry/BR.html.

1981), the tilt aftereffect (Wade and Wenderoth, 1978) and the McCollough effect (White et al., 1978). At least some of these aftereffects exhibit interocular transfer, which itself implies a central, binocular site for adaptation. The failure of suppression to retard the build-up of these aftereffects, then, implies that suppression transpires at a neural locus subsequent to binocular site(s) of adaptation. Whether such a locus is to be construed as "late" is debatable, but adaptation's escape from the effects of suppression limits how early "early" might be.

In fact, suppression's effect on the build-up of visual adaptation aftereffects may need to be reexamined. The studies cited above used relatively high-contrast adaptation patterns, and in several studies the authors compared the magnitude of the resulting aftereffect to conditions where the adapting target was physically turned on and off over time in a sequence mimicking rivalry alternations. This comparison condition, however, may not be appropriate, for it presumes that any effect of rivalry suppression would be equivalent to intermittent physical removal of the adaptation pattern. It is possible that suppression of a given rival figure, although phenomenologically profound, is accomplished through modest neural events involving relatively small shifts in the balance of activity, not wholesale disruption of that activity. Hence a suppressed adaptation pattern may be weakened by suppression, but by an amount insufficient to affect its potency as an adapting pattern. It would be instructive to repeat some of those experiments using weak adapting patterns for which modest reductions in, say, contrast produced significantly weaker aftereffects. In this way one could avoid potential ceiling effects in aftereffect strength. Research along these lines was initiated by Lehky and Blake (1991), and more comprehensive studies are currently underway in our laboratory.

At the same time, it should be noted that suppression does reduce the magnitude of the spiral aftereffect (Wiesenfelder and Blake, 1990), which involves adaptation to rotation/expansion, and the magnitude of the plaid motion aftereffect (van der Zwan et al., 1993), which involves adaptation to motion uniquely specified by several components. Now, it is commonly believed that these more complex forms of optic flow are registered in "higher" visual areas (e.g., Graziano et al, 1994). As pointed out earlier, one contemporary version of "late" rivalry posits that the effects of rivalry are amplified at higher processing stages, which could explain why suppression does retard the buildup of aftereffects induced by more complex forms of motion and, for that matter, aftereffects induced by subjective contours (van der Zwan and Wenderoth, 1994).

6.3.4   *Binocular Rivalry with Rapid Eye Swapping*

Logothetis et al. (1996) created a new type of rival display that completely eliminated the possibility of "eye" rivalry, that is a variant of "early" rivalry. With these displays, two orthogonally oriented gratings were interchanged between the eyes three times each second and, as well, each grating flickered on/off at 18 Hz in the attempt to mask the transients accompanying the eye swaps. If it were the "eyes" that engaged in rivalry, observers should see a grating rapidly flipping back and forth between the two orientations, since this is what each eye viewed. But, in fact, observers experienced alternating periods of prolonged dominance of one orientation and then the other, with the durations of dominance and suppression being equivalent to those measured under conventional rivalry conditions. Logothetis et al. termed this "stimulus" rivalry since a given stimulus remained dominant over multiple eye exchanges. This remarkable observation has been replicated in our laboratory (Lee and Blake, 1999), and there is no doubt that stimulus rivalry represents a potentially important instance of bistable perception. But does it disprove "eye" rivalry?

Several studies have shown that dissimilar monocular targets that are flickered rapidly do not undergo binocular rivalry. Instead, the two targets appear superimposed (Wade, 1973; Wolfe, 1983a). According to Wolfe (1983b), transient signals appear to overwhelm or bypass the rivalry mechanism. Of course, the stimulus conditions created by Logothetis et al. involved a steady train of transients, as the two patterns were swapped back and forth between the eyes at 3 Hz and were flickered on/off at 18 Hz. Commenting on the Logothetis et al. study, Wolfe (1996) speculated that the combination of contrast and flicker in Logothetis et al.'s displays might create a state that "preferentially disrupts the interocular processes." (p. 588). Sang-Hun Lee and I (Lee and Blake, 1999) sought to evaluate the role of sharp transients in the production of stimulus rivalry, by varying the flicker rate and by "smoothing" the transitions over time using a gaussian ramp rather than a square-wave. Figure 6.6 summarizes one of our experiments. We started by identifying the flicker rate, spatial frequency, and contrast values maximizing the incidence of stimulus rivalry with abrupt eye exchange (the condition studied by Logothetis et al.). Using those stimulus values, we then measured the incidence of stimulus rivalry ("slow alternations in dominance") and "eye" rivalry ("regular, rapid alternations in grating orientation") under several different conditions of flicker. Stimulus rivalry was markedly reduced when the eye exchanges were made gradually, not abruptly, or when the 18 Hz flicker was removed. When the gratings were exchanged at 3 Hz between the eyes with ramped contrast and when 18 Hz flicker was eliminated, observers nearly always saw regular, rapid alternations (implying "eye" rivalry). These findings imply, then, that abrupt transients are an important ingredient in the production of stimulus rivalry.

Now, there is no doubt that alternations in dominance can be observed under conditions that preclude "eye" rivalry, and this finding represents an important contribution to the study of perceptual bistability. It remains arguable, however, whether rivalry under conventional conditions also entails alternations in dominance between competing stimuli, not between competing eyes. This question deserves further examination, and it is gratifying to see that other laboratories are already pursuing the answers (Matthews et al., 2000; Bonneh et al., 2001).

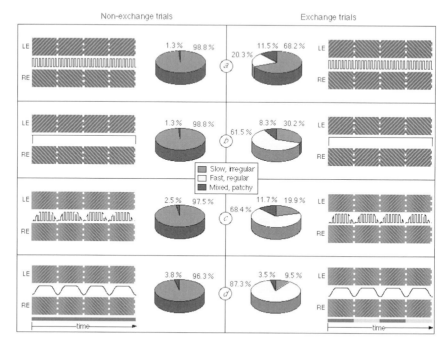

FIGURE 6.6. Schematic of stimulus conditions used to study the incidence of "stimulus" rivalry, "eye" rivalry, and "patchwork" rivalry during "nonexchange trials" (on which a given eye continued to see the same rival target) and "exchange trials" (on which rival targets were exchanged between the eyes at the rate of 3 swaps/second). (a) Rival gratings flickered on/off abruptly at 18 Hz and, for the "exchange" trials, the eye swaps occurred abruptly (these conditions essentially replicate those used by Logothetis et al., 1996); (b) 18 Hz flicker was eliminated, but eye swaps on exchange trials occurred abruptly; (c) gratings flickered at 18 Hz but 3 Hz eye swapping followed a gaussian temporal window that eliminated abrupt transients at the time of swaps (equivalent contrast ramping occurred in the nonexchange trials, but without eyeswapping); and (d) 18 Hz flicker was eliminated and eye-swapping was gradual. Following each 10-sec observation period, observers indicated whether the gratings underwent slow and irregular changes in dominance (indicative of "stimulus" rivalry in the case of "exchange trials"), fast and regular changes (indicative of "eye" rivalry in the case of exchange trials), or mixed and patchy rivalry; histograms show the incidence of these three categories of phenomenal report for nonexchange and exchange trials. Reprinted from *Vis. Res,*, 39: 1447–1454, 1999, Lee, S-H. and Blake, R., Rival ideas about binocular rivalry, with permission from Elsevier Science.

6.3.5   Other Evidence Bearing on the "Early" vs. "Late" Distinction

There are several other lines of investigation that potentially bear on the debate concerning "early" versus "late" rivalry, and I will just mention them in passing.

To the extent that rivalry dominance and suppression are influenced by cognitive factors such as meaning or affective content, we would be inclined to characterize rivalry as a "late" process (e.g., Walker, 1978). But as pointed out earlier, rivalry transpiring at a relatively early level could be modulated by "top-down" influences triggered by more refined analyses of the dominant stimulus at higher stages. In general the existence of feedback from higher to lower visual areas blurs the distinction between these two alternative views of rivalry.

Three other, related lines of evidence that could shed light on this debate are provided by single-cell recordings from alert animals experiencing rivalry (e.g., Leopold and Logothetis, 1996; Sheinberg and Logothetis, 1997), by evoked potential measurements in humans (e.g., Brown and Norcia, 1997; Valle-Inclan et al., 1999) and by brain imaging experiments in humans (Lumer et al., 1998; Tong et al., 1998; Tononi et al., 1998; Tong and Engel, 2001; Lee and Blake, 2001). But from the outset one must be mindful of what potentially constitutes a "neural signature" for rivalry in single cell responses, in evoked potentials, and in brain imaging studies. The answer to this question is not unequivocal. From the perceptual standpoint, rivalry undeniably involves the appearance and disappearance of complex, potentially interesting stimuli for several seconds at a time. But these remarkable fluctuations in dominance and suppression do not necessarily mean that the underlying neural events are comparably remarkable. As suggested earlier, modest shifts in the balance of activity among competing neural representations could be sufficient to trigger alternations in perception. And we must always be mindful of the possible role of feedback projections from higher visual areas onto "early" areas, feedback that could produce modulations in fMRI signals. For that matter, it is conceivable that suppression is occasioned by a temporary disruption in the temporal patterning of activity within populations of neurons, not just by reductions in activity level (e.g., Fries et al., 1997; Lumer, 1998; Engel et al., 1999).

Finally, I wish to acknowledge Pettigrew's very stimulating ideas on binocular rivalry. In a recent essay (Pettigrew, 2001), he argues that individual differences in rivalry rates and the correlation of rate with mood and circadian rhythm implicate brain areas other than V1 in binocular rivalry. Instead, he develops a case for rivalry being driven by a neural oscillator located in subcortical structures. He concludes by characterizing rivalry as a form of perceptual decision making, which — along with his rejection of V1 rivalry — probably qualifies him as a member of the "late" camp. Some of his arguments against V1 are based on circumstantial evidence (e.g., rivalry is abolished by the emotional relief provided by laughter), but considering all his reasons as an aggregate one must acknowledge that it provides a fresh, provocative outlook on rivalry.

6.4   Does the "Early" vs. "Late" Distinction Remain Tenable?

The discussion of possible neural concomitants of rivalry brings me to a final point concerning the "early" versus "late" debate. In the past, students of rivalry, myself included, have tended to speak of "**the** rivalry mechanism" as if rivalry resulted from a single neural process operating at a given site — "early" or "late" — within the visual pathways. This is certainly the way Walker (1978) framed the debate in his important review article, and it is the way I cast my neural model of rivalry a decade later (Blake, 1989). We now have good reason to believe that this view is wrong, for it now appears that rivalry entails multiple neural operations that may well be neurophysiologically and, perhaps, neuroanatomically distinct. Rather than paraphrase to make this point, let me reproduce a passage from my recent *Brain and Mind* (2000) review article on rivalry:

> Binocular rivalry can be characterized in terms of its spatial extent, its temporal dynamics and its generality beyond those conditions triggering rivalry. These aspects of rivalry are not necessarily tied to a single, omnibus process. ... The stimulus determinants of fluctuations in dominance and suppression (i.e., temporal dynamics) are not necessarily those governing the spatial extent of rivalry. Similarly, it is conceivable that the well-established stimulus determinants of suppression phases (i.e., energic variables like luminance and contrast) differ from those controlling dominance durations (e.g., context, meaning). Indeed, to reconcile the diverse findings concerning rivalry — including controversial results on the role of meaning and context in rivalry — it may be important to distinguish between processes responsible for initiation of rivalry and selection of one eye's input for dominance from processes responsible for the implementation and maintenance of suppression. It is entirely plausible that suppression and selection are the result of separate mechanisms operating at different stages in the visual system.

This idea that rivalry entails multiple, distributed neural operations has been elaborated on by Fox (1991), and it is gaining popularity in the contemporary literature (e.g., Logothetis, 1998; Ooi and He, 1999; Bonneh et al., 2001). To the extent that this view takes hold, it will become meaningless to continue the "early" versus "late" debate, for rivalry would comprise "early" and "late" components.

Does this new way of thinking about rivalry undermine its utility as a psychoanatomical tool? The answer is "no" and, in fact, as we gain deeper understanding of rivalry's operations, we may well have in hand a tool with multiple purposes. First, rivalry suppression may indeed provide a paradigm condition for studying the neural correlates of conscious visual awareness (Blake, 1997; Logothetis, 1998), for during rivalry a complex, suprathreshold stimulus can disappear from visual awareness for seconds at a time. When exploiting suppression, however, we need to be mindful of the distinction between "stimulus suppressed" and "stimulus absent." Second, during dominance phases of rivalry locally distributed features within the visual field tend to unite, forming coherent shapes (Kovacs et al., 1996; Alais and Blake, 1999; Whittle et al., 1968; Alais et al., 2000). Because of this tendency, rivalry offers a potentially useful means for studying global perceptual organization and the resolution of perceptual ambiguity (Wolfe, 1996). And third, rivalry involving interocular competition may provide a useful tool for assessing the binocular visual system's tolerance for monocular mismatches. Rivalry, in other words, reveals some of the limits for binocular fusion. At the same time, binocular rivalry may shed light on one important process underlying binocular single vision. It is true that left and right foveae rarely receive dissimilar stimulation for any length of time. The oculomotor system attempts to correct this situation by altering vergence angle until matching features are imaged on the two foveae. There are, however, locations on the two retinae where dissimilar monocular images strike corresponding retinal areas, an inevitable consequence of the geometry of binocular vision. Objects located well in front of or well behind the horopter do cast images on distinctly different areas of the two eyes, and the resulting disparities will be too large for the stereoscopic system to resolve. Yet one ordinarily does not experience the consequences — confusion or diplopia — of this dissimilar monocular stimulation. Binocular single vision, then, may be accomplished by two processes — binocular fusion and binocular suppression (Ono et al., 1977). The potent inhibitory process revealed during rivalry, in other words, may contribute to normal binocular single vision.

In conclusion, as an inferential tool for studying levels of perceptual processing, binocular rivalry remains alive and well. Indeed, its utility may exceed previous expectations, particularly as future work expands our understanding of the workings of binocular rivalry.

6.5   Postscript

Coverage in this review is selective and, consequently, some important recent work on rivalry is not included. Interested readers may want to consult any of several review articles (Blake, 2001) and chapters (Fox, 1991; Howard and Rogers, 1995) for a more complete overview of rivalry. Interested readers should also consult Robert O'Shea's up-to-date bibliography of rivalry articles, available at:

http://psy.otago.ac.nz/r_oshea/br_bibliography.html

Finally, readers interested in experiencing different aspects of binocular rivalry are invited to visit the author's webpage where a number of rivalry phenomena are demonstrated:

http://www.psy.vanderbilt.edu/faculty/blake/rivalry/BR.html

Acknowledgement

Some of the work reported in this chapter was supported by NIH Grant EY13358. I am grateful to Sang-Hun Lee for help in preparing this chapter and to Jeremy Wolfe and Nikos Logothetis for helpful discussion.

References

Abadi, R. (1976). Induction masking – a study of some inhibitory interactions during dichoptic viewing. *Vis. Res.*, 16: 269–275.

Alais, D. and Blake, R. (1999). Grouping visual features during binocular rivalry. *Vis. Res.*, 39: 4341–4353.

Alais, D., O'Shea, R. P., Mesana-Alais, C. and Wilson, I. G. (2000). On binocular alternation. *Percept.*, 29: 1437–1445.

Andrews, T. J. and Blakemore, C. (1999). Form and motion have independent access to consciousness. *Nature Neuroscience*, 2: 405–406.

Andrews, T. J. and Purves, D., (1997). Similarities in normal and binocular rivalrous viewing. *Proc. Nat. Acad. Sci. USA*, 94:9905–9908.

Blake, R. (1988). Dichoptic reading: The role of meaning on binocular rivalry. *Percept. Psychophys.*, 44: 133–141.

Blake, R. (1989). A neural theory of binocular rivalry. *Psyc. Rev.*, 96: 145–167.

Blake, R. (1995). Psychoanatomical strategies for studying human vision. In T. Papathomas, C. Chubb, E. Kowler and A. Gorea (eds), *Early Vision and Beyond*, Cambridge, MA: MIT Press.

Blake, R. (1997). What can be perceived in the absence of visual awareness? *Cur. Dir. Psych. Sci.*, 6: 157–162.

Blake, R. (2001). A primer on binocular rivalry, including current controversies. *Brain and Mind*, 2: 5–38.

Blake, R., Ahlstrom, V. and Alais, D. (1997). Can context boost predominance during binocular rivalry? *Invest. Ophthal. Vis. Sci.*, 38: S908.

Blake, R. and Fox, R. (1974). Adaptation to "invisible" gratings and the site of binocular rivalry suppression. *Nature*, 249: 488–490.

Blake, R., O'Shea, R. P. and Mueller, T. J. (1992). Spatial zones of binocular rivalry in central and peripheral vision. *Vis. Neurosci.*, 8: 469–478.

Blake, R., Westendorf, D. and Overton, R. (1979). What is suppressed during binocular rivalry? *Percept.*, 9: 223–231.

Blake, R., Yu, K., Lokey, M. and Norman, H. (1998). Binocular rivalry and visual motion. *J. Cog. Neurosci.*, 10: 46–60.

Bonneh,Y., Sagi, D. and Karni, A. (2001). A transition between eye and object rivalry determined by stimulus coherence. *Vis. Res.*, 41: 981–989.

Breese, B. B. (1899). On inhibition. *Psycholog. Monograph.*, 3: 1–65.

Brown, R. J. and Norcia, A. M. (1997). A method for investigating binocular rivalry in real-time with the steady-state VEP. *Vis. Res.*, 37: 2401–2408.

Carlson, T. A. and He, S. (2000). Visible binocular beats from invisible monocular stimuli during binocular rivalry. *Cur. Biol.*, 10: 1055–1058.

Carney, T., Shadlen, M. and Switkes, E. (1987). Parallel processing of motion and colour information. *Nature*, 328: 647–649.

Cave, C., Blake, R. and McNamara, T. (1998). Binocular rivalry disrupts visual priming. *Psych. Sci.*, 9: 299–302.

Creed, R. S. (1935). Observations on binocular fusion and rivalry. *J. Physiol. (Lond.)*, 84: 381–392.

Diaz-Caneja, E. (1928). Sur l'alternance binoculaire. *Ann. Oculist*, October: 721–731.

Dörrenhaus, W. (1975). Musterspezifischer visueller Wettstreit. *Naturwissenschaften*, 62: 578–579.

Engel, A. K., Fries, P., Konig, P., Brecht, M. and Singer, W. (1999). Temporal binding, binocular rivalry and consciousness. *Conscious. Cog.*, 8: 128–151.

Fahle, M. (1987). Naso-temporal asymmetry of binocular inhibition. *Invest. Ophthal. Vis. Sci.*, 28: 1016–1017.

Fox, R. (1991). Binocular rivalry. In D. M. Regan (ed.), *Binocular Vision and Psychophysics*. pp. 93–110. London: MacMillan Press.

Fox, R. and Check, R. (1968). Detection of motion during binocular rivalry suppression. *J. Exp. Psych.*, 78: 388–395.

Fox, R. and Check, R. (1972). Independence between binocular rivalry suppression duration and magnitude of suppression. *J. Exp. Psych.*, 93: 283–289.

Fries, P., Roelfsema, P. R., Engel, A. K., Konig, P. and Singer, W. (1997). Synchronization of oscillatory responses in visual cortex correlates with perception in interocular rivalry. *Proc. Nat. Acad. Sci. USA*, 94: 12699–12784.

Graziano, M. S. A., Andersen, R. A. and Snowden, R. J. (1994). Tuning of MST neurons to spiral motions. *J. Neurosci.*, 14: 54–67.

Grossberg, S. (1987). Cortical dynamics of three-dimensional form, color and brightness perception: 2. Binocular theory. *Percept. Psychophys.*, 41: 117–158.

Grossberg, S. (2000). The complementary brain: unifying brain dynamics and modularity. *Trends Cog. Sci.*, 4: 233–246.

Hollins, M. (1980). The effect of contrast on the completeness of binocular rivalry suppression. *Percept. Psychophys.*, 27: 550–556.

Howard, I. P. and Rogers, B. J. (1995). *Binocular Vision and Stereopsis*. New York: Oxford University Press.

Julesz, B. (1971). *Foundations of Cyclopean Perception*. Chicago, IL: University of Chicago Press.

Kanwisher, N. and Wojciulik, E. (2000). Visual attention: insights from brain imaging. *Nature Rev. Neurosci.*, 1: 91–100.

Kovacs, I., Papathomas, T. V., Yang, M. and Feh'r, A. (1996). When the brain changes its mind: Interocular grouping during binocular rivalry. *Proc. Nat. Acad. Sci. USA*, 93: 15508–15511.

Lack, L. (1978). *Selective Attention and the Control of Binocular Rivalry*. The Hague: Mouton.

Lee, S-H. and Blake, R. (1999). Rival ideas about binocular rivalry. *Vis. Res.*, 39: 1447–1454.

Lee, S-H. and Blake, R. (2001). V1 activity is reduced during binocular rivalry. *Vis. Sci.*, Abstract #446.

Lehky S. R. (1988). An astable multivibrator model of binocular rivalry, *Percept.*, 17: 215–228.

Lehky, S. and Blake, R. (1991). Organization of binocular pathways: Modeling and data related to rivalry. *Neural Comp.*, 3: 44–53.

Lehmkuhle, S. and Fox, R. (1976). Effect of binocular rivalry suppression on the motion aftereffect. *Vis. Res.*, 15: 855–859.

Lennie, P. (1998). Single units and visual cortical organization. *Percept.*, 27: 889–935.

Leopold, D. and Logothetis, N. (1996). Activity changes in early visual cortex reflect monkeys' percepts during binocular rivalry. *Nature*, 379: 549–553.

Levelt, W. (1965). *On Binocular Rivalry*. Soesterberg: Institute for Perception RVO-TNO.

Logothetis, N. K. (1998). Single units and conscious vision. *Phil. Trans. Royal Soc. Lond. B*, 353: 1801–1818.

Logothetis, N. K., Leopold, D. A. and Sheinberg, D. L. (1996). What is rivalling during binocular rivalry? *Nature*, 380: 621–624.

Lumer, E. D. (1998). A neural model of binocular integration and rivalry based on the coordination of action-potential timing in primary visual cortex. *Cereb. Cortex*, 8: 553–561.

Lumer, E. D., Friston, K., and Rees, G. (1998). Neural correlates of perceptual rivalry in the human brain. *Science*, 280: 1930-1934.

Mather, G. (2001). Object-oriented models of cognitive processing. *Trends Cogn. Sci.*, 5: 182-184.

Matthews, N., Geesaman, B. J. and Quan, N. (2000). The dependence of motion repulsion and rivalry on the distance between moving elements. *Vis. Res.*, 40: 2025–2036.

Matsuoka, M. (1984). The dynamic model of binocular rivalry. *Biol. Cyber.*, 49: 201–208.

Meenes, M. (1930). A phenomenological description of retinal rivalry. *Am. J. Psych.*, 42: 260–269.

Mueller, T. J. (1990). A physiological model of binocular rivalry. *Vis. Neurosci.*, 4: 63–73.

Ngo, T. T., Miller, S. M., Liu, G. B. and Pettigrew, J. D. (2000). Binocular rivalry and perceptual coherence. *Cur. Biol.*, 10: R134–R136.

Nguyen, V. A., Freeman, A. W. and Wenderoth, P. (2001). The depth and selectivity of suppression in binocular rivalry. *Percept. Psychophys.*, 63: 348–360.

Ono, H., Angus, R. and Gregor, P. (1977). Binocular single vision achieved by fusion and suppression. *Percept. Psychophys.*, 21: 513–521.

Ooi, T. L. and He, Z. J. (1999). Binocular rivalry and visual awareness: the role of attention. *Percept.*, 28: 551–574.

O'Shea, R. P. and Crassini, B. (1981). Interocular transfer of the motion aftereffect is not reduced by binocular rivalry. *Vis. Res.*, 21: 801–804.

Pettigrew, J. D. (2001). Searching for the switch: neural bases for perceptual rivalry alternations. *Brain and Mind*, 2: 85–188.

Sheinberg, D. L. and Logothetis, N. K. (1997). The role of temporal cortical areas in perceptual organization. *Proc. Nat. Acad. Sci. USA*, 94: 3408–3413.

Sherrington, C. S. (1906). *Integrative Action of the Nervous System*. New Haven: Yale University Press.

Sobel, K. and Blake, R. (2001). Does context influence a rival target's escape from suppression? *Vis. Sci.*, #173.

Teller, D. Y. (1984). Linking propositions. *Vis. Res.*, 24: 1233–1246.

Tong, F. and Engel, S. (2001). Interocular rivalry revealed in the cortical blind-spot representation. *Nature*, 411: 195–199.

Tong, F., Nakayama, K., Vaughan, J. T. and Kanwisher, N. (1998). Binocular rivalry and visual awareness in human extrastriate cortex. *Neuron*, 21: 753–759.

Tononi, G., Srinivasan, R., Russell, D. P. and Edelman, G. M. (1998). Investigating neural correlates of conscious perception by frequency-tagged neuromagnetic responses. *Proc. Nat. Acad. Sci. USA*, 95: 3198–3203.

Treisman, A. M. (1962). Binocular rivalry and stereoscopic depth perception. *Q. J. Exp. Psych.*, 14: 23–37.

Valle-Inclan, F., Hackley, S. A., de Labra, C. and Alvarez, A. (1999). Early visual processing during binocular rivalry studied with visual evoked potentials. *NeuroReport*, 10: 21–25.

van der Zwan, R. and Wenderoth, P. (1994). Psychophysical evidence for area V2 involvement in the reduction of subjective contour tilt aftereffects by binocular rivalry. *Vis. Neurosci.*, 11: 823–830.

van der Zwan, R., Wenderoth, P. and Alais, D. (1993). Reduction of a pattern-induced motion aftereffect by binocular rivalry suggests the involvement of extrastriate mechanisms. *Vis. Neurosci.*, 10: 703–709.

Wade, N. J. (1973). Binocular rivalry and binocular fusion of afterimages. *Vis. Res.*, 13: 999–1000.

Wade, N. J. (1974). The effect of orientation in binocular contour rivalry of real images and afterimages. *Percept. Psychophys.*, 15: 227–232.

Wade, N. J. and Wenderoth, P. (1978). The influence of colour and contour rivalry on the magnitude of the tilt after-effect. *Vis. Res.*, 18: 827–836.

Wales, R. and Fox, R. (1970). Increment detection thresholds during binocular rivalry suppression. *Percept. Psychophys.*, 8: 90–94.

Walker, P. (1978). Binocular rivalry: central or peripheral selective processes? *Psych. Bul.*, 85: 376–389.

Walker, P. and Powell, D. J. (1979). The sensitivity of binocular rivalry to changes in the nondominant stimulus. *Vis. Res.*, 19: 247–249.

White, K. D., Petry, H. M., Riggs, L. A. and Miller, J. (1978). Binocular interactions during establishment of McCollough effects. *Vis. Res.*, 18: 1201–1215.

Whittle, P., Bloor, D. and Pocock, S. (1968). Some experiments on figural effects in binocular rivalry. *Percept. Psychophys.*, 4: 183–188.

Wiesenfelder, H. and Blake, R. (1990). The neural site of binocular rivalry relative to the analysis of motion in the human visual system. *J. Neurosci.*, 10: 3880–3888.

Wiesenfelder, H. and Blake, R. (1991). Apparent motion can survive binocular rivalry suppression. *Vis. Res.*, 31: 1589–1600.

Wilson, H. R., Blake, R. and Lee, S-H. (2001). Dynamics of travelling waves in visual perception. *Nature*, 412: 907–910.

Wolfe, J. M. (1983a). Influence of spatial frequency, luminance, and duration on binocular rivalry and abnormal fusion of briefly presented dichoptic stimuli. *Percept.*, 12: 447–456.

Wolfe, J. M. (1983b). Afterimages, binocular rivalry, and the temporal properties of dominance and suppression. *Percept.*, 12: 439–445.

Wolfe, J. M. (1986). Stereopsis and binocular rivalry. *Psych. Rev.*, 93: 269–282.

Wolfe, J. M. (1996). Resolving perceptual ambiguity. *Nature*, 380: 587–588.

7

The Making of a Direction Sensing System for the Howard Eggmobile

Hiroshi Ono, Linda Lillakas, and Alistair P. Mapp

7.1 Preamble

In 2000, the first author gave a talk entitled "The Laws of Visual Direction Require Some Y2K Upgrades" at a meeting of *The Institute of Electronics, Information, and Communication Engineers* (Japan). This provided him with an opportunity to introduce our ideas to an audience unfamiliar with our work and he was looking forward to the meeting. But when the time came, he dropped the ball. After a brief historical introduction, he defined several terms and then gave an axiomatic statement of what Ono and Mapp (1995) have called the Wells–Hering's laws of visual direction and the deductions from these laws. The main message was to be that the laws, as stated, are incomplete. His talk was received with uniform boredom. Since that time we have been wondering about how to make that body of literature more interesting and "understandable," and in this chapter we are going to try a different approach. In tribute to Ian Howard in this Festschrift, we tell a fairy tale about how a group of engineers worked on a system that processes directional information.

7.2 Introduction

Once upon a time (for our Japanese colleagues, "mukashi mukashi"), a group of engineers and sorcerers were asked by their patron the CVR (Centre for Victims of Retirement) at York University to design a direction–processing system to be incorporated into Ian Howard's eggmobile (Figure 7.1). The initial requirement was simple. The system was to calculate the direction of an object without having to identify or recognise it, attend to it, or to compute its distance. Different groups of wizards were to work on those problems. The Direction Group's task was to devise an input—output system capable of calculating two types of direction; the relative and the absolute directions of objects as discussed in Howard's (1982) *Human Visual Orientation*. More specifically, this system was to calculate and

OntarioScienceCentre   August1978

FIGURE 7.1. The poster for the eggmobile contest. In 1978, the Ontario Science Centre in Toronto sponsored a contest to see how far an egg could be propelled using a single rubber band for power. The idea was to design a very energy efficient transportation system. Ian Howard won the Canadian championship with the eggmobile shown here. In the international competition Ian moved the egg 351.21 meters and placed a close second behind Steve Darling, a Rolls-Royce engineer.

output the direction of an object as being (a) to the left or right of another object (relative direction) and by how much, and (b) to the left or right of the eggmobile's median plane (absolute direction) and by how much. These outputs of left or right and of how much to the left or right are not necessarily the same for the two types of directional tasks. This is because the output for each of the two direction types is specified with respect to a different reference axis or fiducial line, namely, (a) an arbitrarily chosen reference stimulus and (b) the median plane of the frontend. The CVR wanted the eggmobile to process information about both relative and absolute direction, because they could foresee attaching a miniature Canadarm in the future. When the eggmobile goes to a pub with Ian, for example, the relative direction of a pitcher and a mug is critical in pouring beer, while the absolute direction of the bull's-eye can be critical when throwing darts.

   In this chapter we tell the tale of how this ragtag group of misfits and wanderers, including Ian and Toni Howard and the authors of this chapter, struggled with the development of this direction-sensing system. What was initially thought to be a simple task became quite complicated as the other groups of engineers and sorcerers demanded modifications to suit their needs, most of which were unrelated to the task at hand for the Direction Group. This competition of needs and specifications demanded by the various groups prevented us from producing a simple, elegant, and straightforward system, which would have served its purpose in most situations. As our story unfolds, it will become clear why the final incarnation of the system attached to the eggmobile does not work perfectly.

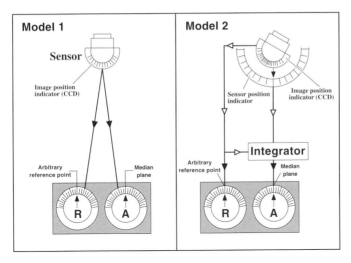

FIGURE 7.2. Schematic representations of Models 1 and 2. Both models are single-sensor systems. In Model 1 the sensor is rigidly fixed in place, whereas in Model 2 the sensor can rotate. In Model 1, the relative and absolute directions are computed from information about the location of images on the sensor's CCD and are registered as outputs in the R (relative) and A (absolute) dials. In Model 2, the relative direction is computed as in Model 1, but the absolute direction is computed by combining the information from the CCD and that from the sensor position indicator. The output dials are the same as in Model 1.

7.3 The Making of Models 1 and 2: A Single, Centrally Located Input Device

The first two systems designed by the Direction Group engineers were modelled after the Schmidt telescope, which "combines large aperture, favourable ratio of aperture to focal length, and extraordinarily wide field of view with a highly efficient photographic process" (Marx and Pfau, 1992, p. 4). We redesigned the optical system of the Schmidt telescope to focus on objects closer than the stars and we replaced the film with the CCD (charged coupled device) of a video system. A schematic representation of our first system, "Model 1," is depicted in Figure 7.2. Since the input device for this model was rigidly fixed onto the eggmobile's frontend, the software required to compute the two possible outputs (i.e., relative and absolute direction) was straightforward. Information about the relative direction of one object with respect to another was provided by the location of the two images on the CCD. The greater the separation between the images, the greater the difference in relative direction. Also, the location of the object's image relative to the centre of the CCD provided complete information about the absolute direction of the object relative to the frontend. If the image was on the centre of the CCD, the object was on the median plane, and the farther the image was from this centre, the more eccentric was the absolute direction of the object.

The CVR was satisfied with Model 1, and all was well until they heard from another group of engineers who were working on a different system. The group working on object and pattern recognition needed an expensive high-resolution imaging system to perform their job, but they were restricted by a CVR-imposed financial constraint. To meet this constraint, the Pattern Recognition Group decided to restrict the high-resolution portion of the CCD to its central region. They reasoned that it would be cheaper to attach a motor to rotate the sensor than to have high resolution over the entire CCD. Their idea was that once an object was detected the sensor could be rotated so that the image of the object would fall on the area of highest resolution. The CVR agreed with this idea and hired another group of engineers to work on the motors to rotate the sensor. This new group was called the Oculomotor Group.

Having the sensor rotate did not create a problem for the Direction Group. The Schmidt telescope was designed to rotate in order to point to different parts of the sky and to track the stars and, therefore, we were familiar with the idea of a rotating sensor. Our relative direction software did not require any modifications because the angular position of the sensor is irrelevant for the computation of relative direction; all that is needed is information about the locations of the images of the reference stimulus and the object of interest, just as in Model 1. For absolute direction, however, the angular position of the sensor is crucial. That is, to compute the absolute direction of an object, information about the location of the object's image on the CCD must be integrated with information about the angular position of the sensor. One consequence of this additional computation is that the output of absolute direction is less precise than that of relative direction, because it has two sources of noise. A schematic representation of this system, "Model 2," is depicted in Figure 7.2.

Our patron was satisfied with our design. This satisfaction, however, was also short-lived as the Distance/Depth Group convinced the CVR that they needed two sensors. The CVR ordered us to create an input–output system consisting of two sensors, each with the same characteristics as the single sensor used in Model 2.

7.4   The Making of Model 3: Two Frontally Located Input Devices

There were lively debates among the members of the Direction Group as to how to proceed. First, we tried to persuade CVR to change its mind. We argued that if the eggmobile were to move leftward and rightward, upward and downward, or some combination of the two, information about distance and depth could be derived from its single sensor (see, e.g., Bourdon, 1902; Heine, 1905; Rogers and Graham, 1979; Ono, Rivest, and Ono, 1986). But the CVR insisted on implementing the Distance/Depth Group's proposal (which was referred to by some as creating a "two-eyed monster"). Our group was of two minds. One subgroup argued for keeping the software from Model 2 intact and writing additional software integrating the outputs from the two units, while the other subgroup argued for writing brand new software. Unable to resolve this disagreement, the Direction Group hired a consultant, a German engineer named Hering (1879/1942, 1868/1977) who, as it turned out, was already a consultant to the newly established Oculomotor Group.

Hering (1879/1942, 1868/1977) sided with the second group. He advised us to treat the two sensors not as two separate units but as two halves of a single unit. He told us that he also advised the Oculomotor Group to treat the two sets of motors, each controlling one sensor, as a single unit. For example, his advice for rotating the two sensors horizontally was to have two subsystems, one to rotate the two sensors rightward or leftward and another to rotate them inward or outward.

What became obvious while discussing the new requirements with Hering was that a centre or origin, like that of a polar coordinate system in plane geometry, would be advantageous. For relative direction, information about where an arbitrary reference line passes through the origin is useful to code whether an object is located to the left or right of that reference line. For absolute direction, information about where the median plane passes through the origin is useful to code whether an object is located to the left or right of it. Without an origin or a reference point the coding would be more complicated. In Models 1 and 2, the nodal point of the single sensor served as the origin of direction. That is, without thinking about this problem, we had placed the origin at the nodal point of the optical system, because we were specifying the direction as an angle. With two sensors, however, the issue of where to place the reference point had to be addressed. What is straight ahead of one sensor is not straight ahead of the other sensor, and two objects that are in the same direction for one sensor are not so for the other sensor. Hering's (1879/1942, 1868/1977) suggestion was to specify the directions of objects from an imaginary sensor positioned midway between the two actual sensors.[1] We will use the term "cyclopean point" when discussing the system we designed for the eggmobile and the term "cyclopean eye" when we discuss what is in the literature with respect to two-eyed humans.

The Direction Group decided to accept all the recommendations made by Hering, but we also had to incorporate the distinction between relative and absolute direction which he did not make in our discussion with him. In addition, we undertook three projects to assess the abilities and limitations of this new two-input direction system. First, we made a schematic representation of the new system as we did for Models 1 and 2; second, we worked out a set of logical rules to compute the output of the system; and third, we created an analytical tool to visualise how the new system would work.

We quickly came up with a schematic representation of the new system: Model 3 (Figure 7.3). Note that, as ordered by CVR, there are two input devices (sensors) and the hardware used to capture the image location on the CCDs and the

---

[1] Since Hering made this suggestion, this reference point has been given many names: binoculus, centre of visual direction, cyclopean eye, double eye, egocentre, and projection centre. The term "cyclopean eye" suggested by another German engineer, Helmholtz (1910/1962), is the most popular one today.

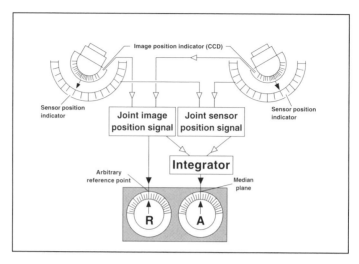

FIGURE 7.3. Schematic representation of Model 3. This model is comprised of two sensors, both of which can rotate. The relative direction is registered as an output in the R dial after the information from the two CCDs is combined. The absolute direction is registered as an output in the A dial after information from the joint image positional signal and that from the joint sensor position signal are combined.

angular position of the sensors are the same as in Model 2. Information about relative direction is output to the R (relative direction) dial after the inputs from the two CCDs are combined. The eggmobile would use this information when performing such tasks as pouring beer from a pitcher to a mug, or making a Nonius or Vernier judgement. This information also goes to the integrator, where it is combined with information about joint sensor position. As in Model 2, information about absolute direction is output to the A (absolute direction) dial after the integrator combines the inputs from the two sources. This information is used in tasks such as throwing a dart, judging whether something is straight-ahead or not, or pointing to a target with an unseen Canadarm. The schematic representation of Hering's idea, with an explicit distinction between relative and absolute directions, makes clear why the output for absolute direction is less precise than that for relative direction. There are two sources of noise (two sets of inputs) for absolute direction, whereas there is only one source (one set) for relative direction. We speculate later as to the consequences of the different noise levels associated with the two outputs.

Finishing the group's second project was easy because of the work done by another consultant, a Scottish engineer named Wells (1792). He made an explicit distinction between coded direction (he called it "visible direction") and coded distance, and he worked out the geometry with clearly defined terminology. The term "coded" acknowledges that the system sometimes makes an error in processing direction or distance. He defined the common axis as a line that passes through the intersection of the two optic axes and the mid-point between the two sensors, which is illustrated in Figure 7.4. Using this concept and assuming that the software written by the Distance/Depth Group correctly calculates the distances of objects, he predicted the coded location of a stimulus using Cartesian coordinates.[2] The Direction Group used Wells's definitions and added the construct of the cyclopean point recommended by Hering to create a set of logical rules. These rules do not make an explicit distinction between relative and absolute direction (i.e., the relative direction is the difference between two absolute directions). Ono and Mapp (1995) labelled these rules the Wells–Hering's laws of visual direction. In this chapter, however, no more is said of this for the reasons stated in the preamble and because the Direction Calculator that we discuss next describes the same predictions.

To devise a magic wand needed to visualise Hering's idea, we relied on two descriptions of his recommendation, one by Julesz (1971) and another by Ono, Mapp, and Howard (2002). The Julesz description reads in part: "This hypothetical eye incorporates the two real eyes into a single entity (with two overlapping retinae) and lies midway between the two real eyes." (Julesz, 1971, p. xi). Ono et al.'s reads: "[T]he vector defined by a visual target and its retinal image (i.e., the visual axis or visual line) . . . transfers to the cyclopean eye by rotating about the point at which [it] intersect[s] the horizontal horopter that includes the intersection of the two visual axes."[3] We realised that this rotation would "overlap" the images on the two CCDs consistent with Julesz's description. Thus, all we had to do was to rotate each optic axis about the point at which it intersected the horopter until it met the common axis. This would bring each CCD to the imaginary CCD. Examples of how this was done and the consequences of doing it are "animated" in the CD-ROM enclosed with this book. We called this wand the Direction Calculator.

First, we used the Direction Calculator to determine the directions of two points on a horopter when the two sensors were directed to the point in the median plane. The Direction Calculator rotated each sensor's field of view as shown in Figure 7.4a. In this situation the system coded the directions correctly. The point to which the two sensors were directed (i.e., the bifixated point[4]) was on the common axis and coded in the correct direction with respect to the cyclopean point. The point on the cyclopean line (i.e., the eccentric point) was also coded in the correct direction. Next, we directed the two sensors to the eccentric stimulus (Figure 7.4b); again we found that both points were coded correctly. Last, we used the Direction Calculator on a more complicated stimulus situation comprised of five points located in different directions, at different distances, and not on the

---

[2]Although Wells does not use the reference point of polar coordinates, his predictions about direction are very close to those made by Hering's recommendation. See Ono (1991) for details.

[3]The horizontal horopter is the circle that passes through the intersection of two optical axes and the nodal points of the two sensors.

[4]The term "bifixation" is used throughout this chapter to mean that the intersection of the two visual axes is on the stimulus as defined in Cline, Hofstetter, and Griffin (1989, page 71), "The fixation of a single object by both eyes simultaneously so that the image on each retina is on the fovea."

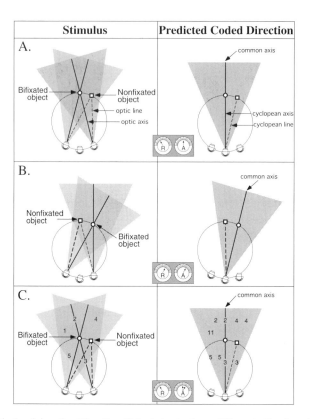

FIGURE 7.4. Applying the Direction Calculator to three different stimulus situations. In the left column, the stimulus situations are depicted and in the right column the predicted coded direction for each stimulus with the output dials R and A are depicted. In row A, the two stimuli are located on the horopter and the one on the median plane is bifixated. Each field of view is rotated at the intersection of the optic axis (or the optic line) and the horopter until the optic axis (or optic line) meets the common axis (the cyclopean line). In row B, the object to the right of the median plane is bifixated, and each field of view is rotated in the same way. In rows A and B, output dial R indicates the relative direction of the bifixated stimulus with respect to the other stimulus, and output dial A indicates the absolute direction of the bifixated stimulus. In row C, five stimuli (1 to 5) are located not on the horopter. When the two fields of view are rotated, each of the five stimuli is coded as occupying two locations. Dial R indicates the relative direction of the diplopic images (the right eye's image relative to that of the left eye) and dial A indicates that each stimulus is coded as having two different directions (dial settings are for stimulus 3). See QuickTime animation of each rotation on the CD-ROM.

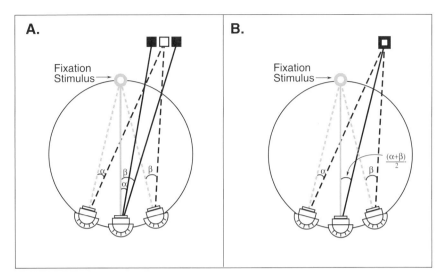

FIGURE 7.5. Coding a stimulus outside the horopter as single. In (a), the stimulus (empty square) outside the horopter subtends different angles with respect to each optic axis (α and β) and the Direction Calculator would place them in two different locations (filled squares). Averaging α and β, as shown in (b), and coding the averaged value as the stimulus's directional value, places the coded direction at the correct direction with respect to the cyclopean point.

horopter. What we found was that all five points were coded as double (in two incorrect directions; Figure 7.4c).

It seemed to us that coding the stimulus as double would be intolerable, behaviourally, for the eggmobile, and we asked our patron, the CVR, to resurrect Model 2 or at least to move the two sensors to each side of the eggmobile so as to eliminate the overlapping inputs. The CVR's answer was firm: they were committed to putting two sensors on the front of the eggmobile. They would instruct the Oculomotor Group to have the two sensors rotate quickly to any object for which direction information was necessary for the well-being of the eggmobile. After all, CVR said, referring to Figures 7.4a and 7.4b, when the two sensors are directed to a point, the direction of that point is correctly coded.

The CVR did take our report about the doubly coded directions seriously, however. They wanted us to do something about the direction of objects that are near the intersection of the two optic axes. We were told that the Distance/Depth Group was working on representing these objects singularly and in depth. Therefore, our assignment was to come up with a way to code the correct direction value for an object near the intersection of the two optic axes. They told us that they would instruct the Attention Group to create a system that would ignore the direction outputs from stimuli that are far away from the intersection. Also, they told us that they would instruct the Oculomotor Group not to use the coded direction to program eye movements, because most stimuli would be mislocalized. The CVR

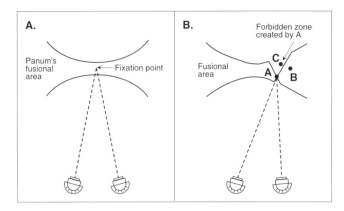

FIGURE 7.6. Area of single vision for a given bifixation. Panum's area of single vision is illustrated in (a). Here the two borders between single and double vision are specified for different eccentricities, one for that of crossed disparity and the other for that of uncrossed disparity. The modified Panum's area proposed by Burt and Julesz (1980a, b) in which there is a forbidden zone for one of two points to fuse, is shown in (b). This zone is defined by the disparity gradient of two points (the difference in their disparities divided by their separation in visual angle) being less than one. If two stimuli are close in direction with respect to the cyclopean point, then one of the two points will not fuse. In (b), point A is bifixated and fused, but point C is not fused, because it is in the forbidden zone, created by point A. Point B is fused, however, because it is not in the zone.

reminded us that humans do not use information about perceived location to program eye movements (Ono and Nakamizo, 1977; Ono and Steinbach, 1983), and there is no reason for the eggmobile to do so. They noted that there is sufficient information on the two CCDs to compute the exact directions and magnitudes of vergence and version required to move the eyes to their destination correctly.

Our task turned out to be simpler than we had anticipated. All that was needed was to average the two direction values with respect to each sensor and, voilà, we had the correct direction with respect to the cyclopean point (Figure 7.5).

The CVR was not explicit on what should be considered "near" or "far" from the intersection of the two optic axes, but we were told about the work started by Panum (1858) in specifying what is coded as single or double. The details of the work Panum started are not yet worked out, but the zone in which a stimulus will be coded single is called Panum's area (Figure 7.6a). According to the plan, the size of the area is dependent upon the eccentricity, the characteristics of the stimulus, and the surrounding stimuli. See Howard and Rogers (1995) for more detail. We were satisfied with this solution, as was the CVR, until *poof!* (or *totsuzen detekita* for our Japanese colleagues) the greatest engineer of all, Leonardo da Vinci, appeared on the scene.

Leonardo told us that this solution could produce a serious problem. If our system is to compute the direction of two objects in Panum's area as having the same direction from the vantage point of the cyclopean point, there is no way to represent what they are — a serious problem for the Pattern Recognition Group. Consider two objects collinear with the cyclopean point. If the objects are opaque, the far one should not be visible because they are both located in the same direction. Leonardo's opinion was that, since he could not solve this problem when he was painting from a "station point," there was no way that we could solve it. According to him, it is a geometrical imperative that if two opaque objects are collinear with respect to a point in space, then the view of the farther object is blocked from this point (see Ono, Wade, and Lillakas (2002) for further discussion).

Our solution to what we now call Leonardo's constraint was to follow Burt and Juleszs's (1980a, b) suggestion and create what is called the "forbidden zone for fusion." In this forbidden zone, only the direction of the bifixated point is coded as single, all other points are coded as double and not in the same direction as the bifixated point (Figure 7.6b). Of course, this is not the best solution because double coding can be bothersome, but the CVR agreed to this because they were committed to using two sensors.

Despite their commitment to the use of a two-sensor system, the CVR were curious about what would happen to the coded absolute directions of targets if one of the sensors was broken or if the eggmobile forgot to remove one of the lens caps. Thus, they asked us to apply the Direction Calculator to Model 3 to see how this model would code absolute direction when only one of the two sensors was operative. The question was whether this new system would code the absolute direction with respect to the cyclopean point correctly, as did Model 2 (the single sensor model). The answer, of course, depends upon the location of the common axis and, thus, we consulted the Oculomotor Group. They reported that they had yoked the distance of the intersection of the two optical axes to the accommodative state of the lenses, but that this yoking was not perfect. That is, when only one sensor is being used the common axis would not pass through the object most of the time, and therefore the coded absolute direction would be inaccurate (Ono and Gonda, 1978; Ono and Weber, 1981). How the incomplete coupling between focusing and the intersection produces inaccurate coding of absolute direction is shown in Figure 7.7 and on the CD-ROM enclosed with this book. When the optic axis of the inactive sensor deviates outward (exophoria), the coded direction is displaced toward that sensor. When that optic axis deviates inward (esophoria), the coded direction is displaced in the opposite direction. (The term "displacement" will be used later to mean a deviation of the coded direction from the veridical direction with respect to the cyclopean point.)

While working on this problem, the Direction Group noticed many stimulus situations in which inaccurate coding occurs, even when both sensors are being used. Before reporting the inaccurate coding of absolute direction when only one sensor is used, we decided to do an archive search on the "problems" encountered by two-eyed humans. Figures 7.8, 7.9, and 7.10 show three kinds of illusions that were uncovered. Figure 7.8 and our animation on the CD-ROM show what is referred to as the cyclopean illusion (e.g., Enright, 1988; Helmholtz, 1910/1962; Hering, 1879/1942; Carpenter, 1988) which has recently received attention in human vision research (e.g., Erkelens, 2000; Erkelens and Van Ee, 2002; Ono, et al., 2002). The illusion consists of an apparent lateral shift in the absolute direction

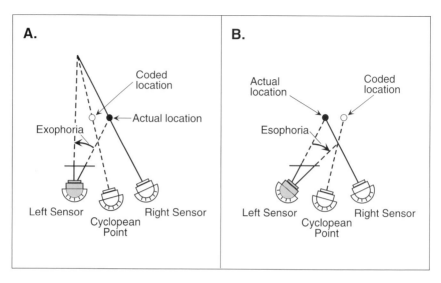

FIGURE 7.7. Coded direction of a stimulus when the left sensor is broken or the lens cap is still on. In (a) exophoria produces apparent displacement to the left, and in (b) esophoria produces apparent displacement to the right. Because the focusing of the lens and the intersection of two optic axes are not completely coupled, using only one sensor produces incorrect absolute direction with respect to the cyclopean point. See the application of the Direction Calculator in the animation on the CD-ROM.

of stimuli positioned on the visual axis of one eye, as one changes fixation from one of the stimuli to another. This occurs because stimuli positioned physically on the visual axis are seen on the common axis, and the angle of the common axis changes as a function of binocular eye position (accommodative-vergence). Figure 7.9 shows four static versions of the cyclopean illusion. In each example, stimuli on the visual axis are incorrectly seen on the common axis. See Alhazen (1083/1989), Hering, (1879/1942), Wells (1792), and Sharpe (1918). Also see Howard (1996) about Alhazen. Figure 7.10 shows the opposite, namely, what is on the common axis is incorrectly seen on the visual axis. See Alhazen (1083/1989), Wells (1792), and Ono and Mapp (1995).

We reported to our patron our results of the inaccurate coding of absolute direction when only one sensor is used and the results of our archive search. We repeated our argument that Model 2 is better than Model 3. The CVR, however, dismissed our argument and responded that they were not surprised with what we reported, saying that they knew from what we had reported earlier that an object that is closer or farther away than the intersection of the optic axes will be coded incorrectly. According to them, our new report just reinforced the idea that the two sensors should be directed toward the object for which direction is of concern. They argued that the situations we found in the archive are not ones likely for the eggmobile to confront and, as such, are only of interest to visual scientists and magicians, and not to real engineers.

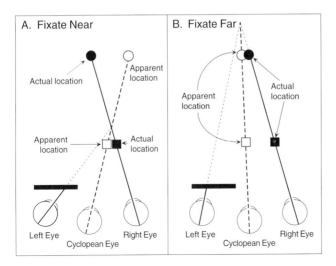

FIGURE 7.8. Illustration of the cyclopean illusion. When fixation changes from the near stimulus (see part a) to the far stimulus (see part b) the absolute visual direction of the far stimulus shifts to the left. The two stimuli on the visual axis of the right eye are seen on the common axis (dashed line from the cyclopean eye). In the cyclopean illusion, the location of the common axis changes with the change in accommodation. See animation on the CD-ROM.

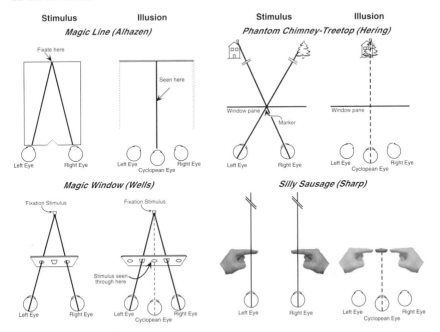

FIGURE 7.9. Illustrations of four static versions of the cyclopean illusion. Each example shows the stimulus and the resultant illusion, that is, a stimulus on the visual axis is seen on the common axis.

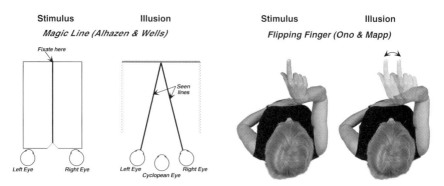

FIGURE 7.10. Illustrations of two illusions that show the opposite of what was shown in Figure 7.9. In the two examples, a stimulus on the common axis is seen on the visual axes.

The CVR was ready to install the hardware along with our software and the software developed by the other groups. However, they realised that the Direction Calculator had only been used to test the coding of direction for pointlike stimuli. They wanted us to further test the system for stimulus situations that would more closely resemble those encountered by the eggmobile i.e., stimuli with surfaces located at different distances.

We applied the Direction Calculator to one of the stimulus situations studied by Leonardo (see Wade, Ono, and Lillakas, 2001, for the other stimulus situations he studied), and found that the software developed for dealing with point stimuli immediately crashed (see animation on the enclosed CD-ROM). It crashed, because there was not enough room on the imaginary CCD at the cyclopean point to insert what was processed by the two sensors. (Leonardo had told us this, but we did not appreciate what he was saying until this point; see Figures 7.11a and b).

One possible "solution" we considered was to exclude half of the area that was visible to only one sensor (the area c to d and the area e to f in Figure 7.11a), but the CVR did not favour this idea. They thought the information in this area might be important for the eggmobile. Given that there was not enough room in the field of view of the station point, or the cyclopean point, the only way for the system to deal with this situation is to compress a part of what is in the imaginary CCD. Moreover, the perceptual displacement that we considered for diplopic images was also necessary to meet the constraint that two opaque objects cannot be seen in the same direction. That is, the area behind the sphere visible to only one sensor must be displaced and compressed (Figure 7.11c).

We proposed this solution to the CVR, and we no longer argued for Model 2, although we felt that compressing and displacing a portion of the field of view seemed unnecessarily complicated. We were tired, and we realised that the CVR was committed to a two-sensor system. The CVR accepted our "solution" saying that these small directional distortions were acceptable since they were only a

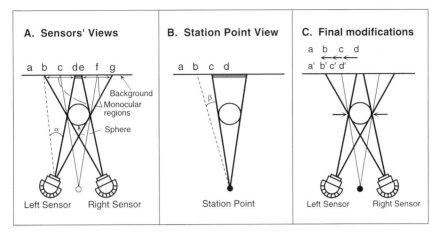

FIGURE 7.11. Illustration of the two sensors' views, the view from the station point, and the final modifications (cyclopean view). Leonardo's constraint that what is seen by two sensors cannot be represented faithfully on a canvas from a single station point is shown in (a) and (b). The field of view represented by angle α cannot fit into the field of view β specified from the station point. In the final modification (c) sent to the CVR, point d is displaced leftward to d', and the area (a) to (d) shrinks to fit into the area a' to d'. This compression is indicated by the distance between a' to d' being smaller than the distance between a to d. Note that similar displacements and compressions occur in the areas seen by the right sensor, but to simplify the figure they are not illustrated.

temporary problem. As the CVR kept reminding us, it had hired the Oculomotor Group to rotate the sensors and to enable the eggmobile to process the direction of a bifixated target correctly. Our job was done as far as the CVR was concerned, and the system we worked on was installed on the eggmobile.

7.5 Conclusion

Not long ago, we had a chance to talk to the eggmobile about how it liked its direction–sensing system. It reported experiencing three illusions, namely the Kanizsa illusion, the Poggendorff illusion, and the moving-moon illusion. Having worked on Models 2 and 3, we offer here our speculations as to why the eggmobile is experiencing these illusions.

We think that the directional distortions caused by including all that is seen from two vantage points into a scene as though from one station point created problems for the Pattern Recognition Group. As illustrated in Figure 7.12b, a square shape will no longer be coded as square nor will a straight oblique line appear straight, when the front surface is bifixated. We think that the CVR asked the Pattern Recognition Group to do something about this directional distortion and that the group succeeded in designing some kind of a correction mechanism. They have not written to us yet to explain what kind of mechanism they made

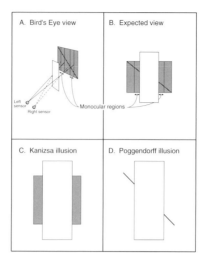

FIGURE 7.12. The expected and coded consequences of the displacement and compression of a portion of the visual field. A bird's-eye view of the stimulus is shown in (a), the expected front view in (b), and the Kanizsa and Poggendorff illusions are shown in (c) and (d) respectively. The effect of compression is illustrated by the smaller width of the monocular regions than the binocular regions in (b), whereas their width is the same in (a). Note that the two illusions in (c) and (d) are opposite to what is expected from the displacement (and compression), as shown in (b). For discussion, see text.

to correct for the directional distortion, but the eggmobile does judge the shape and alignment correctly in three-dimensional space. The studies on humans also show that these two possible consequences are limited to visual direction and do not occur for visual shape and visual alignment in three-dimensional perception (Ohtsuka and Yano, 1994, Ono, Ohtsuka, and Lillakas, 1998, van Ee and Erkelens, 2000). In three-dimensional space, a square behind an occluder is seen as a square (Ohtsuka. 1995a; Ono et al., 1998) and a line behind an occluder is seen aligned (Drobnis and Lawson, 1976; Gyoba, 1978; Liu and Kennedy, 1995; Ohtsuka, 1995b). It appears that the Pattern Recognition Group has modified their software to come up with the correct coding of shape and alignment in three-dimensional space. This correction mechanism is not perfect, however. It causes two-dimensional illusions in shape and alignment (Figures 7.12c and 7.12d). We think that the Pattern Recognition Group used the pictorial cue of occlusion as the trigger for the correction. We are waiting for a fax from this group to quench our curiosity about the correction mechanism they designed.

The eggmobile's experiencing of the moving-moon illusion was a surprise to us, because nothing we worked on in Model 3 should produce the illusion of the moon appearing to move rapidly in the direction opposite to the movement of the clouds. It is as though the information about absolute direction is overridden by the information about relative direction. We speculated that, because of the noise associated with the absolute direction output, the CVR created another group called the Stability Group which worked on location constancy. It is possible that this group decided to use the background as a "frame" to localise an object.[5] We have made several inquiries about this possibility to CVR, but have yet to receive an answer. Perhaps all the members of the CVR were forced to retire at age 65, and there is no one there to answer our questions.

Throughout this fairy tale, we extolled the virtues of Model 2 and pleaded several times with the CVR to reinstate it. We understand why the Distance/Depth Group wanted two frontal sensors, but what we do not understand is why the CVR insisted that we work on a direction-sensing system with two sensors. In retrospect, it seems that when distance/depth information is necessary, the information from two sensors could be used; when direction information is necessary, the information from a single sensor could be used. Perhaps, it is because the Distance/Depth Group decided to discard what visual scientists call "eye signature" after the distance and depth information has been processed, and without it the absolute direction of an object is impossible to determine when only one sensor is used (e.g., Ono and Barbeito, 1985; Steinbach, Howard, and Ono, 1985). Or more likely it is because of some engineering (or alchemy) principles the CVR considered and did not tell us. In any event, the sentiment that Model 2 should be used for direction is reflected by researchers on eye dominance (e.g., Khan and Crawford, 2001; Porac and Coren, 1981; Walls, 1951) and monocular vision (Erkelens, 2000; Erkelens, Muijs, and van Ee, 1996; Erkelens and van Ee 1997, 2002). Some researchers think that the two sensors, each with its own motor unit, work independently, which as we demonstrated with our direction calculator is not correct. This mistake has led to confusion about what the eggmobile codes, which we will not discuss today. We will, instead, refer you to some papers attempting to untangle this confusion (Mapp and Ono, 1999; Ono and Barbeito, 1982; Ono, et al., 2002).

This is the end of the fairy tale of how a group of engineers struggled with making the system to code relative and absolute directions. We hope it is clear that the imperfections of the system are not the fault of the group that worked on it. Rather, they are due to our patron giving in to the demands made by the other groups of engineers working on different systems for the eggmobile. Is the eggmobile happy with our work? We think so, because most of the limitations of our system can be overcome by the system installed by the Oculomotor Group. Yes, indeed the eggmobile is happy enough and so is Ian Howard. The eggmobile now has a machine that registers direction (although not perfectly), distance/depth, pattern, etc, and, perhaps because of these abilities, it wanders off on its own. Ian now spends a fair amount of time looking for it (Figure 7.13).

THE END ("OSHIMAI" for our Japanese colleagues).

---

[5]For a more detailed discussion of the background as a frame for visual stability, see, e.g., Howard, 1991; Ono et al., (2002); Wade and Swanston, 1987.

FIGURE 7.13. Ian is looking for the Howard Eggmobile (photograph by Laurence Harris).

Acknowledgements

This research was supported by Grant A0296 from the Natural Sciences and Engineering Research Council of Canada (H. O.), and a York University Contract Faculty Research Grant (A. P. M.). The authors wish to thank Olivia Espiritu, Esther González, Phil Grove, Miyuki Kamachi, Vladislav Khokhotva, Radha Kohly, Ria Ono, and Krista Phillips for their helpful comments on an earlier version of this chapter.

References

Alhazen, I. (1989). *Book of Optics*, translation by A. I. Sabra, in *The Optics of Ibn al-Haytham*. London: Warburg Institute. (Originally published in 1083.)

Bourdon, B. (1902). *La Perception Visuelle de Lespace*. Paris: Librairie Schleincher Freres.

Burt, P. and Julesz, B. (1980a). Modifications of the classical notion of Panum's fusional area. *Percept.*, 9: 671–682.

Burt, P. and Julesz, B. (1980b). A disparity gradient limit for binocular fusion. *Science*, 208: 615–617.

Carpenter, R. H. S. (1988). *Movement of the Eyes*. London: Pion.

Cline, D., Hofstetter, H. W. and Griffin, J. R., eds. (1989) *Dictionary of Visual Science 4th ed*. Radnor: Chilton Trade Book Publishing, p. 71.

Drobnis, B. J. and Lawson, R. D. B. (1976). The Poggendorff illusion in stereoscopic space. *Percept. and Motor Skills*, 42: 15–18.

Enright, J. T. (1988). The cyclopean eye and its implications: vergence state and visual direction. *Vis. Res.*, 28: 925–930.

Erkelens, C. J. (2000). Perceived direction during monocular viewing is based on signals of the viewing eye only. *Vis. Res.*, 40: 2411–2419.

Erkelens, C. J., Muijs, A. J. M. and van Ee, R. (1996). Binocular alignment in different depth planes. *Vis. Res.*, 36: 2141–2147.

Erkelens, C. J. and van Ee, R. (1997). Capture of the visual direction of monocular objects by adjacent binocular objects. *Vis. Res.*, 37: 1735–1745.

Erkelens, C. J. and van Ee, R. (2002). The role of the cyclopean eye in vision: sometimes inappropriate, always irrelevant. *Vis. Res.*, 42: 1157–1163..

Gyoba, J. (1978). The Poggendorff illusion under stereopsis. *Tohoku Psychologica Folia*, 37: 94–101.

Heine, L. (1905). On perception and conception of distance differences. *Albrecht von Graefes Archiv für Klinische und Experimentelle Opthamologie*, 61: 484–498.

Helmholtz, H. (1962). *Helmholtz's Treatise on Physiological Optics. Vol. 3*, 3rd German ed. and translation by J. P. C. Southall, New York: Dover. (Originally published in 1910.)

Hering, E. (1942). *Spatial Sense and Movements of the Eye*, translation by C. A. Radde, Baltimore: American Academy of Optometry. (Originally published in 1879.)

Hering, E. (1977). In B. Bridgeman (trans.), B. Bridgeman and L. Stark (eds.), *The Theory of Binocular Vision*. New York: Plenum Press. (Originally published in 1868.)

Howard, I. P. (1982). *Human Visual Orientation*. Chichester: John Wiley and Sons.

Howard, I. P. (1991). Spatial vision within egocentric and exocentric frames of reference. In S. R. Ellis (ed.), *Pictorial Communication in Virtual and Real Environment*. New York: Taylor and Francis.

Howard, I. P. (1996). Alhazen's neglected discoveries of visual phenomena. *Percept.*, 25: 1203–1217.

Howard, I. P. and Rogers, B. (1995). *Binocular Vision and Stereopsis*. Oxford: Oxford University Press.

Julesz, B. (1971). *Foundations of Cyclopean Perception*. Chicago: University of Chicago Press.

Khan, A. Z. and Crawford, J. D. (2001). Ocular dominance reverses as a function of horizontal gaze angle. *Vis. Res,*, 41: 1743–1748.

Liu, C. H. and Kennedy, J. M. (1995). Misalignment effects in 3-D versions of Poggendorff displays. *Percept. and Psychophys.*, 57: 409–415.

Mapp, A. P. and Ono, H. (1999). Wondering about the wandering cyclopean eye. *Vis. Res.*, 39: 2381–2386.

Marx, S. and Pfau, W. (1992). *Astrophotography with the Schmidt Telescope*. (Translated by P. Lamble.) New York: Cambridge University Press. (Originally published in 1990.)

Ohtsuka, S. (1995a). Perception of direction in three-dimensional space with occlusion. *The Institute of Electronics, Information, and Communication Engineers Technical Report*, 95: 31–36. (Abstract in English.)

Ohtsuka, S. (1995b). Relationship between error in inclination perception in observing Poggendorff figures and stereopsis. *The Institute of Electronics, Information, and Communication Engineers Technical Report*, 95: 24–26. (Abstract in English.)

Ohtsuka, S. and Yano, S. (1994). The phenomenon causing the Poggendorff illusion compensates geometrical error in reconstructed 2D image from stereopsis. *The Institute of Television Engineers of Japan (ITE) Technical Report*, 18–60, 25–30. (Abstract in English.)

Ono, H. (1991). Binocular visual directions of an object when seen as single or double. In D. Regan (ed.), *Vision and Visual Dysfunction*, Vol 9, *Binocular Vision* (pp. 1-18). London: MacMillan.

Ono, H. and Barbeito, R. (1982). The cyclopean eye vs. the sighting dominant eye as the center of visual direction. *Percept. and Psychophys.*, 32: 201–210.

Ono, H. and Barbeito, R. (1985). Utrocular discrimination is not sufficient for utrocular identification. *Vis. Res.*, 25: 289–299.

Ono, H. and Gonda, G. (1978). Apparent movement, eye movement and phoria when two eyes alternate in viewing a stimulus. *Percept.*, 7: 75–83. Also reprinted in I. Rock (1997) (ed.) *Indirect Perception* [Title changed to: Apparent motion based on changing phoria] (pp. 265-276). Cambridge: MIT Press.

Ono, H. and Mapp, A. P. (1995). A restatement and modification of Wells-Hering's laws of visual direction. *Percept.* 24: 237–252.

Ono, H., Mapp, A. P. and Howard, I. P. (2002). The cyclopean eye in vision: the new and old data continue to hit you right between the eyes. *Vis. Res.*, 42: 1307–1324.

Ono, H. and Nakamizo, S. (1977). Saccadic eye movements during changes in fixation to stimuli at different distances. *Vis. Res.*, 17: 233–238.

Ono, H., Ohtsuka S. and Lillakas, L. (1998). The visual system's solution to Leonardo da Vinci's paradox and to the problems created by the solution. *Proceeding for The Workshop on Visual Cognition*, Tsukuba, Japan: Science and Technology Association and National Institute of Bioscience and Human-Technology, 125–136.

Ono, M. E., Rivest, J. and Ono, H. (1986). Depth perception as a function of motion parallax and absolute-distance information. *J. Exp, Psych.: Hum. Percept. and Perf.*, 12: 331–337.

Ono, H. and Steinbach, M. J. (1983). The Pulfrich phenomenon with eye movement. *Vis. Res.*, 23: 1735–1737.

Ono, H., Wade, N. J. and Lillakas, L. (2002). The pursuit of Leonardo's constraint. *Percept.*, 31: 83–102.

Ono, H. and Weber, E. V. (1981). Nonveridical visual direction produced by monocular viewing. *J. Exp. Psych.: Hum. Percept. and Perf.*, 7: 937–947.

Panum, P. L. (1858). *Physiological Investigations Concerning Vision with Two Eyes.* Translated by C. Hubscher, 1940 (Hanover, NH: Dartmouth Eye Institute).

Porac, C. and Coren, S. (1981). *Lateral Preferences and Human Behavior*. New York: Springer-Verlag.

Rogers, B. J. and Graham, M. E. (1979). Motion parallax as an independent cue for depth perception. *Percept.*, 8: 125–134.

Sharpe, W. L. (1918). The floating-finger illusion. *Psych. Rev.*, 35: 171–173.

Steinbach, M. J., Howard, I. P. and Ono, H. (1985). Monocular asymmetries in vision: we don't see eye to eye. *Can. J. Psych.*, 39: 476–478

van Ee, R. and Erkelens, C. J. (2000). Is there an interaction between perceived direction and perceived aspect ratio in stereoscopic vision? *Percept. and Psychophys.*, 62: 910-926.

Wade, N. J, Ono, H. and Lillakas, L. (2001). Leonardo's struggle with representations of reality. *Leonardo*, 34: 231–235.

Wade, N. J. and Swanston, M. T. (1987). The representation of nonuniform motion: induced motion. *Percept.*, 16: 555–571.

Walls, G. L. (1951). A theory of ocular dominance. *Arch. Ophthalmol.*, 45: 387–412.

Wells, W. C. (1792). *An Essay Upon Single Vision with Two Eyes: Together with Experiments and Observations on Several Other Subjects in Optics*. London: Cadell.

8

Levels of Processing in the Size-Distance Paradox

Helen E. Ross

Phenomena such as the moon illusion and accommodation–convergence micropsia pose a paradox. Visual cues and oculomotor changes normally associated with a far distance may make objects appear both large and near, while those associated with a near distance have the opposite effect. The reported distance effects are paradoxical because they contradict the classical theory of size–distance invariance, in which objects of the same angular size appear larger in linear size because they appear further away. One recent solution is to say that certain cues produce a change in perceived angular size, which in turn influences perceived distance and perceived linear size. An older solution is the 'further–larger–nearer' account, in which distance is first processed at a subconscious (automatic) level, and determines the perceived linear size; the perceived linear size is then inappropriately used as a cue for the conscious judgement of distance. Some authors dissociate size and distance processing, and some do not accept more than one type of perceived size. Several authors distinguish between cognitive judgements of distance (which may be biased by perceived size) and the automatic use of distance cues to scale size. Various experiments show that conflicting spatial values can sometimes be held at the same time in the visual and tactile-kinaesthetic systems. Tactile spatial judgements cannot, therefore, provide a reliable measure of visual perception. It remains difficult to find an empirical distinction between cognitive and automatic processes.

> "Paradoxes have no place in science. Their removal is the substitution
> of true for false statements and thoughts." William Thomson, Baron
> Kelvin of Largs. Address to the Royal Institution (1887).

8.1   The Size–Distance Paradox

The size-distance paradox refers to those perceptual phenomena where objects appear both larger and nearer, or smaller and further, than would be predicted on the basis of size–distance invariance (SDI). The SDI hypothesis (see Sedgwick, 1986) is that an object's perceived linear size is determined in a geometrical manner by its true angular size and its perceived distance (Figure 8.1). On that basis,

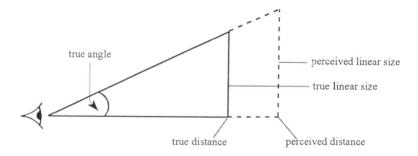

FIGURE 8.1. Classical size–distance invariance. Angular size is correctly known. Errors in perceived distance cause errors in perceived linear size.

many size illusions can be explained by the misperception of distance: objects appear larger in a mist because they appear further away (e.g., Ross, 1974, p. 54); and people appear different sizes in an Ames room because they appear to be at the same distance when they are actually at different distances (e.g., Gregory, 1998, p. 186). When disparate images of equal size are fused in binocular stereopsis, apparently nearer images appear smaller and apparently farther images appear larger (e.g., Kaufman and Kaufman, 2000).

There are several phenomena that contradict this account. Perhaps the most famous is the moon illusion: the moon (or sun) appears to be both larger and nearer on the horizon than it does when high in the sky (e.g., Hershenson, 1989). Similar phenomena occur when looking up or down from a great height, or when looking horizontally through one's legs: objects in these circumstances appear small and usually far (e.g., Ross, 1974, pp. 63–65). Some brightness and colour effects (known as advancing and retreating colours — Pillsbury and Schaefer, 1937) also disobey SDI: bright and red objects appear larger and nearer than dim or blue objects. Geometrical illusions are another example: certain parts of figures appear larger than other parts of the same size, while appearing to be at the same distance on the plane of the paper (see Robinson, 1972). The size–distance paradox has also sometimes been found with changes in the vergence angle of the two eyes when stimuli of the same angular size are viewed in reduced cue situations: close vergence makes an object appear smaller (known as accommodation–convergence [A/C] micropsia) but also as further away, while far vergence has the opposite effect. Effects of this type were reported by Wheatstone (1852) and by several later authors (see Mon-Williams and Tresilian, 1999). Similarly–objects viewed in dense fog are sometimes reported as "looming large" and also appearing near (e.g., Myers, 1911). In all the examples given here the size effects are agreed, but the distance effects are more disputable.

Many attempts have been made to explain these paradoxes while preserving some aspects of SDI, and they often involve different levels of processing. Before considering these explanations, I will first consider the levels of processing proposed for classical (nonparadoxical) SDI, that is, the order in which processes are supposed to occur and their level of consciousness.

8.2 Historical Background to Levels of Processing in SDI

It has often been asked whether size and distance are directly perceived, or whether they are calculated at a conscious or unconscious level. There is little agreement on these issues. An early view was that of the atomists or Epicureans, who believed that perception was directly impressed upon the senses by means of images or replicas that were given off by objects (Bailey, 1926; Siegel, 1970). Sensations were veridical, and any errors were those of interpretation. Thus, sizes (at least at close distances) should be correctly perceived, unless a cognitive mistake is made. However, according to Epicurean physics, an object's size appears to decrease with distance because the image gets worn down by other atomic bodies: in that case the misperception of size is not a cognitive error, but occurs at the level of the image.

The Stoic philosophers opposed the Epicurean beliefs. They held mixed theories about how images entered the eye. They believed in a continuous outflow of pneuma or flux, which touched the object and brought back an image to the eye. Some Stoics used the metaphor of a stick or ray, while others thought of the pneuma as cone-shaped. Cones and rays formed the basis of the geometrical approach to size and distance perception.

The Greek and Latin languages make a clear distinction between appearance and reality, and "apparent" usually corresponds to the modern psychological meaning of "perceived." It is a mistake to assume that apparent size always meant true angular size (as it does in astronomy). Sometimes it corresponds to perceived linear size, and sometimes the meaning is debatable. Euclid (c. 300 BC) dealt with the question of perceived size in a geometrical manner, and stated that perceived size followed the visual angle subtended at the eye. He wrote in his *Optics* (Theorem 5): "Objects of equal size unequally distant appear unequal and the one lying nearer to the eye always appears larger" (Trans. Burton, 1945, p. 358). In this passage Euclid used the language of appearances. In another passage (Theorem 21, "To know how great is a given length") he argued that linear size (true object size) could be calculated in a geometrical manner from the angular size and from the distance: but in this passage he used the language of calculation rather than appearances.

Ptolemy (a Stoic of the second century AD) often used the language of appearances. His writings contain one of the earliest statements that the same angular size gives rise to different perceived linear sizes at different perceived distances (Ross and Plug, 1998). However, in some other passages he hints at calculation, as when using the word "assumed" in describing aerial perspective (*Optics* II.126): "The same illusion also stems from differences in colors, for an object whose color is dimmer seems farther away and is therefore immediately assumed to be larger, just as happens with objects that actually are; that is, when objects are seen under equal angles while some of them lie at a greater distance." (Trans. Smith, 1996, p. 121.) In other passages Ptolemy implies that, at the most primitive level, the perception of size depends only on visual angle: distance and slant must be taken into account at a higher level, or later stage, to give true size. Smith (1996, p. 28) describes the process thus: "From such determinations, finally, perceptual judgments or inferences are drawn during the ... concluding phase of the process. The result is a sort of conceptual conclusion about the object as it actually exists in physical space."

Another Stoic writer who used the language of appearances was the Greek astronomer Cleomedes (c. first–third century AD). He attempted to explain the sun illusion by SDI, and wrote (*Meteora* II.1): "However, the sun's distance seems to us to vary. The sun appears to us nearest in the meridian, and further away when rising and setting. ... Whenever it appears near it appears very small; and whenever its distance appears greater, its size also appears larger." (Trans. Ross, 2000).

Ptolemy and Cleomedes described sight in terms of a visual cone with outgoing visual rays, but neither author explained how the length of the visual ray was known — an essential procedure if angular size is to be scaled for distance to give linear size. They acknowledged that the procedure was subject to error, from causes such as refraction and aerial perspective. Another cause — peculiar to the concept of outgoing rays — was the difficulty of looking upward. This idea was suggested by Vitruvius (III 5.9) in the first century BC, who wrote: "For the higher the glance of the eye rises, it pierces with the more difficulty the denseness of the air; therefore it fails owing to the amount and power of the height, and reports to the senses the assemblage of an uncertain quantity of the modules." (Granger, 1970, Vol.1, p. 191). Ptolemy made similar comments concerning the sun illusion (*Optics*, III 59). He stated that the visual ray travelling upward had a reduced sensation of distance and other characteristics, and concluded that "objects high in the sky seem small because of the unusual conditions and the difficulty of the action." (Trans. Ross and Ross, 1976). In these examples the error occurs at the sensory level, before any cognitive processes can take place. Ptolemy was well aware that some illusions occurred at the sensory level, while others were misjudgements at a cognitive level.

The Arab mathematican and physicist Alhazen (or Ibn al-Haytham) was clear in his *Optics* (c. 1030) that the visual rays were incoming rather than outgoing, and that distance was judged by cues other than ray length. His language often suggests some cognitive calculations, or perhaps an "unconscious inference" (Howard, 1996). He wrote (Book VII): "When human vision perceives the size of visible objects, it perceives it from the size of the angles that visible objects project to the centre of vision, and from the degree of intervening space, and by comparing the angles with the intervening space." (Trans. Ross and Plug, 1998)

The idea of a conscious calculation was expressed more strongly by Descartes in his *Sixth Discourse* (1637, 1965 p. 107): "Their size is estimated according to the knowledge, or the opinion, that we have of their distance, compared with the size of the images that they imprint on the back of the eye."

Many variations on these ideas were written by later scholars and scientists, some stressing the conscious and some the unconscious nature of the calculations and the perceptions. The later history is well described elsewhere (e.g., Epstein, 1977; Rock, 1977). The idea of "unconscious inference" is sometimes attributed to Helmholtz, but it obviously goes back much earlier than the nineteenth century. For example, John Locke held a theory of unconscious perception in the seventeenth century: "The ideas we receive by sensation are often altered by the judgement, without our taking notice of it" (cited by Morgan, 1977, p. 77).

8.3   Modern Approaches to Classical SDI

It has often been pointed out that there are several objections to the classical form of SDI. The most serious objection is that judgements of distance and linear size are not strictly related to true angular size in the required geometrical manner (see Sedgwick, 1986). Another difficulty is the distinction between size and distance. The difference is fairly clear in the usual diagram (as in Figure 8.1) where a (small) upright object is viewed horizontally and only one distance is involved. If the object is oriented at a slant in relation to the observer (Figure 8.2), the distinction becomes less clear. Several distances or angles could be involved; and if one distance and one angle are known, in addition to the subtended angle at the eye, object size can be calculated (see Sedgwick, 1986; Schwartz, 1994, p. 69ff; Baird et al., 1990). This makes it more difficult to determine which are the primary sources of information and which are secondary calculations. However, the examples of the size–distance paradox usually refer to the simpler case of Figure 8.1. For classical SDI, most authors assume that distance and angular size are primary sources, and that linear size is a secondary calculation. Schwartz (p. 65ff) calls this the TAD approach (taking-account-of-distance) to emphasise the mechanism by which SDI is supposed to occur. He makes the point that SDI could hold empirically, but for reasons other than TAD.

Attempts have been made to preserve the TAD approach even when the size and distance judgements are not strictly in accordance with geometry. Koffka (1936, p. 229) described the arguments of Eissler, Klimpfinger and Holaday (all in 1933) concerning irregularities of shape or size constancy, saying (with some disapproval): "All three authors . . . [maintain] that the functionally effective depth data need not become conscious as such, so that the 'mediation of perceptual things' takes place on a level lower than that which carries conscious processes." This line of arguing had in fact been used in the nineteenth century to account for the paradoxical breakdowns of SDI (see Section 8.5). The stages and levels of processing in SDI are not usually made explicit, but modern proponents of the

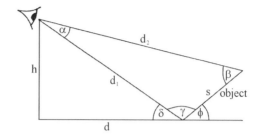

FIGURE 8.2. Sources of information for a general solution to the computation of size (s) by triangulation. The subtended angle (α) is assumed to be known. Other sources are the line-of-sight distances d_1 and d_2 and the interior angles β and γ; or the observer's eye height (h), the ground distance to the base of the target, and the target orientation (ϕ).

theory seem to hold some of the following opinions:

(a) Retinal image size is correctly processed to give the angular size of an object. Angular size may or may not be open to consciousness.

(b) Object distance is computed, sometimes incorrectly, from information within the retinal image and from other ocular and kinaesthetic sources. Retinal image size may be a distance cue: it might operate through relative angular size; or, in the case of a familiar object, it might be compared with the remembered image size at a remembered distance. The perceived distance is open to consciousness, but not the process of computation.

(c) Angular size and perceived distance are used to compute the perceived linear size, but the process is not open to consciousness.

(d) The perceived (or calculated) linear size is open to consciousness.

These ideas are very difficult to test, partly because it is hard to separate retinal/angular size from distance cues. It is also hard to determine what is held in consciousness, and what stages come first. However, various attempts have been made to determine whether retinal image size (or angular size) is correctly known. Most experiments have used size matching techniques, but with "retinal image size" instructions. In conditions of totally reduced cues, correct matches usually occur (see Baird, 1970). However, if any distance cues or other size cues are present, there is a large overestimation of image size, which increases with viewing distance (e.g., Holway and Boring, 1941; Gilinsky, 1955; Leibowitz and Harvey, 1967, 1969). The same is true of numerical estimates of angular size (Higashiyama, 1992). These findings imply that true image size is not normally open to consciousness, and that some scaling occurs at a preconscious level.

A different approach to the problem is to compare the variability of judgements of angular and linear size. If a judgement requires more levels of processing, there is more opportunity for "noise" to enter the system, and discrimination is likely to be poorer. McKee and Smallman (1998) reviewed the relevant literature and

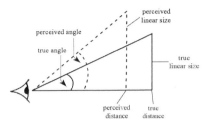

FIGURE 8.3. Errors in perceived angular size contribute to errors in perceived distance and in perceived linear size.

concluded that two routes can be used for judging objective size. The direct (and usual) route is the relative size of an object to its surroundings, which can give very precise objective matches at different distances without the need for TAD; the precision is much greater than would be predicted from the relatively high variability of distance judgements. The indirect route is to use distance information, which normally plays only a "supporting role" in size constancy. The authors argued that linear size was normally more open to consciousness than angular size.

The levels of processing seem very uncertain in classical SDI, so cannot be used as a guide to what might happen in paradoxical cases. The evidence suggests that true angular size is not normally open to consciousness, and that perceived angular size is not an accurate rendering of true angular size. Regardless of the levels of processing, the simple classical account of SDI is inconsistent with examples of the size–distance paradox.

8.4   Perceptual SDI: Misperceiving Angular Size

A different approach to SDI is to say that angular size can be misperceived independently of any misperception of distance. Perceived angular size and perceived distance can then be combined to give perceived linear size in a geometrical manner (Figure 8.3). This account was made explicit by McCready (1965, 1985, 1986).

Wheatstone (1852) was one of the first to hint at such an approach. He investigated A/C micropsia and found consistent changes in perceived size, but very variable effects on perceived distance. He denied that convergence first determined perceived distance, and that perceived size was scaled according to this distance. Instead, he argued that convergence had a direct effect on perceived size, which in turn had an uncertain effect on perceived distance:

> From the experiments I have brought forward, it rather appears to me that what the sensation [of convergence] immediately suggests is a correction of the retinal magnitude to make it agree with the real

magnitude of the object, and that distance, instead of being a simple perception, is a judgement arising from a comparison of the retinal and perceived magnitudes. However this may be, unless other signs accompany this sensation the notion of distance we thence derive is uncertain and obscure, whereas the perception of the change of magnitude it occasions is obvious and unmistakable.

Wheatstone's "perceived magnitude" here seems to be equivalent to perceived angular size, though he obviously thought the true retinal size remained available as a source of information. Von Kreis (1910/1962, pp. 601–602) made some equally ambiguous remarks about 'the impression of absolute size', when discussing A-C micropsia, the geometrical illusions, and the moon illusion. In the first sentence quoted below the phrase seems to mean "perceived angular size," but in the last sentence it must mean "perceived linear size."

The impressions we have of absolute size evidently belong also to the judgments that are immediately connected with physiological processes... It is this that enters the consciousness immediately and is retained in the memory, when we are unable either to say what the angular size is or to recall it ... Our immediate impressions of the absolute dimensions of observed objects are sometimes related to each other in ways that are mathematically impossible... This is the case chiefly with reference to the relations between distance, angular size, and the impression of absolute size.

This type of model is particularly popular for oculomotor effects, where perceived angular size is said to change for reasons other than perceived distance (e.g., McCready, 1965; Roscoe, 1989). The moon illusion is similarly explained (McCready, 1986) by saying that its angular size appears larger on the horizon for oculomotor or other reasons, and that its apparently enlarged angular size causes it to appear closer. The perceived linear size of the moon is then proportional to its perceived angular size and its perceived distance. Geometrical illusions receive similar treatment (McCready, 1985): size contrast makes parts of the figure appear angularly enlarged, and thus linearly enlarged on the plane of the paper or slightly in front of it. This interpretation is quite different from the perspective account of the geometrical illusions (Sections 8.5 and 8.7), in which the apparently enlarged part is supposed to appear either on the plane of the paper or slightly behind it.

The stages and levels in McCready's approach seem to be:

(a) Retinal image size is processed, sometimes incorrectly, to give the angular size of an object. Perceived angular size is open to consciousness, but true angular size is not.

(b) Object distance is computed, sometimes incorrectly, from various sources of information. One source is perceived angular size. The perceived distance is open to consciousness.

(c) Perceived angular size and perceived distance are used to compute the perceived linear size, which is open to consciousness.

This approach might seem to avoid different levels of processing, because all perceptual variables are open to consciousness. However, angular size affects distance, and both of these affect linear size, while linear size does not affect the other two. Linear size is therefore at a higher level of processing. The theory proposes only one type of distance perception but two types of size perception (angular and linear). It is harder to test than classical SDI because there are two possible sources of error (perceived angular size and perceived distance) rather than just one (perceived distance). It is also very difficult to devise an adequate measure of perceived angular size.

The next class of explanation differs in that it makes a hidden appeal to perceived distance.

8.5   The Further-Larger-Nearer Hypothesis and Classical SDI

The classical SDI hypothesis held such sway that attempts were made to save it while explaining the paradox. A solution concerning different levels of consciousness was elaborated in the nineteenth century. The idea was credited to Aubert in 1876 by James (1890/1931, p. 235); and to Brentano in 1892 and Thiéry in 1895 by Sanford (1898). It has also been taken up by several modern authors (reviewed by Robinson, 1972; Gillam, 1998). The basic idea is that objects can be perceived as both far and near at the same time (or perhaps in sequence), but at different levels of consciousness. An object may appear far and therefore enlarged at a low level of consciousness, but large and therefore near at a higher level of consciousness.

Several authors advanced this type of explanation for the geometrical illusions. They argued that some geometrical patterns contain depth cues, which alter the (unconsciously) perceived depth and thus the consciously perceived (linear) size of certain parts of the figures. However, conventional size–distance invariance breaks down in these illusions, because the perceptually enlarged parts appear to be either at the same distance or nearer than the perceptually diminished parts. To counteract this criticism most authors stressed that the perceived depth was unconscious. Sanford (p. 222) commented:

> Even in casual observation the figure is seen perspectively — not consciously, but in effect... That no such inclination of the planes is seen by the observer until it is suggested to him is fully recognized and even insisted upon by Thiéry. It is no more necessary, however, that the perspective factor should be conscious in order that it may influence the final form of the perception than that the partial tones

in a note on a violin should be consciously recognized before it can be distinguished from a note of the same pitch on a flute.

Myers (1911, pp. 282–283) wrote in similar terms about geometrical illusions and size–distance effects in a fog:

> In the latter case [a foggy atmosphere]... and in the suggestions of perspective in a drawing — it is the apparent size which determines the apparent distance. Yet primarily, the apparent size must be dependent on some unconscious influence of distance. Possibly we have here a schema..., or unconscious disposition, in regard to the distance of objects; and when this schema undergoes change, it manifests itself in consciousness by effecting a change in apparent size, whereupon the apparent size determines our awareness of the distance of the object.

The further-larger-nearer solution has also been applied to the moon illusion. Reimann (1902a, b) stated that the flattened appearance of the sky causes the horizon moon to be seen as more distant and therefore as larger; this greater perceived size then causes the horizon moon to be consciously perceived as closer — "floating" before the sky. Claparède (1906, p. 132) suggested that the two size–distance judgements were simultaneous: the horizon moon is subconsciously judged to be further and therefore larger; and the enlarged moon is consciously judged to be closer. Dees (1966) argued that the horizon moon appears further away through monocular depth cues such as linear perspective and interposition; size–distance scaling makes it appear larger; then the larger size makes it appear closer. Gogel (1974) and Gogel and Mertz (1989) put forward a more detailed version of the theory. They argued that the "equidistance tendency" mainly determines the perceived distance of the horizon moon, so that it is seen to be close to the horizon; but the "egocentric reference distance" mainly determines that of the zenith moon, so that it is seen to be very close. The "reported distance" of the moon is then cognitively determined by the size judgement.

The argument, as put forward by many of these authors, attempts to preserve perceived distance as an explanation by proposing two different levels of consciousness for perceived distance. The idea of two levels of consciousness may be defensible; but the attempt to preserve SDI is unacceptable because it confounds linear and angular size. In the first judgement perceived size is analogous to linear size, and in the second to angular size.

8.6   Independence of Size and Distance

An apparently clean solution to the size–distance paradox is to abolish it by denying any necessary connection between size and distance. Berkeley (1709, Section 53) took this line. He inveighed against the dominant geometric approach of the "optic writers," and argued that size and distance judgements were separately conditioned to various cues. So dominant is the belief in SDI that Berkeley is widely misquoted as supporting it (e.g., Boring, 1942, pp. 223 and 298). Berkeley rejected SDI firmly:

> What inclines men to this mistake (beside the humour of making one see by geometry) is, that the same perceptions or ideas which suggest distance, do also suggest magnitude...I say they do not first suggest distance, and then leave it to the judgement to use that as a medium, whereby to collect the magnitude; but they have as close and immediate a connexion with the magnitude, as with the distance; and suggest magnitude as independently of distance, as they do distance independently of magnitude.

An excellent discussion of Berkeley and Gibson, and a critique of SDI, can be found in Schwartz (1994). Gibson's approach is ambiguous, and changed over time, but for him the connections between size and distance information were meaningful and necessary. For Berkeley, the connections were arbitrary — anything could be learned. In these approaches different levels of processing are not required, because size and distance judgements can be made simultaneously. Both are conscious.

Several modern authors support independent processing of size and distance (e.g., Day and Parks, 1989), or multiple levels of processing for both size and distance (e.g., Coren, 1989). Such statements are interesting as a denial of SDI, but they tend to lack explanatory detail. As Carr (1935, p. 396) wrote when criticising the perspective explanation of the geometrical illusions: "The most probable explanation of the correlation between perceptible distance and size is the assumption that both are the effects of the same causes, and hence we are still confronted with the problem of explaining how these accessory lines affect our judgments of both size and distance."

Few authors would go so far as to deny any link at all between size and distance processing. The problem remains as to what the links are. Some attempts at expounding the links are discussed below.

8.7   Half-Way Houses: Automatic and Cognitive Perceptions

Several recent authors have put forward arguments similar to the further-larger-nearer hypothesis, but without claiming to maintain SDI. I will try to describe the views of Kaufman and Rock, and of Gregory, because they have made serious attempts to reconcile the issues, and they illustrate most of the options and difficulties.

In their work on the moon illusion Kaufman and Rock (1962a, b, 1989) and Rock and Kaufman (1962) distinguished between two types of distance perception — the "registered" distance, and the judged or apparent distance. Distance cues are "registered" by the nervous system, and are used, together with angular size, to compute linear size. Judgements of distance and size are influenced by many factors such as the psychophysical procedures and instructions, and the observer's knowledge and biases. Observers may or may not be aware of the available cues and their meaning; consequently their verbal judgements may sometimes conflict with what the cues indicate, resulting in a size–distance paradox. In their earlier work, Kaufman and Rock suggested that the registered distance of the horizon was unconscious, and that the judged or apparent distance of the moon was consciously perceived; whereas in their 1989 account they argued that the registered distance of the horizon moon coincided with the perceived distance of the horizon, and that the judgement of the moon's distance was cognitive rather than perceptual and was distorted by the moon's perceived size. In this later account the levels of consciousness for distance judgements seem to be reversed. In their most recent account (based on perceived distance settings for stereoscopically adjusted artificial moons projected to the horizon or high in the sky) Kaufman and Kaufman (2000) argued that "the horizon moon is seen as larger because the perceptual system treats it as though it is much further away." However, they qualified this by saying: "Thus the term apparent in so-called apparent-distance theories is inappropriate. Rather, we suggest that the physical cues to distance affect both perceived distance and perceived size."

Language of this sort was also used by Rock in various publications, causing Schwartz (1994, pp. 64–65) to comment:

> Rock waffles somewhat when it comes to spelling out what registered distance amounts to. On one reading it is an unconscious representation of a specific distance value. Often, though, he talks as if what are registered are only the (distance) cues themselves, and that they directly influence size. But if it is registered cues about distance, not a distance value itself, that plays a role, it would seem that Rock has gone a long way towards accepting one of Berkeley's central criticisms of the TAD model.

Gregory (1963, 1968, 1998) runs into similar difficulties. He produced an account of the geometrical illusions like the nineteenth century perspective accounts. However, he elaborated it more formally. He distinguished between two types of scaling of perceived size: "primary" or "bottom up" scaling, in which those perspective cues indicating distance automatically enhance perceived size; and "secondary" or "top down" scaling, in which perceived size is scaled according to perceived distance, in a manner similar to SDI. However, he did not distinguish between angular and linear size. Primary size cues are similar to distance cues, but need not result in consciously perceived distance. The paradox of geometrical illusions can then be explained because the perspective cues within the figures automatically enlarge the typically more distant parts, while other cues keep the perceived distance on the plane of the page. Gregory (1998, p. 226) wrote: "All perspective pictures have a curious depth paradox: they represent depth, with their perspective and other depth cues; yet as objects the pictures are flat and their textured surfaces provide depth cues showing that they are flat." As an empirical test of the perspective theory, Gregory claims that that when suitable illusion figures are made luminous, and are viewed at an appropriate distance in the dark, they do appear in the required depth. However, some authors dispute this result, and argue about what constitutes a typical perspective (see Robinson, 1972; Gillam, 1998). A more difficult point to interpret is Gregory's reliance on the evidence from luminous figures. The argument might imply that size scaling depends on unconsciously registered depth — and that is how many of his critics have interpreted him. However, he does not appeal to unconscious depth in most of his writings; instead he follows Helmholtz and compares perception to an "unconscious inference." Unfortunately, there is no empirical test to demonstrate the existence of unconscious inferences.

Gregory (1998) explained the moon illusion slightly differently from geometrical illusions: certain ground cues automatically enhance the perceived size of the horizon moon in a bottom-up manner, but cannot affect the raised moon which retains a "default" size (undefined). Thus the moon's size is enhanced for the same reason that certain lines or other elements are enlarged in geometrical illusions. The perceptually enlarged horizon moon then acts as a source of information (unspecified as to whether top-down or bottom-up) leading to a relatively close perceived distance. This is an additional perceptual stage — one that Gregory does not propose for 2-D geometrical illusions (though McCready does so).

There are similarities between the positions of Gregory and Kaufman in that they both propose an automatic scaling of size by distance cues. Both also appeal to evidence about perceived depth to support this idea — Gregory to the perceived depth of luminous figures, and Kaufman to perceived depth measured in a stereoscopic experiment. This leaves unresolved the question raised by Schwartz and others as to whether the scaling depends on an unconsciously registered depth or operates directly from the distance cues.

8.8   Space Perception in Vision and Touch

Up till now the discussion has been restricted to levels of processing within the visual system. However, different levels of processing can occur between vision and touch. Tactile measures of size and distance are sometimes used as an arbiter of measures within the visual system, so it is necessary to consider the rationale behind such comparisons.

8.8.1   Undistorted Vision

Many authors believed that the touch sense was less fallible than vision, and that touch educates vision. Plotinus wrote in the third century: "Touch conveys a direct impression of a visible object" (*Second Ennead*, VIII; Trans. MacKenna and Page, 1952, pp. 64–65). The question of whether tactile and visual size were innately linked was discussed by Molyneux in the seventeenth century, and many inconclusive investigations were made of formerly blind people who recovered their sight (Morgan, 1977). Berkeley continued the debate and argued that tactile experience enables us to interpret the visual image size and perceive the true linear size, even for objects beyond the reach of touch. He argued that both the visible size (image size) and the tangible size (physical object size, whether tactile or visual) were open to perception, but especially the latter as it was of more practical importance:

> Hence it is, that when we look at an object, the tangible figure and extension thereof are principally attended to; whilst there is small heed taken of the visible figure and magnitude, which, though more immediately perceived, do less concern us, and are not fitted to produce any alteration in our bodies. (1709, Section LIX).

What exactly Berkeley meant by this remains obscure. He failed to abolish the problem of two types of perceived size — he merely moved one of them from the visual realm to some intersensory realm.

Modern research confirms that vision and touch interact when reaching for close objects. Visual size and distance judgements guide the hand to grasp an object correctly (e.g., von Hofsten and Ronnqvist, 1988; Jeannerod, 1981); and tactile judgements serve to correct vision when it errs through optical distortion (Welch, 1986). Different spatial values can therefore be held simultaneously in the two perceptual systems. It has recently been shown that the brain has two visual pathways leading from the primary visual cortex — a ventral stream serving perception and cognition and a dorsal stream serving action (see review by Milner, 1997). Size perception and object identification are usually thought to belong to the ventral stream; while the localization of targets in space, and the guidance of movement, belong to the dorsal stream. There must be some interaction between visual and motor areas to account for the interaction of vision and touch. Usually the two systems compute size fairly consistently, but discrepancies may sometimes arise even when distortions are not present (see reviews by Marks, 1978; Seizovacajic, 1998).

The fact that grasp size is normally set for linear size can be taken as evidence that (at least at close distances) the visual system perceives linear size at a primary level, rather than having to calculate it at a secondary level.

8.8.2   *Optical Distortion*

In the twentieth century there was much interest in the perceptual effects of optical distortion, for example, when the normal relation between size and distance is altered by magnification or minification. Experimental studies disproved the inherent dominance of touch, and showed instead the phenomenon of "visual capture": the hand feels to be the size it looks. However, after some time, both vision and touch are modified, and a new compromise between the senses is reached (reviewed by Welch, 1986). It is usually argued that the discrepancy between the senses is what drives the adaptation process. However, adaptation occurs even though the observer is not consciously aware of the discrepancy, which implies that different spatial values can be held at different levels of consciousness in the visual and tactile perceptual systems. On the other hand, many researchers use unseen tactile adjustments as a way of measuring visual perception, which suggests that they believe that the two systems share the same spatial values.

Evidence that the two systems need not share the same distance values was provided by Mon-Williams and Tresilian (1999). They altered vergence by placing a prism in front of one eye. When the observers converged to a closer distance than the target, they reported that the target appeared smaller and further; and the opposite effects held for far vergence. However, the distance effects only occurred in verbal reports: when the observers reached with their finger (hidden from view) to the perceived location of the target, they reached closer for near vergence and further for far vergence. The authors interpret this to mean that the motor system receives accurate but "cognitively impenetrable" visual information about distance, while the cognitive visual system relies on various cues including perceived size (type unspecified). They concluded that the size–distance paradox is cognitive rather than perceptual. A difficulty with this analysis is to define precisely which methods involve "cognitive" judgements and which are more direct. Verbal judgements may be defined as cognitive — but are all motor judgements noncognitive? The verbal reports that overhead luminous targets appear nearer than horizontal targets of the same size have been confirmed in studies in which the distance of the targets is adjusted manually (e.g., Wood et al., 1968, Zinkus and Mountjoy, 1969). It could be argued that these motor adjustments are cognitive because they rely on memory. But unless the criteria are made clear there is a danger of circularity: any judgements involving the size–distance paradox could be called cognitive.

8.8.3   Perceptual Distortion

Distortions of visual perception can arise for reasons other than optical distortion, as in the geometrical illusions. It has been claimed that size–contrast illusions affect visually perceived size but not manually adjusted grip size (e.g., Aglioti et al., 1995); but other authors dispute this and claim that visual illusions can affect dorsal stream activities (e.g., Daprati and Gentilucci, 1997). Recent reviews suggest that the size of the effects varies with the method of measurement and the nature of the task (Bingham et al., 2000; DeLucia et al., 2000; Mon-Williams and Bull, 2000). It would indeed be strange if the primary visual cortex always fed accurate size and distance information to the dorsal stream but misleading information to the ventral stream. Given the controversial status of these studies, they cannot shed much light on whether perceptually conflicting size or distance values can be held within the ventral stream.

8.9   Conclusions

In this chapter I have described what various authors have said about levels of processing in size constancy and the size–distance paradox, and I have attempted to classify their positions. This has not been easy because most early authors and some modern authors are inexplicit on the crucial issues. There is not much evidence to help us decide between the different descriptions of the paradox. It is clear that classical SDI does not hold, and in any case it cannot explain the paradox. The available evidence does suggest that true angular size is not normally open to consciousness, and that some scaling in the direction of size constancy takes place at a preconscious level. Further scaling may take place at a higher level. Interactions between size and distance perception occur at all levels, so that it does not make much sense to speak of discrete stages of processing.

Different spatial values can be held simultaneously in vision and touch, which suggests that these senses are not equally open to consciousness. Size and distance perception are like many other aspects of perception, in that observers are unaware of the cues they use to produce conscious judgements. Verbal and other "cognitive" judgements about visual appearances may conflict with "direct" tactile judgements - but it remains difficult to find an objective test of what is cognitive and what is direct.

Acknowledgements

I should like to thank Lloyd Kaufman and Robert P. O'Shea for helpful comments on a draft of this chapter.

References

Aglioti, S., Goodale, M. A. and DeSouza, J. F. X. (1995). Size–contrast illusions deceive the eye but not the hand. *Current Biol.*, 5: 679–685.

Bailey, C. (ed.). (1926) *Epicurus. The Extant Remains.* : Oxford, UK: Clarendon Press.

Baird, J. C. (1970). *Psychophysical Analysis of Visual Space.* Oxford, UK: Pergamon Press.

Baird, J. C., Wagner, M. and Fuld, K. (1990). A simple but powerful theory of the moon illusion. *J. Exp. Psych.: Hum. Percept. Perf.*, 16: 675–677.

Berkeley, G. (1709/1972). *A New Theory of Vision.* London, UK: Dent. .

Bingham G. P., Zaal, F., Robin, D. and Shull, J. A. (2000). Distortions in definite distance and shape perception as measured by reaching without and with haptic feedback. *J. Exp. Psych.: Hum. Percept. and Perf.*, 26: 1436–1460.

Boring, E. G. (1942). *Sensation and Perception in the History of Experimental Psychology.* : New York: Appleton-Century-Crofts.

Burton, H. E. (1945). The Optics of Euclid. *J. Opt. Soc. Am.*, 35: 357–372.

Carr, H. A. (1935). *An Introduction to Space Perception.* New York: Longmans, Green and Co.

Claparède, E. (1906). L'agrandissement et la proximité apparente de la lune à l'horizon. *Arch. de Psych.*, 5: 121–148.

Coren, S. (1989). The many moon illusions: an integration through analysis. In M. Hershenson (ed.), *The Moon Illusion*, pp. 351–370, Hillsdale, NY: Erlbaum.

Daprati, E. and Gentilucci, M. (1997). Grasping an illusion. *Neuropsychol.*, 35: 1577–1582.

Day, R. H. and Parks, T. E. (1989). To exorcize a ghost from the perceptual machine, In M. Hershenson (ed.), *The Moon Illusion*, pp. 343–350, Hillsdale, NY: Erlbaum.

Dees, J. W. (1966). Moon illusion and size–distance invariance: an explanation based upon an experimental artifact. *Percept. and Motor Skill*, 23: 629–630.

DeLucia, P. R., Tresilian, J. R. and Meyer, L. E. (2000). Geometrical illusions can affect time-to-contact estimation and mimed prehension. *J. Exp. Psych.: Hum. Percept. and Perf.*, 26: 552–567.

Descartes, R. (1637/1965). *Discourse on Method, Optics, Geometry, and Meteorology.* Trans. P. J. Olscamp. Indianapolis, IL: Bobbs-Merrill.

Epstein, W. (1977). Historical introduction to the constancies. In W. Epstein (ed.), *Stability and Constancy in Visual Perception*, pp. 1-22, : New York, NY: Wiley.

Gillam, B. (1998). Illusions at century's end. In J. Hochberg (ed.) *Handbook of Perception and Cognition*, 2nd Edition, pp. 95-136, London, UK: Academic Press.

Gilinsky, A. S. (1955). The effect of attitude upon the perception of size. *Am. J. Psych.*, 68: 173–192.

Gogel, W. C. (1974). Cognitive factors in spatial response. *Psycholog.*, 17: 213–225.

Gogel, W. C. and Mertz, D. L. (1989). The contribution of heuristic processes to the moon illusion. In M. Hershenson (ed.), *The Moon Illusion*, pp. 235-258, Hillsdale: Erlbaum.

Granger, F. (1970). *Vitruvius: On Architecture.* Cambridge, MA: Harvard University Press.

Gregory, R. L. (1963). Distortion of visual space as inappropriate constancy scaling. *Nature*, 203: 1407.

Gregory, R. L. (1968). Perceptual illusions and brain models. *Proc. Roy. Soc. Lond. B*, 71: 279–296.

Gregory, R. L. (1998). *Eye and Brain*, 5th edition, Oxford: Oxford University Press.

Hershenson, M. (1989). That most puzzling illusion. In M. Hershenson (ed.), *The Moon Illusion*, pp. 1-3, Hillsdale: Erlbaum.

Higashiyama, A. (1992). Anisotropic perception of visual angle: Implications for the horizontal–vertical illusion, over constancy of size, and the moon illusion. *Percept. Psychophys.*, 51: 218–230.

Holway, A. H. and Boring, E. G. (1941). Determinants of apparent visual size with distance variant. *Am. J. Psych.*, 54: 21–37.

Howard, I. P. (1996). Alhazen's neglected discoveries of visual phenomena. *Percept.*, 25: 1203–1207.

James, W. (1890/1931) *The Principles of Psychology*, Vol.II, New York: Holt.

Jeannerod, M. (1981). Intersegmental coordination during reaching at natural objects. In J. Long and A. Baddeley (eds.), *Attention and Performance IX*, pp. 153–168, Hillsdale: Erlbaum.

Kaufman, L. and Kaufman, J. H. (2000). Explaining the moon illusion. *Proc. Nat. Acad. Sci.*, 97: 500–505.

Kaufman, L. and Rock, I. (1962a). The moon illusion, I. *Science*, 136: 953–961.

Kaufman, L. and Rock, I. (1962b). The moon illusion. *Sci. Am.*, 207: 120–130.

Kaufman, L. and Rock, I. (1989). The moon illusion thirty years later. In M. Hershenson (Ed.), *The Moon Illusion*, pp. 193–234, Hillsdale: Erlbaum.

Koffka, K. (1936). *Principles of Gestalt Psychology*. London: Kegan Paul.

Leibowitz, H. W. and Harvey, L. O. (1967). Size matching as a function of instructions in a naturalistic environment. *J. Exp. Psych.*, 74: 378–382.

Leibowitz, H. W. and Harvey, L. O. (1969). Effect of instructions, environment, and type of test object on matched size. *J. Exp. Psych.*, 81: 36–43.

MacKenna, S. and Page, B. S. (1952). Plotinus: The six Enneads. *Encyclopaedia Britannica, Great Books of the Western World*, Vol.17. Chicago: Benton.

Marks, L. E. (1978). *The Unity of the Senses: Interrelations Among the Modalities*, New York: Academic Press.

McCready, D. (1965). Size–distance perception and accommodation-convergence micropsia — a critique. *Vis. Res.*, 5: 189–206.

McCready, D. (1985). On size, distance, and visual angle perception. *Percept. Psychophys.*, 37: 323–334.

McCready, D. (1986). Moon illusions redescribed. *Percept. Psychophys.*, 39: 64–72.

McKee, S. P. and Smallman, H. S. (1998). Size and speed constancy. In V. Walsh and J. J. Kulikowski (eds.), *Perceptual Constancy: Why Things Look As They Do*, pp. 373–408, Cambridge: Cambridge University Press.

Milner, A. D. (1997). Vision without knowledge. *Phil. Trans. Roy. Soc. Lond. B*, 352: 1249–1256.

Mon-Williams, M. and Bull, R. (2000). The Judd illusion: evidence for two visual streams or two experimental conditions? *Exp. Brain Res.*, 130: 273–276.

Mon-Williams, M. and Tresilian, J. R. (1999). The size–distance paradox is a cognitive phenomenon. *Exp. Brain Res.*, 126: 578–582.

Morgan, M. J. (1977). *Molyneux's Question: Vision, Touch and the Philosophy of Perception*, Cambridge: Cambridge University Press.

Myers, C. S. (1911). *A Textbook of Experimental Psychology. Part 1 — Text-book. 2nd edition*, Cambridge: Cambridge University Press.

Pillsbury, W. B. and Schaefer, B. R. (1937). A note on "advancing and retreating" colors. *Am. J. Psych.*, 33: 150–161.

Reimann, E. (1902a). Die scheinbare Vergrösserung der Sonne und des Mondes am Horizont, I. Geschichte des Problems. *Zeitschrift für Psychologie*, 30: 1–38.

Reimann, E. (1902b). Die scheinbare Vergrösserung der Sonne und des Mondes am Horizont, II. Beobachtungen und Theorie. *Zeitschrift für Psychologie*, 30: 161–195.

Robinson, J. O. (1972). *The Psychology of Visual Illusion*, London: Hutchinson University Library.

Rock, I. (1977). In defense of unconscious inference. In W. Epstein (ed.), *Stability and Constancy in Visual Perception*, pp. 321–373, New York: Wiley.

Rock, I. and Kaufman, L. (1962). The moon illusion, II. *Science*, 136: 1023–1031.

Roscoe, S. N. (1989). The zoom-lens hypothesis. In M. Hershenson (ed.), *The Moon Illusion*, pp. 31–57, Hillsdale: Erlbaum.

Ross, H. E. (1974). *Behaviour and Perception in Strange Environments*, London: Allen and Unwin.

Ross, H. E. (2000). Cleomedes (c. 1st century AD) on the celestial illusion, atmospheric enlargement and size–distance invariance. *Percept.*, 29: 853–861.

Ross, H. E. and Plug, C. (1998). The history of size constancy and size illusions. In V. Walsh and J. J. Kulikowski (eds.), *Perceptual Constancy: Why Things Look As They Do*, pp. 499–528, Cambridge: Cambridge University Press.

Ross, H. E. and Ross, G. M. (1976). Did Ptolemy understand the moon illusion?. *Percept.*, 5: 377–385.

Sanford, E. C. (1898). *A Course in Experimental Psychology. Part I: Sensation and Perception*, Boston: Heath and Co.

Schwartz, R. (1994). *Vision: Variations on Some Berkeleian Themes*, Oxford: Blackwell.

Sedgwick, H. A. (1986). Space perception. In K. R. Boff, L. Kaufman and J. P. Thomas (eds.) *Handbook of Perception and Human Performance. Vol.I. Sensory Processes and Perception*, pp. 21.1–21.57, New York: Wiley.

Seizovacajic, T. (1998). Size perception by vision and kinaesthesia. *Percept. and Psychophys.*, 60: 705–718.

Siegel, R. E. (1970). *Galen On Sense Perception*. Basel: Karger.

Smith, A. M. (1996). *Ptolemy's Theory of Visual Perception: An English Translation of the Optics With Introduction and Commentary*.Philadelphia, PA: The American Philosophical Society.

von Hofsten, C. and Ronnqvist, L. (1988). Preparation for grasping an object: A developmental study. *J. Exp. Psych.: Hum. Percept. and Perf.*, 14: 610–621.

von Kries, C. (1910/1962). Notes. In H. von Helmholtz, *Handbook of Physiological Optics*, Vol.III, (Trans. 1925 by J. P. C. Southall). New York: Dover.

Welch, R. B. (1986). Adaptation of space perception. In K. R. Boff, L. Kaufman and J. P. Thomas (eds.), *Handbook of Perception and Human Performance. Vol.I. Sensory Processes and Perception*, Chapter 24, New York: Wiley.

Wheatstone, C. (1852). Contributions to the physiology of vision - Part the second. On some remarkable, and hitherto unobserved, phenomena of binocular vision. *Phil. Trans. Roy. Soc. Lond.*, 142: 1–17.

Wood, R. J., Zinkus, P. W. and Mountjoy, P. T. (1968). The vestibular hypothesis of the moon illusion. *Psychonom. Sci.*, 11: 356.

Zinkus, P. W. and Mountjoy, P. T. (1969). The effect of head position on size discrimination. *Psychonom. Sci.*, 14: 80.

9

The Level of Attention: Mediating Between the Stimulus and Perception

Jeremy M. Wolfe

Current conceptions of visual processing make good use of the metaphor of levels of vision. At the very least, there are meaningful distinctions to be made between early vision, mid-level vision, and high-level vision. Without getting too committed to the details, early vision is a level of local processing of simple stimulus attributes like the orientation and motion of line segments. Perhaps these can be considered to be the atoms of vision. If so, then mid-level vision is concerned with the molecules — larger pieces put together out of the early vision atoms. Like the "wetness" of water, these mid-level molecules may have properties that are not easy to predict from their early vision precursors. Examples might include Gestalt observations about the whole being greater than the sum of its parts (Kellman, 1998; Rock and Palmer, 1990) or the work of Adelson and Gilchrist (this volume) and others on the apparent brightness of surfaces. At a still higher level, the molecules of mid-level vision give rise to recognizable objects.

Once upon a time, perhaps in the first flush of excitement about single-unit recordings in the visual system (Barlow, 1995), we might have thought of these levels in a fairly straightforward, hierarchical manner. The atoms (lines) made the molecules (corners, junctions) that made the object (grandmother) (Lettvin, 1995). Even then, it was a bit difficult to believe in this effectively unidirectional story given evidence for massive feedback from apparently higher levels in the visual nervous system to earlier stages (reviewed in Lamme, Super, and Spekreijse, 1998). Moreover, even if everything was feeding forward, it became clear that not everything was being allowed to pass unimpeded to the higher levels. You could not simultaneously recognize all of the objects in the visual field. Somewhere there was a gate or a filter or bottleneck that was permitting some information to flow to higher levels while other information was blocked, perhaps lost. Attention altered the filter or moved the bottleneck around the visual field. Reading provides a clear example. Even if letters of text are made big enough to be read without eye movements, reading proceeds in a serial fashion as one word after another is somehow selected and processed. The stimulus on the retina might remain constant, but the contents of the later stages had to be changed by an act of selecting some stimuli and ignoring others. Neisser (1967) distinguished between levels of processing that were preattentive in which all input could be processed in parallel and attentive levels that processed only a selection of the available input at any one time.

The purpose of this chapter is to revisit the role of attention in mediating between levels of visual processing. The heart of the argument will be that attention governs the surprisingly narrow gate between visual perception and the visual stimuli that give rise to that perception. To make that argument, this discussion will review several lines of research, drawn largely from the visual search literature. This evidence will be seen to back us into a corner. The data can be used to make a good case for the argument that we only "see" one object at a time (c.f. Mack and Rock, 1998a, b; O'Regan, 1992; Rensink, 2000; Simons and Levin, 1997). How can we reconcile this evidence with our subjective impression of a rich visual world, populated with many objects? We will assume that the there is a rich physical world out there to be seen. We will argue, against some current thinking, (e.g., O'Regan and Noë, 2001), that perceivers experience a rich perceptual representation of that world. The curious, perhaps counterintuitive argument of this chapter is that the representation is in rather limited contact with the stimuli that give rise to that representation. Thus, the cat you "see" may not be quite the same as the cat in the world. Indeed, that physical cat might be gone from the physical scene. It is through the narrow gate of attention that the perceived cat is linked to the cat in the world. This is a version of the venerable thought that we infer the visual world (Brainard, Wandell, and Chichilnisky, 1993; Freeman, 1994; von Helmholtz, 1924; Nakayama and Shimojo, 1992). We will examine the role of attention in the maintenance of that inference.

9.1   Evidence for the Parallel Processing of Visual Features

Visual search experiments provide one line of evidence for the parallel processing of some visual features. In a typical visual search experiment, subjects look for a target among a variable number of distractors. A useful dependent measure is the reaction time (RT), the time to respond that, "yes", a target is present or "no", it is not. The slope of the function relating RT to the number of items (set size) is a measure of the efficiency of a search task. For a limited set of stimulus attributes, that slope is near zero when the target is defined by the presence of a feature. There are, perhaps, a dozen dimensions that will support this sort of highly efficient search (reviewed in Wolfe, 1998). Figure 9.1 shows some examples: top left — luminance polarity (Enns and Kingstone, 1995; Gilchrist, Jane, and Heiko, 1997; O'Connell and Treisman, 1990) which may or may not be the same dimension as color (Carter, 1982; D'Zmura, 1991; Green and Anderson, 1956; Treisman and Gormican, 1988); top middle — orientation (Foster and Ward, 1991; Moraglia, 1989; Nothdurft, 1993); top right — size (Aks and Enns, 1996; Cavanagh, Arguin, and Treisman, 1990; Williams, 1966), bottom left — line termination (Treisman

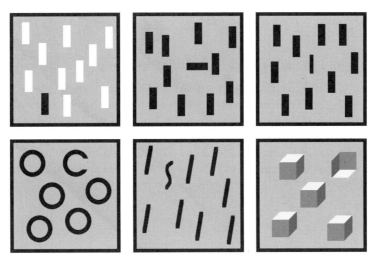

FIGURE 9.1. Examples of simple searches for "basic features."

and Gormican, 1988) or, conversely, closure (Elder and Zucker, 1993), bottom middle — curvature (Wolfe, Yee, and Friedman-Hill, 1992) (see also Kristjansson and Tse, 2001); and bottom right — various aspects of 3D structure (Enns and Rensink, 1990a, b, 1992; Kleffner and Ramachandran, 1992; Sun and Perona, 1996a, b).

It is not entirely trivial to determine what is and is not a basic feature in visual search. For example, it is sometimes possible to search for conjunctions of two or more features very efficiently (Theeuwes and Kooi, 1994; Wolfe, 1992). It is unlikely that this means that the visual system possesses a mechanism for parallel processing of conjunctions of orientation and luminance polarity, for instance. More likely, the system can guide attention simultaneously toward items with the target orientation and the target polarity (Wolfe, Cave, and Franzel, 1989). The claim that any particular attribute is a basic feature is strengthened when there is converging evidence, notably from visual search asymmetries (Treisman and Gormican, 1988; Treisman and Souther, 1985; Wolfe, 2001) and texture segmentation (Beck, 1966; Julesz and Bergen, 1983; Wolfe, 1992). For present purposes, the important point is that there is a limited set of features that appear to be processed in parallel, across the visual field and that appear to be available to guide the deployment of attention.

9.2   Basic Features and Early Vision

It is important not to confuse "basic features," as defined within the visual search literature, with early vision features as assessed with classical psychophysical and electrophysiological methods. There are many important features that make it clear that basic features in visual search are not properties of cells in early stages of visual cortical processing (e.g., V1).

9.2.1   The Lists of Features Are Different

There are a number of candidates for basic feature status in visual search that are not generally found on lists of early vision features. These might include lighting direction and shading (Braun, 1993; Enns and Rensink, 1990a; Kleffner, Polichar, and Ramachandran, 1990 ; Ramachandran, 1988; Rensink and Cavanagh, 1993; Sun and Perona, 1996a, b), binocular lustre (shininess) (Wolfe and Franzel, 1988), and a variety of depth cues (Enns and Rensink, 1990a, b; Previc and Naegele, 2001; Rensink and Cavanagh, 1994; Sun and Perona, 1996a).

9.2.2   Preattentive Basic Features Can Be Created as "Second Order" Stimuli

For example, it is easy to find a vertical target among horizontal distractors. These vertical and horizontal stimuli can be created by simple luminance differences between stimulus and background, but they can also be based on texture or grouping of other elements (even other oriented elements: Bravo and Blake, 1990). The oriented regions can be defined by attributes such as motion, color, or stereopsis (Cavanagh et al., 1990).

9.2.3   Coding of Preattentive Basic Features Appears to Be Quite Coarse

One can measure "just noticeable differences" (jnd) with standard psychophysical methods in order to determine when two stimuli can be discriminated at some threshold level. One can measure a different sort of jnd in visual search by measuring the slope of the RT × set size function for a range of differences between target and distractors. A somewhat arbitrary slope threshold can define a preattentive jnd just as a somewhat arbitrary discrimination threshold defines the classic jnd. When such experiments have been done in color (Nagy and Sanchez, 1990) and orientation (Foster and Westland, 1992, 1998), we find that preattentive jnds are much larger than classical jnds.

The effects of distractor heterogeneity reveal another sense in which preattentive basic features are coarsely coded. It is generally true that increasing distractor heterogeneity decreases search efficiency (Duncan and Humphreys, 1989). More specifically, search is particularly difficult when the distractors flank the targets in feature space. This is demonstrated in Figure 9.2.

In the first row of the figure, the distractors flank the target in feature space (e.g., in the size panel, the target is intermediate in size between the large and small distractors.) In the second row, one of the distractor types is changed so that it lies between the target and the other distractor in the feature space. This makes the distractors, on average, more similar to the target but the task becomes easier.

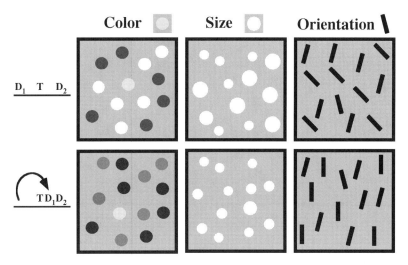

FIGURE 9.2. It is relatively hard to find targets that are flanked by distractors in feature space (e.g., medium-size target among big and small distractors). It is easier to find targets that are to one side of the distractors in feature space.

This may be clearest in the orientation example, where it is easier to find the line tilted 15 deg left (-15) among 0 and 15 deg distractors (bottom row) than among flanking -45 and 15 deg distractors (top row). Extensive experimental support for this claim has been obtained for color (Bauer, Jolicoeur, and Cowan, 1996a, b, 1998; D'Zmura, 1991) and orientation (Wolfe and Friedman-Hill, 1992a, 1992b; Wolfe et al., 1992).

9.2.4   Coding of Preattentive Features May Be Categorical

In Figure 9.3, the target orientation on the left is -10 deg and the distractors are $+30$ and -70. If we add 20 deg to all orientations, we get 10 deg among $+50$ and -50 deg. In this latter case, the target is categorically unique. It is the only "steep" item. The stimuli on the right yield search that is markedly more efficient that the stimuli on the left (Wolfe, Friedman-Hill et al., 1992). In orientation, the preattentive categories seem to be "steep," "shallow," "left," and "right." In size, the categories are probably merely "big" and "small"; in depth, "near" and "far," and so forth.

9.2.5   Reverse Hierarchy

Given this set of experimental findings, it is quite clear that preattentive vision is a relatively late abstraction of the visual input. Its sensitivity to second-order stimuli such as texture boundaries and its coarse, categorical nature point to a locus beyond the detailed local processing of V1. An interesting problem is posed by this fact. A lot of information seems to have been lost on the way from early

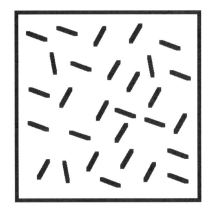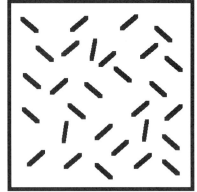

FIGURE 9.3. The targets, tilted 10 deg off vertical, are easier to find on the right where they are the only "steep" items than on the left, where they are not categorically unique.

vision to the preattentive representation of the visual information that supports efficient visual search. When attention is directed to a stimulus, that information can be recovered. How is that done? In their Reverse Hierarchy Theory, Ahissar and Hochstein (1997) revive a thought that had been discussed earlier, notably in the physiological literature and in some computational models (notably Tsotsos, 1988, 1993; Tsotsos et al., 1995). Perhaps the role of selective attention is to allow the perceiver to reach back into earlier stages of visual processing and to recover for specific items the details that had been lost in general. In this view, visual attention mediates between levels of visual processing (c.f. re-entrant processes: Di Lollo, Enns, and Rensink, 2000).

9.3   The Nature of Preattentive Objects

To understand the implications of this view for an understanding of visual perception, a few more facts are needed. First, let us consider the nature of the preattentive representation in some more detail. One more piece of evidence that preattentive vision comes after early vision is that, in the preattentive representation, the visual scene has been parsed into some sort of objects. In contrast, early vision seems to be concerned with local features and rather minimally with whether features are part of the same or neighboring objects. The primary body of evidence supporting the idea that the preattentive representation contains objects is the evidence that attention tends to select objects (reviewed in Goldsmith, 1998; Tipper and Weaver, 1998). Attention to one part of an object seems to "flow" to other parts of objects (Baylis and Driver, 1993; Egly, Driver, and Rafal, 1994; Tipper, Weaver, Jerreat, and Burak, 1994) (for a physiological analog see Roelfsema, Lamme, and Spekreijse, 1998). Some aspects of object structure seem to be available preattentively (e.g., occlusion Rensink and Enns, 1995). New objects capture attention while equivalent changes in low-level features do not (Yantis

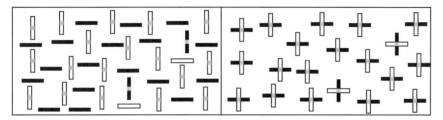

FIGURE 9.4. In the left and right panels, search for the black vertical lines (two in each panel). It is much easier on the left because preattentive feature information can be used to guide attention to objects that are just black and vertical. On the right, all objects have the attributes "black" and "vertical." It requires attention to determine where those features are bound to the same part of the object.

and Hillstrom, 1994; Yantis and Jonides, 1996).

A precise definition of preattentive "object" has eluded the field to date. In part because it seems clear that the same entity (e.g., "nose") can be an object at one moment and a part of an object at another. Further complicating matters, items that might each be considered to be an object may be grouped to form different objects for the purpose of visual search (Bravo and Blake, 1990) (see also the "anthill" phenomenon of Nelson, 1974).

While the nature of an object is not entirely clear, it does seem clear that basic features are rather loosely attached to preattentive objects. Treisman (1982) originally proposed that, prior to the arrival of attention, features were in some sense "free floating" and able to migrate quite widely in a scene, forming "illusory conjunctions" with other features (Treisman and Schmidt, 1982). The notion of completely free-floating features seems to overstate the case (Cohen and Ivry, 1989). Treisman (personal communication) argues that the rapid action of attention would provide some coarse location coding for features (see also Cohen and Ivry, 1991). Another hypothesis, tied to the idea that attention is directed to some sort of preattentive objects, holds that features like the color, size, and orientation of an object are loosely "bundled" with the object prior to the arrival of attention (Wolfe and Bennett, 1997). If an object possesses multiple examples of a single type of feature on a single object (e.g., two or more colors or orientations), the relationship of those features to the object and to each other would not be made explicit until attention permitted the accurate "binding" of those features (Treisman, 1996). As an example, consider Figure 9.4.

On the left of Figure 9.4, it is quite easy to find black vertical targets because, even if the features are not bound to each other preattentively, attention can be guided to the preattentive bundle that includes the features "black" and "vertical". A version of this conjunction search experiment, using red and green rather than black and white stimuli, yield a target present slope of 5.9 msec/item. In contrast, it is harder to find the black verticals in the right-hand panel of Figure 9.4 (there are two of them). The vertical and horizontal items have been combined into "plusses." The resulting preattentive objects all have the features "black, white, vertical, and horizontal." These objects only differ when correctly bound. As a result, search is much less efficient: 47.2 msec/item in a red-green version (Wolfe and Bennett, 1997).

To summarize, visual search data suggest that, prior to the arrival of attention, the visual system codes about a dozen basic features in parallel and represents these as bundles of features, loosely aggregated into preattentive objects. Attention to one of these bundles allows the features to be properly bound, making explicit their relationship to one another. This explicit relationship, in turn, makes object recognition possible. There is some evidence for what could be considered "implicit binding" (e.g., Houck and Hoffman, 1986) and even for implicit recognition (e.g., Tipper and Weaver, 1998) but the ability is fairly limited (e.g., Neumann and DeSchepper, 1992). It seems reasonably clear that attention to specific objects in a scene is needed to recognize those specific objects.

9.4   Post-attentive Vision

What happens after attention has been used to select, bind, and recognize an object? Does its status in visual search change? This can be called the problem of "post-attentive vision" (Wolfe, Klempen, and Dahlen, 2000). In order to address this question, we have performed an extensive series of "repeated search" experiments. These are somewhat different from standard search tasks. In a standard search task, the subject knows that she is looking for a specific target. On each trial, a new set of items appears. It may or may not contain the target. In a repeated search task, the situation is reversed. For a block of trials (several hundred trials in some experiments), all the items in the display remain static, in fixed locations. On each trial, the subject is asked about the presence or absence of a randomly chosen item. Half the time it is an item from the display. Half the time it is absent from the display. The subject simply responds "present" or "absent" in the usual manner.

We have used many different types of stimuli for this task: Letters, novel objects, realistic objects, sometimes embedded in naturalistic scenes. For the example presented here, the stimuli were photorealistic objects provided by Michael Tarr: http://www.cog.brown.edu/~tarr. A sample display is shown in Figure 9.5. The actual items were colored. Subjects were taught the specific names used for each object, and all objects were tested for recognizability at all eccentricities used when subjects were fixating a central spot.

In prior studies, we had presented a visual probe or a word on the screen to inform subjects of the target identity for each trial. In this experiment, we used an auditory probe. This allowed the visual stimulus to remain utterly unchanged for a block of trials. Subjects looked at a display like the one shown in Figure 9.5 and heard, for example, "anchor," to which they would have responded in the affirmative, or "cow," to which they would have responded in the negative.

Ten subjects were tested for two blocks of 50 practice and 100 experimental

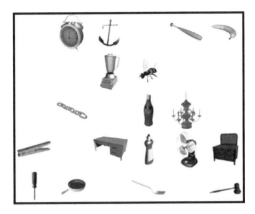

FIGURE 9.5. Sample stimuli from a repeated search experiment using realistic objects and auditory probes. The stimuli would remain static across a block of hundreds of trials. On each trial, subjects would hear the name of a target to be searched for (e.g., "bee," present; "apple," absent).

trials at each of four set sizes (4, 6, 10, and 20). Average RTs are shown in Figure 9.6. These results are typical. There is little or no improvement over, in this case, 150 trials. The differences seen here are not significant. In many other experiments, we have found little or no improvement in the efficiency of search, as measured by the slope of the RT × set size function. There is no increase in efficiency over the first few trials nor over blocks of up to 350 trials (Wolfe et al., 2000). Apparently, even when only a few items are clearly visible in the field, some bottleneck prevents simultaneous access to all items. You believe that you can see N items. You have memorized N items (at least, if N is relatively small.). Yet, if asked to confirm that you can see one of those items, your behavior depends on the number of items in the display just as it would if you were searching a completely new display.

If the visual stimuli are removed, search actually does become more efficient. The task becomes a memory search task. It is known that memory search can become "automatic" — independent of set size — with several hundred trials of practice (Logan, 1992). In the visual case, however, search does not become automatic. Prolonged exposure to the same, unvarying stimulus does not remove the capacity limitation on search.

9.5   One Object at a Time?

How should we understand the failure of repeated search to become efficient? It may be useful to think about the requirements for object recognition more generally. In order for a visual stimulus to be recognized, its features must be bound (as illustrated by the "plus" example shown in Figure 9.4). That bound representation must then be linked to an identifying representation in memory. Ian Howard

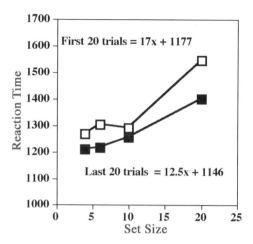

FIGURE 9.6. Results of a repeated search experiment showing little (statistically insignif-icant) change in search efficiency after 100 trials.

is not recognized as Ian Howard until you have bound the Howard features and linked that bound representation to the Howard representation in your memory. Without that link, the bound representation is something, but it is not recogniz-ably Ian. The repeated search experiments suggest that either the binding or the linking operations — or both — are limited to a single object at a time. This is shown in cartoon form in Figure 9.7. Like most cartoons, this one is not intended to be taken too seriously or too literally. It simply illustrates the idea that there is a stimulus containing a vast amount of information. We each possess a memory containing a vast amount of information. In addition, we seem to have a repre-sentation of the stimulus that is, itself, rich with information. The reality of that internal representation is discussed in the next section. For the present, the central point is that contact between levels is very restricted. (Note that even though the "stimulus" and "representation" are identical here, this should not be taken as a claim that we create a faithful representation of the outside world.)

The cartoon shows the fan as the attended item. If the bee was replaced by a button while the fan was attended, the "change blindness" literature tells us that an observer would not notice until attention happened to be directed to the object that was previously the bee (Rensink, 2000; Simons and Levin, 1997). In a change blindness experiment, observers see one version of a scene change into another and report on the nature of the change. If something is done to hide low-level transients that would cue the location of a change, observers prove to be very bad at reporting the change. Objects can appear and disappear and, yet, the change can go unreported until attention happens to be directed to the relevant object. Apparently, attentional bottlenecks keep observers from noticing changes that occur, literally, right before their eyes.

Returning to the cartoon, one can speculate that these bottlenecks serve to pre-vent the confusion that might arise from crosstalk if there were multiple links

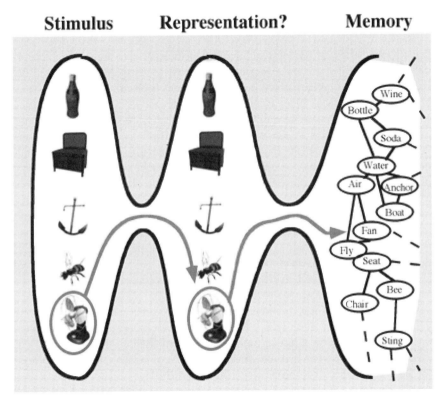

FIGURE 9.7. A cartoon illustrating three broad levels in perception. First, the stimulus; second, a hypothetical internal visual representation and third, memory. Bottlenecks, governed by attention, limit contact between levels.

between levels. Without a bottleneck, an observer might not be sure which object was being recognized as a fan and which as a bee. Note that, though the cartoon shows bottlenecks between stimulus and representation and between representation and memory, current data are not adequate to determine the number of bottlenecks, only to assert that there must be bottlenecks.

9.6   What Do We Actually See?

9.6.1   The Argument for Not Much

Data from a variety of different paradigms have been used to argue that the internal representation of the visual world is, at best, very sparse. We cannot faithfully integrate information across saccades (Blackmore, Brelstaff, Nelson, and Troscianko, 1995; Grimes, 1996; Henderson, 1997; Irwin, 1996; Irwin, Yantis, and Jonides, 1983; Irwin, Zacks, and Brown, 1990) though some memory can guide eye movements (e.g., Carlson-Radvansky, 1999; Hollingworth and Henderson, 2001; Karn and Hayhoe, 2000). Even without eye movements, the change blindness literature shows that we are very poor at detecting changes in images if the transients produced by those changes are hidden (e.g., Rensink, O'Regan, and Clark, 1996; Rensink, 2000; Simons and Levin, 1997; Simons, 2000). During on-line tasks, we seem to continually go back to acquire information (Ballard, Hayhoe, and Pelz, 1995; Ballard, Hayhoe, and Pook, 1995; Hayhoe, Bensinger, and Ballard, 1998). We fail to report very basic properties of the visual scene if they are unexpected and unattended (Mack and Rock, 1998a, b). Finally, as noted above, the repeated search data suggest that we do not have simultaneous access to multiple bound, linked, and recognized objects.

Given this body of information, it has been argued that we only see currently attended items (Mack and Rock, 1998a, b) or that we preserve only the minimal information needed for a just-in-time visual system (Ballard, Hayhoe, and Pelz, 1995; Rensink, 2000) or that the world itself is the representation with no need for an internal representation (O'Regan and Noë, 2001).

9.6.2   The Argument for a Rich Representation

The difficulty is that people think that they see something and, if one wants to understand visual perception, it is ultimately unsatisfying to say that the experience does not exist. So, what do we see? On the one hand, we do not see the stimulus in any very direct form. Even if we did not concern ourselves with attention, phenomena like binocular rivalry (Blake, 1989; Breese, 1909; Wolfe, 1986; Blake, this volume) make clear the disconnection between what is on the retina and what is perceived. It is also worth noting that, while acuity and other visual functions fall off rapidly with distance from fixation, our perception does not seem comparably degraded. Until we fixate and direct attention to the periphery, it is usually not obvious that very little of the visual field contains well-focused detail. This suggests that what we see can be built up over multiple fixations (e.g., Noton and Stark, 1971) or deployments of attention even if precise transaccadic integration cannot be found. On the other hand it is obvious, but worth noting, that we don't just see some memory of the prior objects of attention. Closing the eyes fundamentally changes the experience. The experience of seeing is ultimately based on visual input. When it is not we call it a dream or a hallucination.

There is a richness to visual experience that is at odds with the data reviewed above. We attend to one or, perhaps, a few objects at a time. Yet the visual world appears to contain many objects. Even in a brief exposure, too brief to permit eye movements, we seem to see a complete visual field filled with the qualities of visual experience at all points. Where does this richness come from?

There are at least three potential sources of perceptual richness. First, even if we can only attend to a single object at a time, this does not mean that the consequences of attention are entirely lost when attention moves elsewhere (no matter what I have previously argued: Wolfe, 2000; Wolfe et al., 2000). Returning to Figure 9.7, it seems possible that successive deployments of attention might serve to populate the representation with objects. If you want to check if an object in the representation still corresponds with an object in the stimulus or, perhaps, if you want to identify a specific object in the representation, you would need to pass through an attentional bottleneck. However, multiple objects of attention might persist. (We could call this "persistence" but the term is already used in a more basic sense, e.g., Di Lollo, Lark, and Hogben, 1988 ; Francis, Grossberg, and Mingolla, 1994.) One way to think about this possibility is to ask what it means to say that only one object can be attended at a time. While only one object may be attended at a specific instant, you do not experience the present time as an instant. As James (1890) and many others recognized, the "psychological" (or "sensible") present has a duration. Various methods of measuring this duration yield estimates of a "present" that is hundreds of msec in length. If we assume that attention can be deployed at a rate of 20-40 Hz (for a discussion, see Moore and Wolfe, 2000), this would yield a perceptual experience of many objects even if only one were actually attended in the physical, instantaneous present.

The second source of a rich and spatially continuous perception is the preattentive visual information discussed at the start of this chapter. It seems clear that we are consciously aware of visual "stuff" (Adelson and Bergen, 1991) throughout the visual field without the need for that "stuff" to be selected by attention. It might be a good idea to consider this awareness of some visual input across the field to be evidence of diffuse attention since there are circumstances of attentional tunnel vision where this awareness seems to be lost (Williams, 1985). Whether it is diffuse attention or preattention, under normal circumstances, we seem to be aware of something like the texture of preattentive basic features throughout the visual field. As with the apparent multiplicity of objects, one must be cautious about declaring that subjects see this preattentive "stuff" because attention is required any time one wishes to get a subject to make an explicit response to the presence of a stimulus. Thus, subjects might see color at all locations but they can only respond to color, one object or location at a time. Implicit measures can show that features were registered (e.g., Houck and Hoffman, 1986) but they cannot demonstrate perception. The fact that the representation (if any) is walled off from stimulus and/or from response by attentional bottlenecks seems likely to render the experimental evidence forever ambiguous. Thus, even if you are convinced that you have a rich perceptual life and even if you are willing to assume that others do too, it will be difficult, if not impossible to prove the point.

A third factor that could contribute to the creation of a rich perceptual representation is the ability to extract some meaning from unbound, minimally attended stimuli. It is striking that people appear to be able to gain some understanding of the meaning of a scene very quickly (Intraub, 1980; Thorpe, Fize, and Marlot, 1996; Van Rullen and Thorpe, 2001). The times required to show evidence of semantic processing are short enough that is difficult to imagine that the meaning is extracted by a succession of attentional deployments to a succession of objects. Recently, Oliva and Torralba (2001) have shown that it possible, in principle, to extract meaning from nonlocalized structural information encoded in the frequency spectrum. They devised a series of simple, linear, feed-forward filters that can be used to classify scenes on axes such as natural/artificial, rough/smooth, and open/closed. In the space defined by these axes, scenes naturally cluster into semantically meaningful categories like mountain scenes, beach scenes, street scenes, and so forth. We could call this the "unbound semantics" of the image.

Given these sources of information it should be possible for an observer to quickly develop a theory about the stimulus. That theory will be modulated by the observer's current biases and predispositions. The idea that the contents of the observer's mind might make a difference is an old one, enshrined in Shakespeare ("In the night, imagining some fear, how easy is a bush supposed a bear," *Midsummer Night's Dream* 5:1:21–22). In vision research it is perhaps best known in Helmholtz's (1924) discussion of "unconscious inference." Bayesian theories are the modern incarnation of this thought (Brainard and Freeman, 1997; Freeman, 1994; Lee, 1995). In all of these varying degrees of sophistication, the core idea is that what we experience as seeing is a theory. Work on attention adds to this idea by emphasizing the tenuous nature of the link between the theory that we see and the stimulus on the retina.

These ideas can be summarized by referring to another cartoon (Figure 9.8). Imagine that you are looking at a farm scene with a few chickens in the yard. The farm scene is projected onto your retina and fed forward to early vision levels of analysis (1). The box at the center of the cartoon is the hypothetical level of the perceptual representation. Massively parallel, feed-forward pathways (2) create the coarse preattentive aspects of perception (3) and provide the data for the unbound semantic analysis that might identify this as a farm scene (4). In the cartoon, attention is directed to one of the chickens (5). This puts the perceived chicken in contact with the stimulus (6) and makes it possible to link the chicken to the relevant node in memory (7). Other, previously attended chickens (8) remain part of the perceptual experience even though the perceived chickens are not in current contact with the stimulus. Indeed, in the cartoon, the second

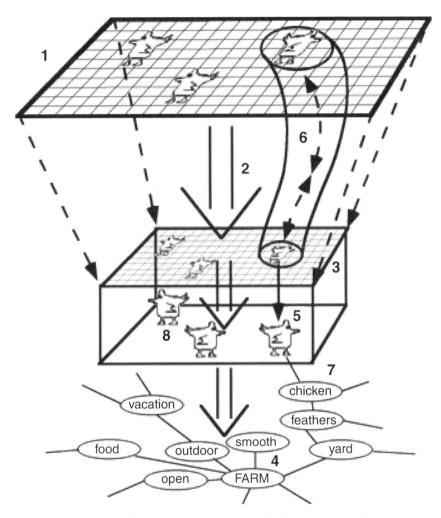

FIGURE 9.8. A cartoon illustrating how the stimulus (1) might be processed in a massively parallel fashion (2) to yield a coarse, pre-attentive representation of the world (3). Further parallel processing might yield the unbound semantic notion that this was a farm scene (4). Representation of individual attended objects (like the chicken 5) can be bound via feedback to ealry levels (6) and linked to the relevant nodes in memory (7). Objects recently attended may be post-attentively represented in perception (8) even though they are not currently linked to their counterparts in the stimulus (note the mismatch in orientation of the "real" and "seen" middle chicken).

chicken is facing different directions in the "world" and in the perceptual representation. This mismatch would not be noted or corrected until attention was redeployed to the relevant chicken.

9.7   Conclusions

It should be obvious that this chapter has moved from the concrete to the speculative. There are some things that we know:

1. Some features are processed in parallel across the entire visual field.

2. The resulting preattentive representation of these features is coarse, perhaps categorical. It is cruder than the analysis that is performed by early visual processing stages (e.g., primary visual cortex).

3. Preattentive vision appears to parse the input into candidate objects.

4. In the absence of attention, features appear to be bundled only loosely with their object.

5. Attention can be directed to those objects. This permits more accurate binding of features to objects.

6. Attention permits recovery of detailed information that is not available preattentively. This may involve feedback/re-entrant pathways.

7. Attention permits linkage between an object and its representation in memory, thus allowing the object to be recognized.

8. Only one (or perhaps a few) objects can be attended at one time.

9. If attention has been deployed elsewhere, it must be redirected back to an object in order to confirm that object's presence (e.g., in repeated search tasks) or to detect a change (e.g., change blindness).

10. Finally, these facts must somehow be reconcilable with the overarching fact that observers think that they are perceiving a visual world that extends across the visual field and is richly populated with coherent objects.

The proposal sketched above tries to reconcile #1–9 with #10 by accepting the old idea that we see our current "theory" about the external world and adding the notion that this theory is in very limited contact with the world at any moment in time.

Acknowledgements

I thank Jennifer DiMase and Aude Oliva for comments on drafts of this chapter. This research was supported by grants from the National Institutes of Health: National Eye Institute and National Institute for Mental Health, National Science Foundation, and the Air Force Office of Scientific Research.

References

Adelson, E. H. and Bergen, J. R. (1991). The plenoptic function and the elements of early vision. In M. Landy and J. A. Movshon (eds.), *Computational Models of Visual Processing*, pp. 3–20. Cambridge: MIT Press.

Ahissar, M. and Hochstein, S. (1997). Task difficulty and visual hierarchy: Counter-streams in sensory processing and perceptual learning. *Nature*, 387: 401–406.

Aks, D. J. and Enns, J. T. (1996). Visual search for size is influenced by a background texture gradient. *J. Exp. Psych.: Hum. Percept. and Perf.*, 22: 1467–1481.

Ballard, D., Hayhoe, M. and Pelz, J. (1995). Memory representations in natural tasks. *J. Cogn. Neurosci.*, 7: 66–80.

Ballard, D., Hayhoe, M. and Pook, P. (1995). *Deitic Codes for the Embodiment of Cognition*. Technical Report 95.1, University of Rochester.

Barlow, H. (1995). The neuron doctrine in perception. In M. S. Gazzaniga (ed.), *The Cognitive Neurosciences*, pp. 415-435. Cambridge: MIT Press.

Bauer, B., Jolicoeur, P. and Cowan, W. B. (1996a). Distractor heterogeneity versus linear separability in colour visual search. *Percept.*, 25: 1281–1294.

Bauer, B., Jolicoeur, P. and Cowan, W. B. (1996b). Visual search for colour targets that are or are not linearly-separable from distractors. *Vis. Res.*, 36: 1439–1466.

Bauer, B., Jolicoeur, P. and Cowan, W. B. (1998). The linear separability effect in color visual search: ruling out the additive color hypothesis. *Percept. and Psychophys.*, 60: 1083–1093.

Baylis, G. C. and Driver, J. (1993). Visual attention and objects: evidence for hierarchical coding of location. *J. Exp. Psychol. Hum. Percept. and Perf.*, 19: 451–470.

Beck, J. (1966). Perceptual grouping produced by changes in orientation and shape. *Science*, 154: 538–540.

Blackmore, S. J., Brelstaff, G., Nelson, K. and Troscianko, T. (1995). Is the richness of our visual world an illusion? Transsaccadic memory for complex scenes. *Percept.*, 24: 10750-1081.

Blake, R. (1989). A neural theory of binocular rivalry. *Psych. Rev.*, 96: 145–167.

Brainard, D. H. and Freeman, W. T. (1997). Bayesian color constancy. *J. Opt. Soc. Am. A*, 14: 1393–1411.

Brainard, D. H., Wandell, B. A. and Chichilnisky, E.-J. (1993). Color constancy: from physics to appearance. *Cur. Dir. in Psych. Sci.*, 2: 165–170.

Braun, J. (1993). Shape-from-shading is independent of visual attention and may be a texton. *Spatial Vis.*, 7: 311–322.

Bravo, M. and Blake, R. (1990). Preattentive vision and perceptual groups. *Percept.*, 19: 515–522.

Breese, B. B. (1909). Binocular rivalry. *Psych. Rev.*, 16: 410–415.

Carlson-Radvansky, L. A. (1999). Memory for relational information across eye movements. *Percept. and Psychophys.*, 61: 919–934.

Carter, R. C. (1982). Visual search with color. *J. Exp. Psych.: Hum. Percept. and Perf.*, 8: 127–136.

Cavanagh, P., Arguin, M. and Treisman, A. (1990). Effect of surface medium on visual search for orientation and size features. *J. Exp. Psych.: Hum. Percept. and Perf.*, 16: 479–492.

Cohen, A., and Ivry, R. B. (1989). Illusory conjunction inside and outside the focus of attention. *J. Exp. Psychol. Hum. Percept. and Perf.*, 15: 650–663.

Cohen, A., and Ivry, R. B. (1991). Density effects in conjunction search: evidence for coarse location mechanism of feature integration. *J. Exp. Psychol. Hum. Percept. and Perf.*, 17: 891–901.

Di Lollo, V., Enns, J. T. and Rensink, R. A. (2000). Competition for consciousness among visual events: the psychophysics of reentrant visual processes. *J. Exp. Psych.: Gen.*, 129: 481–507.

Di Lollo, V., Lark, C. D. and Hogben, J. H. (1988). Separating visible persistence from retinal afterimages. *Percept. and Psychophys.*, 44: 363–368.

Duncan, J. and Humphreys, G. W. (1989). Visual search and stimulus similarity. *Psych. Rev.*, 96: 433–458.

D'Zmura, M. (1991). Color in visual search. *Vis. Res.*, 31: 951–966.

Egly, R., Driver, J. and Rafal, R. D. (1994). Shifting attention between objects and loctions: evidence from normal and parietal lesion subjects. *J. Exp. Psych.: Gen.*, 123: 161–177.

Elder, J. and Zucker, S. (1993). The effect of contour closure on the rapid discrimination of two-dimensional shapes. *Vis. Res.*, 33: 981–991. Enns, J. T. (1992). Sensitivity of early human vision to 3-D orientation in line drawings. *Can. J. Psychol.*, 46: 143–169.

Enns, J. T. and Kingstone, A. (1995). Access to global and local properties in visual search for compound stimuli. *Psych. Sci.*, 6: 283–291.

Enns, J. T. and Rensink, R. A. (1990a). Scene based properties influence visual search. *Science*, 247: 721–723.

Enns, J. T. and Rensink, R. A. (1990b). Sensitivity to three-dimensional orientation in visual search. *Psych. Sci.*, 1: 323–326.

Foster, D. H. and Ward, P. A. (1991). Asymmetries in oriented-line detection indicate two orthogonal filters in early vision. *Proc. Roy. Soc. Lond. B*, 243: 75–81.

Foster, D. H. and Westland, S. (1992). Fine structure in the orientation threshold function for preattentive line-target detection. *Percept.*, 22 (Supp. 2 ECVP — Pisa): 6.

Foster, D. H. and Westland, S. (1998). Multiple groups of orientation-selective visual mechanisms underlying rapid oriented-line detection. *Proc. R. Soc. Lond. B*, 265: 1605–1613.

Francis, G., Grossberg, S. and Mingolla, E. (1994). Cortical dynamics of feature binding and reset: control of visual persistence. *Vis. Res.*, 34: 1089–1104.

Freeman, W. T. (1994). The generic viewpoint assumption in a framework for visual perception. *Nature*, 368: 542–545.

Gilchrist, I. D., Jane, R. M. and Heiko, N. (1997). Luminance and edge information in grouping: a study using visual search. *J. Exp. Psych.: Hum. Percept. and Perform.*, 23: 464–480.

Goldsmith, M. (1998). What's in a location? Comparing object-based and space-based models of feature integration in visual search. *J. Exp. Psych.: Gen.*, 127: 189–219.

Green, B. F. and Anderson, L. K. (1956). Color coding in a visual search task. *J. Exp. Psychol.*, 51: 19–24.

Grimes, J. (1996). On the failure to detect changes in scenes across saccades. In K. Akins (ed.) *Perception*, pp. 89–110. New York: Oxford University Press.

Hayhoe, M. M., Bensinger, D. G. and Ballard, D. H. (1998). Task constraints in visual working memory. *Vis. Res.*, 38: 125–137.

von Helmholtz, H. (1924). *Treatise on Physiological Optics*, J. P. C. Southall, Trans. Trans. from 3rd German ed. of 1909, ed.). Rochester, NY: The Optical Society of America.

Henderson, J. M. (1997). Transsaccadic memory and integration during real-world object perception. *Psych. Sci.*, 8: 51–55.

Hollingworth, A. and Henderson, J. M. (2001). Accurate visual memory for previously attended objects in natural scenes. *J. Exp. Psychol: Hum. Percept. and Perf.*, in press.

Houck, M. R. and Hoffman, J. E. (1986). Conjunction of color and form without attention. Evidence from an orientation-contingent color aftereffect. *J. Exp. Psych.: Hum. Percept. and Perf.*, 12: 186–199.

Intraub, H. (1980). Presentation rate and the representation of briefly glimpsed pictures in memory. *J. Exp. Psych.: Hum. Learn. and Mem.*, 6: 1–12.

Irwin, D. E. (1996). Integrating information across saccadic eye movements. *Cur. Dir. in Psych. Sci.*, 5: 94–100.

Irwin, D. E., Yantis, S. and Jonides, J. (1983). Evidence against visual integration across saccadic eye movements. *Percept. Psychophys.*, 34: 49–57.

Irwin, D. E., Zacks, J. L. and Brown, J. S. (1990). Visual memory and the perception of a stable visual environment. *Percept. Psychophys.*, 47: 35–46.

James, W. (1890). *The Principles of Psychology*. New York: Henry Holt and Co.

Julesz, B. and Bergen, J. R. (1983). Textons, the fundamental elements in preattentive vision and perceptions of textures. *Bell Sys. Tech. J.*, 62: 1619–1646.

Karn, K. and Hayhoe, M. (2000). Memory representations guide targeting eye movements in a natural task. *Vis. Cog.*, 7: 673–703.

Kellman, P. (1998). *An Update on Gestalt Psychology, Perception, Cognition, and Language: Essays in Honor of Henry and Lila Gleitman*, Cambridge: MIT Press.

Kleffner, D., Polichar, V. E. and Ramachandran, V. S. (1990). Shape from shading affects motion perception and brightness constancy. *Invest. Ophthalmol. Vis. Sci.*, 31: 524.

Kleffner, D. A. and Ramachandran, V. S. (1992). On the perception of shape from shading. *Percept. Psychophys.*, 52: 18–36.

Kristjansson, A. and Tse, P. U. (2001). Curvature discontinuities are cues for rapid shape analysis. *Percept. Psychophys.*, 41: 390-403.

Lamme, V. A. F., Super, H. and Spekreijse, H. (1998). Feedforward, horizontal, and feedback processing in the visual cortex. *Cur. Opinion in Neurobiol.*, 8: 529–535.

Lee, T. S. (1995). A Bayesian framework for understanding texture segmentation in the primary visual cortex. *Vis. Res.*, 35: 2643–2657.

Lettvin, J. (1995). J. Y. Lettvin on grandmother cells. In M. S. Gazzaniga (ed.), *The Cognitive Neurosciences*, pp. 434–435. Cambridge: MIT Press.

Logan, G. (1992). Attention and preattention in theories of automaticity. *Am. J. Psych.*, 105: 317–339.

Mack, A. and Rock, I. (1998a). *Inattentional Blindness*. Cambridge: MIT Press.

Mack, A. and Rock, I. (1998b). Inattentional blindness: perception without attention. In R. D. Wright (ed.) *Visual Attention*, pp. 55-76. New York: Oxford University Press.

Moore, C. M. and Wolfe, J. M. (2000). Getting beyond the serial/parallel debate in visual search: a hybrid approach. In K. Shapiro (ed.) *The Limits of Attention: Temporal Constraints on Human Information Processing*. Oxford: Oxford University Press.

Moraglia, G. (1989). Display organization and the detection of horizontal lines segments. *Percept. Psychophys.*, 45: 265–272.

Nagy, A. L. and Sanchez, R. R. (1990). Critical color differences determined with a visual search task. *J. Opt. Soc. Am. A*, 7: 1209–1217.

Nakayama, K. and Shimojo, S. (1992). Experiencing and perceiving visual surfaces. *Science*, 257: 1357–1363.

Neisser, U. (1967). *Cognitive Psychology*. New York: Appleton, Century, Crofts.

Nelson, J. I. (1974). Motion sensitivity in peripheral vision. *Percept.*, 3: 151–152.

Neumann, E. and DeSchepper, B. G. (1992). An inhibition-based fan effect: evidence for an active suppression mechanism in selective attention. *Canad. J. Psych.*, 46: 1–40.

Nothdurft, H. C. (1993). Saliency effects across dimensions in visual search, *Vis. Res.*, 33: 839–844.

Noton, D. and Stark, L. (1971). Eye movements and visual perception. *Sci. Am.*, 224: 35–43.

O'Connell, K. M. and Treisman, A. M. (1990). Is all orientation created equal? *Invest. Ophth. Vis. Sci.*, 31: 106.

Oliva, A. and Torralba, A. (2001). Modeling the shape of the scene: a holistic representation of the spatial envelope. *Int. J. Comp. Vis.*, 42: 145–175.

O'Regan, J. K. and Noë, A. (2001). A sensorimotor account of vision and visual consciousness. *Behav. and Brain Sci.*, 24:.

O'Regan, K. (1992). Solving the "real" mysteries of visual perception. The world as an outside memory. *Canad. J. Psych.*, 46: 461–488.

Previc, F. H. and Naegele, P. D. (2001). Target-tilt and vertical-hemifield asymmetries in free-scan search for 3-D targets. *Percept. Psychophys.*, 41: 445–457.

Ramachandran, V. S. (1988). Perception of shape from shading. *Nature*, 331: 163–165.

Rensink, R. A. (2000). Seeing, sensing, and scrutinizing. *Vis. Res.*, 40: 1469–1487.

Rensink, R. and Cavanagh, P. (1993). Processing of shadows at preattentive levels. *Invest. Ophthal. Vis. Sci.*, 34: 1288.

Rensink, R. and Cavanagh, P. (1994). Identification of highlights in early vision. *Invest. Ophthal. Vis. Sci.*, 35: 1623.

Rensink, R. A. and Enns, J. T. (1995). Pre-emption effects in visual search: evidence for low-level grouping. *Psych. Rev.*, 102: 101–130.

Rensink, R., O'Regan, J. K. and Clark, J. J. (1996). To see or not to see: The need for attention to perceive changes in scenes. *Psych. Sci.*, 8: 368–373.

Rock, I. and Palmer, S. (1990). The legacy of Gestalt psychology. *Sci. Am.*, 263: 84–90.

Roelfsema, P. R., Lamme, V. A. F. and Spekreijse, H. (1998). Object-based attention in the primary visual cortex of the macaque monkey. *Nature*, 395: 376.

Simons, D. J. (2000). *Change Blindness and Visual Memory*. A special issue of *Vis. Cog.*, Philadelphia: Psychology Press.

Simons, D. J. and Levin, D. T. (1997). Change blindness. *Trends in Cog. Sci.*, 1: 261–267.

Sun, J. and Perona, P. (1996a). Early computation of shape and reflectance in the visual system. *Nature*, 379: 165–168.

Sun, J. and Perona, P. (1996b). Where is the sun? *Invest. Ophth. Vis. Sci.*, 37: S935.

Theeuwes, J. and Kooi, J. L. (1994). Parallel search for a conjunction of shape and contrast polarity. *Vis. Res.*, 34:3013–3016.

Thorpe, S., Fize, D. and Marlot, C. (1996). Speed of processing in the human visual system. *Nature*, 381: 520–552.

Tipper, S. P. and Weaver, B. (1998). The medium of attention: location-based, object-based, or scene-based? In R. D. Wright (ed.) *Visual Attention*, pp. 77–107. Oxford: Oxford University Press.

Tipper, S. P., Weaver, B., Jerreat, L. M. and Burak, A. L. (1994). Object-based and environment-based inhibition of return of visual attention. *J. Exp. Psych.: Hum. Percep. Perf.*, 20: 478–499.

Treisman, A. (1982). Perceptual grouping and attention in visual search for features and for objects. *J. Exp. Psych.: Hum. Percep. Perf.*, 8: 194–214.

Treisman, A. (1996). The binding problem. *Cur. Op. Neurobiol.*, 6: 171–178.

Treisman, A. and Gormican, S. (1988). Feature analysis in early vision: Evidence from search asymmetries. *Psych. Rev.*, 95: 15–48.

Treisman, A. and Souther, J. (1985). Search asymmetry: A diagnostic for preattentive processing of seperable features. *J. Exp. Psychol. Gen.*, 114: 285–310.

Treisman, A. M. and Schmidt, H. (1982). Illusory conjunctions in the perception of objects. *Cogn. Psych.*, 14: 107–141.

Tsotsos, J. K. (1988). A "complexity level" analysis of immediate vision. *Int. J. Comp. Vis.*, 2: 303–320.

Tsotsos, J. K. (1993). An inhibitory beam for attentional selection. In L. Harris and M. Jenkin (eds.) *Spatial Vision in Humans and Robots*, pp. 313–331). : Cambridge: Cambridge University Press.

Tsotsos, J. K., Culhane, S., Wai, W. Y. K., Lai, Y., Davis, N. and Nuflo, F. (1995). Modeling visual attention via selective tuning. *Artif. Intel.*, 78: 507–545.

Van Rullen, R. and Thorpe, S. J. (2001). Is it a bird? Is it a plane? Ultra-rapid visual categorisation of natural and artifactual objects. *Percept.*, 30: 655–668.

Williams, L. (1966). The effect of target specification on objects fixed during visual search. *Percept. Psychophys.*, 1: 315–318.

Williams, L. (1985). Tunnel vision induced by a foveal load manipulation. *Hum. Fact.*, 27: 221–227.

Wolfe, J. M. (1986). Stereopsis and binocular rivalry. *Psych. Rev.*, 93: 269-282.

Wolfe, J. M. (1992). "Effortless" texture segmentation and "parallel" visual search are not the same thing. *Vis. Res.*, 32: 757–763.

Wolfe, J. M. (1998). Visual search. In H. Pashler (ed.) *Attention*, pp. 13–74. Hove, East Sussex, UK: Psychology Press Ltd.

Wolfe, J. M. (2000). Post-attentive vision and the illusion of perception. Paper presented at "Toward a Science of Consciousness", Tuscon, AZ.

Wolfe, J. M. (2001). Asymmetries in visual search: an introduction. *Percept. Psychophys.*, 63: 381–389.

Wolfe, J. M. and Bennett, S. C. (1997). Preattentive object files: shapeless bundles of basic features. *Vis. Res.*, 37: 25–44.

Wolfe, J. M., Cave, K. R. and Franzel, S. L. (1989). Guided search: an alternative to the feature integration model for visual search. *J. Exp. Psychol. Hum. Percept. Perf.*, 15: 419–433.

Wolfe, J. M. and Franzel, S. L. (1988). Binocularity and visual search. *Percept. Psychophys.*, 44: 81–93.

Wolfe, J. M. and Friedman-Hill, S. R. (1992a). On the role of symmetry in visual search. *Psych. Sci.*, 3: 194–198.

Wolfe, J. M. and Friedman-Hill, S. R. (1992b). Visual search for orientation: the role of angular relations between targets and distractors. *Spat. Vis.*, 6: 199–208.

Wolfe, J. M., Friedman-Hill, S. R., Stewart, M. I. and O'Connell, K. M. (1992). The role of categorization in visual search for orientation. *J. Exp. Psych.: Hum. Percept. Perf.*, 18: 34–49.

Wolfe, J. M., Klempen, N. and Dahlen, K. (2000). Post-attentive vision. *J. Exp. Psych.: Hum. Percept. Perf.*, 26: 693–716.

Wolfe, J. M., Yee, A. and Friedman-Hill, S. R. (1992). Curvature is a basic feature for visual search. *Percept.*, 21: 465–480.

Yantis, S. and Hillstrom, A. P. (1994). Stimulus-driven attentional capture: Evidence from equiluminant visual objects. *J. Exp. Psych.: Hum. Percept. Perf.*, 20: 95–107.

Yantis, S. and Jonides, J. (1996). Attentional capture by abrupt onsets: New perceptual objects or visual masking. *J. Exp. Psych.: Hum. Percept. Perf.*, 22: 1505–1513.

10

Single Cells to Cellular Networks

Robert F. Hess

The early stages of visual processing involve cells tuned to elementary stimulus properties such as spatial frequency, orientation, and so on, while the later stages involve cells tuned to complex object features such as circular stimuli, faces, and so on. We are essentially ignorant of how the outputs of the initial filters are combined to process these more complex object features. An elementary example is that of contour integration. Here I review recent experiments that shed light on how the outputs of cells in the early stages of cortical processing are combined to extract elementary contours. I examine the regional specialization of these network operations, their site in the visual pathway, and their possible neural code.

10.1   Introduction

Our knowledge of the neural basis of visual perception is understandably dominated by the wealth of single-cell neurophysiology that has emerged since Hubel and Wiesel's first recording in cat cortex (Hubel and Wiesel, 1959). The initial stages of visual processing are dominated by the role of different types of filters; spatial, temporal, chromatic, motion, disparity, etc. Initially, the visual system's preoccupation is with throwing away as much information as possible, only transmitting the essentials to higher stages in the pathway. Neurons acting as filters (e.g., spatial, temporal, chromatic) are designed to accomplish this. Our understanding of how we detect relatively simple spatial targets confined to local regions of the field has been successfully based on our knowledge of the filtering properties of single cells. When stimuli are suprathreshold, involve extended regions of the visual field, or are of a more complex form (e.g., spatially, temporally, chromatically), single-cell explanations fall short of the mark. This is especially true for so-called "global" tasks where information in different parts of the field needs to be integrated to solve the task.

Parallel processing by independent banks of neural filters is well and good for solving the transmission problem, but it has contributed little to the problem of analysing the many types of correlations that characterise natural images and which are essential to visual perception (Field, 1987). This information can only be extracted by comparing the outputs of different neural filters in different parts

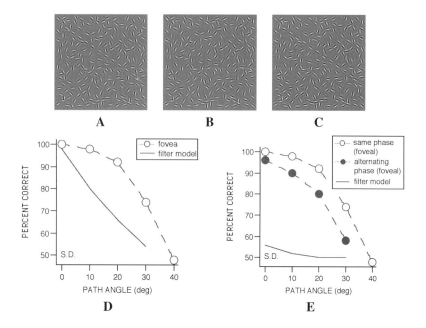

FIGURE 10.1. At the top, straight paths (path angle = 0) composed of aligned (a), orthogonal (b), and phase-alternating (c) elements are embedded in a background field of
identical, randomly oriented elements. In (d) and (e), performance is plotted as a function
of path angle. In each frame human performance is compared with the performance of a
model (solid line) in which there is no integration across filters tuned to different orientations here referred to as a simple filter model (see Hess and Dakin, 1999, for details). In
(d), foveal performance (symbols) is compared with that of the simple filter model (solid
line). In (e), foveal performance (filled symbols) is compared for elements having alternating spatial phase (see (c)), with that of the filtering model (solid line). For comparison,
human performance for elements having the same spatial phase is shown by open symbols.
From Hess and Field (2000). *Trends in Cog. Sci.*, 3: 480–486, with permission.

of the field, and this can only be done within cellular networks. The previously
all-important concept of a receptive field gives way to the just as important concept of the association field of the network as a whole. In this review, I will use the
example of contour integration as a means of showing the importance of network
processing and in particular, the relationship between the receptive fields of single
cortical cells and the association field of their cellular networks.

10.2   The Nuts and Bolts

Contour integration has been of interest to visual psychophysicists for some time
fueled by the Gestalt psychologists who suggested rules by which such complex
relationships could be represented — their rules of good continuation (Gibson,

1950; see also Kovacs, 1996 for review). More recently attempts (Uttal, 1983; Beck, Rosenfeld and Ivry, 1989; Moulden, 1994) have been made to understand these, though it has been difficult to see what the relationship might be between the tuning properties of single cells and the network operations that describe how their outputs are combined. This has been mainly because the elements used in such studies have been spatially broadband (dots and lines) and the contours investigated have been straight. This always leaves open the possibility of an explanation in terms of a single, broadband detector.

David Field, Tony Hayes, and I set out to look at network operations more directly by using spatial frequency narrowband elements and contours of varying curvature. The stimulus is shown in Figure 10.1. Within a field of randomly oriented Gabor elements, a subset of the elements are aligned along a notional contour (Figure 10.1a). This stimulus has the important advantage that we can limit the visual system to one scale of analysis, that of the individual Gabor, while asking questions at the scale of the contour. Subjects were asked to discriminate between images such as that shown in Figure 10.1a and equivalent ones where no contour is present. We ensured that there were no local or global density cues to aid this discrimination. The first indication that our ability to detect the contour displayed in Figure 10.1a was telling us something about visual processing was the finding that contours composed of elements whose local orientation was orthogonal to the contour are less detectable. Such a contour is illustrated in Figure 10.1b. It is only after scrutinizing the stimulus that its presence is revealed. From an informational point of view, Figures 10.1 a and b are equivalent, so any difference in their detectability reflects constraints imposed by the visual system.

One interesting result concerned our ability to detect curved contours. The results displayed in Figure 10.1d plots percent correct against the curvature of the contour expressed in terms of the angular difference between segments of an invisible backbone on which the individual Gabors comprising the contour are aligned (Field, Hayes and Hess, 1993). Performance (unfilled symbols) is good even for quite curved contours suggesting that the output from cells with different orientational preferences are being integrated (rather than summation of cells with the same orientational preference). The solid curve in Figure 10.1d represents the performance of a multichannel, linear filtering model (Hess and Dakin, 1999) in which only the information from single orientation bands is used. As expected, it shows a stronger dependence on contour curvature than that observed psychophysically. Figure 10.1c shows another important stimulus manipulation that reinforces the notion that this task reflects the action of a network rather than that of single neurons. Here we flip the polarity of every other Gabor element. The contour (and background) is now composed of Gabor elements alternating in their contrast polarity. The visibility of the contour in Figure 10.1a and c is similar. Psychophysical measurement shows that although there is a small decrement in performance in the alternating polarity condition (compare filled and unfilled symbols in Figure 10.1e), curved contours are still detectable when composed of elements of alternating polarity. This would not be expected of any single detector with an elongated receptive field since summing over more than one element

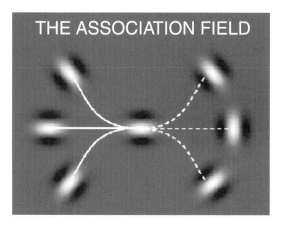

Contour integration only occurs when:

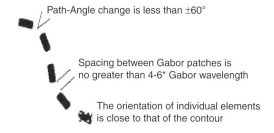

Path-Angle change is less than ±60°

Spacing between Gabor patches is
no greater than 4-6* Gabor wavelength

The orientation of individual elements
is close to that of the contour

Other Variables:
The phase of the Gabor patch was found to be irrelevant
Detection improves as the number of elements increases towards 12

FIGURE 10.2. The "association field."

would be detrimental. This is shown by the performance of the linear filtering model (solid line in Figure 10.1e) being at chance.

The results of these different manipulations suggest that our detection of these extended contours when density cues are removed is due to selective integration of the outputs of cells at different spatial locations with different orientational preferences. This can be summarized in terms of a notional "association field" in much the same way as we have traditionally done for a receptive field. This is depicted in Figure 10.2. The linking strength depend on the orientation and spatial position of individual cells so as to optimize their encoding of simple first order curves. Weakest linking occurs between cells with inappropriate orientation or spatial locations. We proposed (Field, Hayes and Hess, 1993) that the underlying mechanism of this grouping may be the lateral connections between cortical V1 neurons described by a number of laboratories (Gilbert and Wiesel, 1979; Rockland and Lund, 1982). Recent anatomical and neurophysiological studies in the cat (Schmidt et al., 1997), tree shrew (Bosking et al., 1997) and monkey

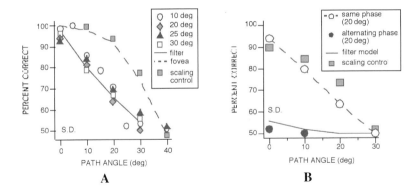

FIGURE 10.3. Detection of paths embedded in a background field of randomly oriented elements. Performance (percent correct) is plotted as a function of path angle for stimuli presented at a range of different retinal eccentricities. In each frame human performance is compared with the performance of a model in which there is no integration across filters tuned to different orientations here referred. From Hess and Dakin (1997), *Nature*, 390: 602–604, with permission.

(Malach et al., 1993) have demonstrated that this behaviourally defined association field maps well onto the sorts of lateral connections that have been found to occur between cells of similar orientations. Using optical imaging, both Malach et al. (1993) and Bosking et al. (1997) for example, determined the orientation columns across the cortex, then used biochemical tracers to track where the horizontal projections of a particular cell project. The results show that the long-range connections primarily project to orientation columns with a similar orientation preference. Furthermore, Bosking et al. have shown that the labeled axons extend for longer distance along the axis of the receptive field than orthogonal to it. This agrees well with the psychophysical results demonstrating much better grouping when the elements are aligned along the axis than orthogonal to the axis (see Figures 10.1a and b). Some recent results have even suggested that the off-axis projections appear to project to off-axis orientations such as that shown in Figure 10.2 (Blasdel, personal communication).

10.3   Regional Specialization

Not all parts of the visual field are endowed with these network interactions. More peripheral parts of the visual field appear to use a different, more rudimentary, but more economical way of processing contours. Performance for contour detection falls off abruptly with eccentricity out to about 10 degree after which performance falls off much more gradually. This result is shown in Figure 10.3a. Furthermore, contours composed of elements of alternating polarity are much less visible in the periphery (Figure 10.3b). Performance under these conditions can be explained simply by summation between cells of similar orientation preference; the linear filtering model can predict performance for both same and alternating polarity condition (solid curve prediction in Figures 10.3a and b) and subjects report seeing only isolated segments of curved contours when viewing peripherally. If this is so it would represent an important economy in the processing of contours; the more computation-intensive operations involving the combination of the outputs of cells with different orientational preferences being confined to the fovea.

10.4   Processing Level

Our use of Gabor elements rests on an assumption that these operations are carried out at a relatively low level in the visual process where the outputs of spatial frequency and orientationally tuned cells have not yet been combined. However, it is quite possible that the linking process that we revealed is a general-purpose one, operating at a level where features constructed from combining spatial frequency and orientation information are linked. It may occur at a stage where cue-invariant operations occur (Albright, 1992). To address this, Steven Dakin and I (Dakin and Hess, 1999) investigated whether contour integration occurred prior to the point where spatial frequency and orientation information is combined in the visual pathway. Georgeson and Meese (1997) had previously shown that the perception of horizontal and vertical structure occurred when two oblique grating were added together. We wondered whether linking would occur between the perceived horizontal and vertical features (after the point where orientation information had been combined) or the oblique components (before the point where orientation information had been combined). The stimuli are shown in Figure 10.4 where now the elements are patches of plaids rather than Gabors. In Figure 10.4a, the horizontal and vertical perceived features of an oblique plaid are illustrated. In b, the contour is constructed from the alignment of the component orientations of the plaids, whereas in c the contour is composed of the aligned features of the plaid. The contour in a is much more visible than it is in b. For an 8-element path, average performance was at 75% correct for a straight path in the case of a and 49% in the case of b (Dakin and Hess, 1999), suggesting that the linking occurs prior to the point in the pathway where orientation information is combined.

A similar question can be ask in terms of spatial scale; does this linking occur at or after the point where information is collapsed across spatial scale? To address this question Steven Dakin and I (Dakin and Hess, 1999) investigated whether linking occurs between elements with very different features but a common spatial scale. We used contours composed of alternate Gabors and phase-scrambled edge elements (fractal noise). The phase-scrambled edge elements have no features in common with the Gabors, but they do contain a common spatial scale (Figure 10.5b). In Figures 10.5c and d, performance, for two subjects, in terms of percent correct is plotted against the spatial frequency of the Gabor element. Performance (triangles) is good and invariant with the Gabor spatial frequency, suggesting that linking occurs within individual spatial scales with equal efficiency.

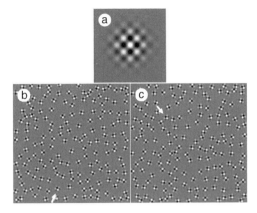

FIGURE 10.4. (a) A Gaussian windowed plaid which, although composed of the sum of two oblique gratings, produces a percept of horizontal and vertical "checkerboard" structure. Contours composed of plaid elements with local contour orientation aligned with either the perceived checkerboard structure (c) or one of the grating components (b). The detectability of contours composed of plaids depends on the orientation of the gratings making up the plaids and not the appearance of the composite. From Dakin and Hess, 1999. *Spatial Vision*, 12: 309–327, with permission.

This compliments a previous result that Steven Dakin and I had obtained concerning the spatial frequency tuning of contour integration (Dakin and Hess, 1998). By changing the spatial frequency of each alternate path (and background) element we were able to show that Gabor elements that were more than an octave different in their spatial frequency did not link together. All of these results (Figures 10.4 and 10.5) when taken together provide support for the notion that contour integration occurs at a relatively early stage in the visual process before information is combined across spatial frequency and orientationally tuned mechanisms to represent more complex features.

10.4.1   The Linking Code

It is presently not known what neural code is used to integrate cellular responses to allow the detection of the type of contours described so far. Two immediate possibilities are the average neuronal activity or firing rate and the synchronicity of neuronal firing. Let us examine each of these in turn.

A number of current models of contour integration (Field, Hayes and Hess, 1993; Kovacs and Julesz, 1993; Yen and Finkel, 1996a, b; Pettet, McKee and Grzywacz, 1996; Polat and Sagi, 1993, 1994; Grossberg, Mingolla and Ross, 1997) rely on the activity level or average firing of neurons for linking operations. The obvious problem with this is that contrast is also conveyed by the same means, suggesting that there should be strong contrast-facilitatory effects during linking. Hess, Dakin and Field (1998) examined whether this was the case using two different approaches. First, they assessed how important relative contrast was

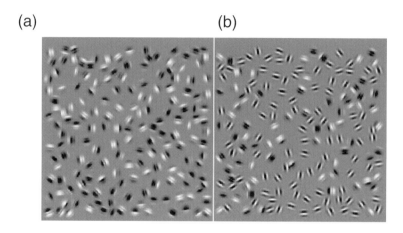

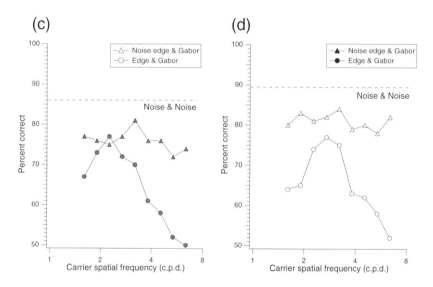

FIGURE 10.5. (a) Path stimulus composed of phase randomised "f+3f+5f" compound Gabor micro-patterns, (b) intermixed with Gabors at 3.2 c.p.d. (c) Detection of "noise-edge" paths for RFH and (d) SCD. Notice in the case of the noise edge and Gabor elements (triangles) that performance is good for all spatial scales that are common to the two stimuli. From Dakin and Hess, 1999. *Spatial Vision*, 12: 309–327, with permission.

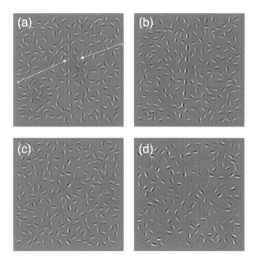

FIGURE 10.6. A shows a 0 deg path where the contrast of all the elements (all 50%) are the same as those of the background elements. B shows a 0 deg path whose elements have 40% higher contrast that those of the background. C shows a random or misaligned path whose elements are 40% higher in contrast to those of the background. D shows a straight path sloping 45 deg to the right in which all elements are uniformly random in contrast between the range 10-90%. Reprinted from *Vis. Res.*, 38: 783–787, 1998, Hess, R. F. et al., The role of "contrast enhancement" in the detection and appearance of visual contours, with permission from Elsevier Science.

for contour integration. Second, they assessed whether the perceived contrast of elements that were part of a contour was different from that of similar elements not part of a contour. The stimuli for these two tasks are displayed in Figure 10.6. Figure 10.7 shows results for two observers for path detection where the contrast of all stimuli (path as well as background) was randomly varied between 0, 10, 20, and 40% about a base level of 50% (see Figure 10.6d for example). Performance is seen to be largely independent of this contrast variation for the different path angles investigated (0, 10, 20, 30 deg). In the extreme case (i.e., 40%), the Gabors varied between 10% and 90% without any loss of performance. The detection of the contour does not appear to be contingent on the average level of neuronal firing because adding noise to this does not affect performance.

The second test of this "average neuronal firing" hypothesis was to assess whether the perceived contrast of elements that were part of a contour was in anyway different to that of similar elements that were not part of a contour. Using a simple matching paradigm, the perceived contrast of contour elements was judged relative to identical elements that were not part of a contour. The stimuli are shown in Figures 10.6a, b and c. There were two conditions; a path condition and a no-path condition. In the path condition, the contrast of elements aligned along the contour (target elements) was judged relative to that of the background elements (Figures 10.6a and b). In the no-path condition, the orientation of the

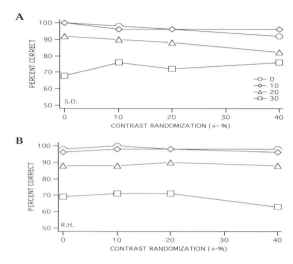

FIGURE 10.7. Path detection for two observers at four path angles (0, 10, 20, 30 deg) for elements of 50% base contrast whose contrast is varied between different ranges (0, 10, 20, 40%). Reprinted from *Vis. Res.*, 38: 783–787, 1998, Hess, R. F. et al., The role of "contrast enhancement" in the detection and appearance of visual contours, with permission from Elsevier Science.

target elements was random (Figure 10.6c). Both path and no-path conditions were run with the background elements set to 25% and 50%. The mean contrast errors (Table 10.1) were found to be low across all conditions (all around 1.5% with s.d. of 1%). There was no systematic bias found in the estimated contrast of "path" compared to "non-path" stimuli. The null hypothesis, that the mean contrast error equals zero, cannot be rejected by a t-test ($p < 0.05$; 15 d.f.) for any of these results (t-values ranged from 0.0 to 0.37). A similar null hypothesis, that data from the jittered and aligned path conditions are significantly different, can also be rejected ($p < 0.05$; 15 d.f.).

Another current idea is that the synchronicity of neuronal activity could provide a linking code that was unaffected by the contrast conveyed by the average neuronal activity. In particular, neurons exhibit high-frequency (40–80Hz) oscillatory behaviour that has been shown to become synchronised when common objects stimulate different individual neurons (Singer and Gray, 1995). On the basis of this, one would expect the linking operation to have good temporal dynamics. Hess et al. (2001) have recently investigated the dynamics of linking using a technique where individual Gabor elements are rotated in and out of alignment along a contour. We did this for both a transient, single-shot and a steady-state, multishot display. The stimulus is diagrammatically displayed in Figure 10.8.

In the transient condition, the test stimulus is presented sandwiched between masks to prevent processing outside the test presentation. In the steady-state condition, the transient stimulus is displayed for many cycles. The results for both types of display are quite similar (Figures 10.9 and 10.10).

	Jitt. path 25%	**Align path** 25%	**Jitt. path** 50%	**Align path** 50%
DF	−0.002 (0.022)	0.003 (0.026)	0.014 (0.038)	−0.016 (0.015)
SCD	−0.02 (0.03)	−0.012 (0.017)	−0.013 (0.05)	0.00 (0.029)
YW	−0.003 (0.02)	−0.017 (0.016)	−0.012 (0.042)	−0.04 (0.033)
IM	0.004 (0.014)	0.00 (0.085)	0.066 (0.033)	0.04 (0.018)
RFH	0.016 (0.027)	0.045 (0.022)	0.025 (0.032)	0.053 (0.02)
AH	0.000 (0.016)	−0.003 (0.014)	0.016 (0.02)	0.014 (0.017)
RD	−0.017 (0.019)	0.001 (0.015)	−0.025 (0.029)	−0.020 (0.04)
CW	0.015 (0.022)	0.001 (0.021)	0.009 (0.027)	0.007 (0.019)

TABLE 10.1. Mean contrast errors, expressed in fractional units (0.0 to 1.0), for matching between path and background elements in aligned and random paths (see Figures 10.6 b and c) of 25% and 50% contrast. Values in parentheses are the estimated standard deviation of errors. From Hess et al., 1998, *Vis. Res.*, 38: 783–787, with permission.

STIMULI SEQUENCE

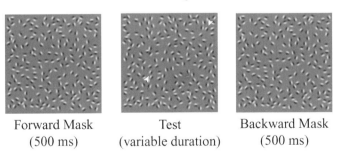

Forward Mask
(500 ms)

Test
(variable duration)

Backward Mask
(500 ms)

FIGURE 10.8. Sequence of stimuli presented in a trial of the single-shot masked condition, and repeated cyclically in the steady-state experiments. Forward mask and backward mask are identical, and built from the orientation randomization of each Gabor element composing the test stimulus. Extremities of the 20 degree path are denoted by white arrows in the test stimulus. Reprinted from *Vis. Res.*, 41: 1023–1037, 2001, Hess, R. F. et al., Dynamics of controu integration, with permission of Elsevier Science.

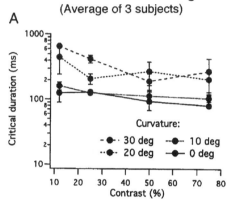

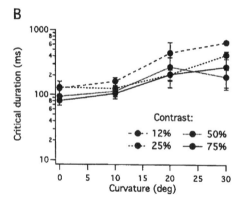

FIGURE 10.9. Critical durations derived from fits to psychometric data. Time constants: (a) as a function of contrast for each curvature; (b) as a function of curvature for each contrast. The dynamics are slow and depend on contour curvature, not contour contrast. Reprinted from *Vi s. Res.* 41: 1023–1037, 2001, Hess, R. F. et al., Dynamics of contour integration, with permission of Elsevier Science.

In Figure 10.9 for the single-shot transient condition, we are plotting critical duration (a measure derived from the psychometric data relating to the time constant of the visual response) against either the curvature of the contour or the contrast of the elements. It can be seen that there is very little dependence on contrast but a clear dependence on contour curvature. In Figure 10.10 for the steady state, multi-shot condition we are plotting critical temporal frequency (the reciprocal of the critical duration) against either the curvature of the contour or the contrast of the elements. Note however for both types of presentation that the dynamics, contrary to what might be expected from the synchronized oscillation hypothesis, are very sluggish. Straight contours whose orientation linking is modulated above 10 Hz cannot be detected, and curved contours whose orientation linking is modulated above 3 Hz can not be detected. The sluggish nature of the linking and its dependence on curvature would not be expected if the linking code is due to high frequency synchronized activity.

Another possibility that we (Hess, Dakin and Field, 1998) favour and are currently exploring involves the relative activity of neurons at different times in their spike trains. Recent work looking at the effects of surround context on the behaviour of visual neurons has shown that individual neurons can independently vary in spike rate at different times following the response of the neuron (Zipser et al., 1996). Figure 10.11, for example, shows this result (from Zipser, Lamme and Schiller, 1996) for a V1 neuron under conditions where the context (well outside of the classical receptive field) was altered. As one can see, the initial transient burst of activity in response to the two stimuli was essentially the same under the two conditions. However, the sustained response to the stimuli showed significant changes. Zipser et al. (1996) interpret their results as suggesting that feedback from higher levels of the visual system alter the response at this later point in the response of the cell.

The temporal coding strategy illustrated in Figure 10.11 may also occur for contrast and connectivity. Whether the difference in sustained response is due to feedback from later stages or lateral connections, these data lead us to the possibility that the initial burst is due to feedforward activity and may provide information regarding contrast, while the variations in the later sustained components of the response provide information regarding the context (e.g., connectivity) of the stimulus. In our stimuli, this would suggest that the initial response of the cells would code for contrast and be the same whether or not the elements were part of the path or the background. However, for the cells responding to the elements along the path, there would be differential activity in the later sustained part of the response.

10.5   Conclusion

At present, neurophysiological approaches to understanding visual perception are stuck at the single-cell level due to technical reasons, yet it is clear that our un-

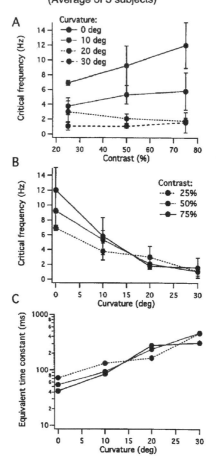

FIGURE 10.10. Critical temporal frequency for orientation modulation derived from psychometric data. Critical frequency: (a) as a function of contrast for each curvature; (b) as a function of curvature for each contrast; (c) expressed as a critical duration (see Figure 10.9). The dynamics are slow and depend on contour curvature not contour contrast. Reprinted from *Vis. Res.*, 41: 1023–1037, 2001, Hess, R. F. et al., Dynamics of controu integration, with permission of Elsevier Science.

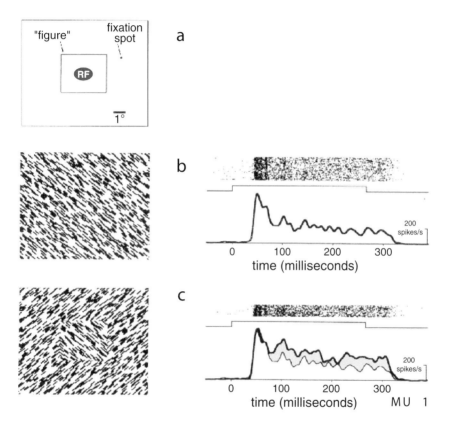

FIGURE 10.11. Illustration of the response of a multi-unit site to stimulation with a texture defined figure. In (a), the stimulus configuration relative to the conventional receptive field. In (b), the response to a homogeneous texture. In (c), the response to a texture defined figure. Note that the initial response is identical, but the tonic phase (shaded region) is elevated in the figure condition. (This figure is reproduced with permission from Zipser et al., 1996, *J. Neurophysiol.*, 16: 7376–7389).

derstanding of even the most rudimentary aspects of visual processing requires a knowledge of how the outputs of cells are combined. Contour integration is a good example. Its investigation has reinforced the importance of understanding network operations, and the concept of an "association field" is a first attempt at this. The rules governing integration will depend on what tuning feature is being integrated (spatial, motion, chromaticity, disparity, etc). For spatial contours there appears to be a specialization for central vision. Linking appears to occur at a level where spatial information has not yet been combined and to possibly involve a temporal code. Whether this occurs in V1 or an extra-striate area is currently unknown. We are only at the very beginning of understanding what appears to be the next important step in going from the filtering /transmission stage to the combination/analysis stage of visual processing.

Acknowledgements

All of the above work was done in collaboration. I am grateful to the intellectual contribution of the following; David Field, Tony Hayes, Steven Dakin, William Beaudot, Kathy Mullen and Tim Ledgeway.

References

Albright, T. D. (1992). Form-cue invariant motion processing in primate visual cortex. *Science*, 255: 1141–1143.

Beck, J., Rosenfeld, A. and Ivry, R. (1989). Line segmentation. *Spatial Vis.*, 42: 75-101.

Bosking, W. H., Zhang, Y., Schofield, B. and Fitzpatrick, D. (1997). Orientation selectivity and the arrangement of horizontal connections in the tree shrew striate cortex. *J. Neurosci.*, 17: 2112–2127.

Dakin, S. C. and Hess, R. F. (1998). Spatial-frequency tuning of visual contour integration. *J. Opt. Soc. Am. A*, 15: 1486–1499.

Dakin, S. C. and Hess, R. F. (1999). Contour integration and scale combination processes in visual edge detection. *Spatial Vis.*, 12: 309–327.

Field, D. J. (1987). Relations between the statistics of natural images and the response properties of cortical cells. *J. Opt. Soc. Am. A*, 4: 2379–2394.

Field, D. J., Hayes, A. and Hess, R. F. (1993). Contour integration by the human visual system: evidence for a local "association field," *Vis. Res.*, 33: 173–193.

Georgeson, M. A. and Meese, T. S. (1997). Perception of stationary plaids: the role of spatial filters in edge analysis. *Vis. Res.*, 37: 3255–3271.

Gibson, J. (1950). *The Perception of the Visual World*, Boston: Houghton Mifflin.

Gilbert, C. D. and Wiesel, T. N. (1979). Morphology and intracortical connections of functionally characterised neurones in the cat visual cortex. *Nature*, 280: 120–125.

Grossberg, S., Mingolla, E. and Ross, W. D. (1997). visual brain and visual perception: how does the cortex do perceptual grouping? *Trends in Neurosci.*, 20: 106–111.

Hess, R. F., Beaudot, W. H. A. and Mullen, K. T. (2001). Dynamics of contour integration, *Vis. Res.*, 41: 1023–1037.

Hess, R. F. and Dakin, S. C. (1997). Absence of contour linking in peripheral vision. *Nature*, 390: 602–604.

Hess, R. F. and Dakin, S. C. (1999). Contour integration in the peripheral field, *Vis. Res.*, 39: 947–959.

Hess, R. F., Dakin, S. C. and Field, D. J. (1998). The role of "contrast enhancement" in the detection and appearance of visual contours. *Vis. Res.*, 38: 783–787.

Hess, R. F. and Field, D. J. (2000). Integration of contours: new insights. *Trends in Cog. Sci.*, 3: 480–486.

Hubel, D. H. and Wiesel, T. N. (1959). Receptive fields of single neurons in the cat's striate cortex. *J. Physiol. (Lond.)*, 148: 574–591.

Kovacs, I. (1996). Gestalten of today: early processing of visual contours and surfaces. *Behav. Brain Res.*, 82: 1–11.

Kovacs, I. and Julesz, B. (1993). A closed curve is much more than an incomplete one: effect of closure in figure-ground segmentation. *Proc. Nat. Acad. Sci. USA*, 90: 7495–7497.

Malach, R., Amir, Y., Harel, H. and Grinvald, A. (1993). Relationship between intrinsic connections and functional architecture revealed by optical imaging and *in vivo* targeted biocytin injections in primary striate cortex. *Proc Natl. Acad. Sci. USA*, 90: 10469–10473.

Moulden, B. (1994). Collator units: second-stage orientational filters. In M. J. Morgan (ed.), *Higher-Order Processing in the Visual System: CIBA Foundation Symposium 184*, pp. 170–184, Chichester: John Wiley and Sons.

Pettet, M. W., McKee, S. P. and Grzywacz, N. M. (1996). Smoothness constrains long-range interactions mediating contour-detection. *Invest. Ophthal. Vis. Sci.*, 37: 4368.

Polat, U. and Sagi, D. (1993). Lateral interactions between spatial channels: suppression and facilitation revealed by lateral masking experiments. *Vis. Res.*, 33: 993–999.

Polat, U. and Sagi, D. (1994). The architecture of perceptual spatial interactions. *Vis. Res.*, 34: 73–78.

Rockland, K. S. and Lund, J. S. (1982). Widespread periodic intrinsic connections in the tree shrew visual cortex. *Science*, 215: 1532–1534.

Schmidt, K. E., Goebel, R., Lowel, S. and Singer, W. (1997). The perceptual grouping criterion of collinearity is reflected by anisotropies of connections in the primary visual cortex., *J. Eur. Neurosci.*, 9: 1083–1089.

Singer, W. and Gray, C. M. (1995). Visual feature integration and the temporal correlation hypothesis. *Ann. Rev. Neurosci.*, 18: 555–586.

Smit, J. T. S., Vos, P. G. and Van Oeffelen, M. P. (1985). The perception of a dotted line in noise: a model of good continuation and some experimental results. *Spatial Vis.*, 12: 163–177.

Uttal, W. R. (1983). *Visual Form Detection in 3-Dimensional Space*. Hillsdale: Lawrence Erlbaum.

Yen, S.-C. and Finkel, L. H. (1996a). Salient contour extraction by temporal binding in a cortically-based network. In D. S. Touretzky, M. C. Mozer and M. E. Hasselmo (eds.) *Advances in Neural Information Processing Systems*, Boston: MIT Press.

Yen, S.-C. and Finkel, L. H. (1996b). "Pop-out" of salient contours in a network based on striate cortical connectivity. *Invest. Ophthal. Vis. Sci.*, 37S: 297.

Zipser, K., Lamme, V. A. F. and Schiller, P. H. (1996). Contextural modulation in primary visual cortex. *J. Neurophysiol.*, 16: 7376–7389.

Part III

Eye Movements and Perception

11

Levels of Fixation

Richard V. Abadi, Richard Clement, and Emma Gowen

11.1 Introduction

Objects are best seen when their images fall on the fovea and are held relatively steady. If retinal image slip velocities exceed 4 deg/sec, blur and oscillopsia occur. On the other hand, if the image velocities are dramatically reduced or even stabilized, then there is fragmentation and the eventual perceptual loss of the object of regard. Consequently, the phrase "relatively steady" is usefully defined by a range of retinal slip velocities.

The search for how fixation is kept in check has led to the finding that there are a number of important control systems in operation. Their number and operation depend greatly on the nature of gaze (primary, secondary, or tertiary) and also whether the individual is seated in a laboratory with teeth embedded in a slab of wax viewing a single target or is moving freely in a multi-textured natural environment. For example, retinal image slip velocities when the head is stabilized using a bite-bar are usually less than 0.25 deg/sec, but these can rise to several degrees per second when the head is free to move (Figure 11.1). Sometimes these control systems do not function efficiently and on occasion can even fail catastrophically. In this case, steady fixation breaks down and a variety of ocular intrusions or oscillations occur. The purpose of this chapter is to first review the underlying mechanisms that are responsible for steady gaze and second to describe the levels of fixation instabilities that can occur when things go wrong.

11.2 Control Mechanisms for Holding the Eyes Steady on Primary Gaze

There are three main control mechanisms for maintaining steady gaze: fixation, the vestibulo-ocular reflex, and a gaze-holding system (the neural integrator) that operates whenever the eyes are required to hold an eccentric gaze position. In addition, slow control also contributes to gaze holding.

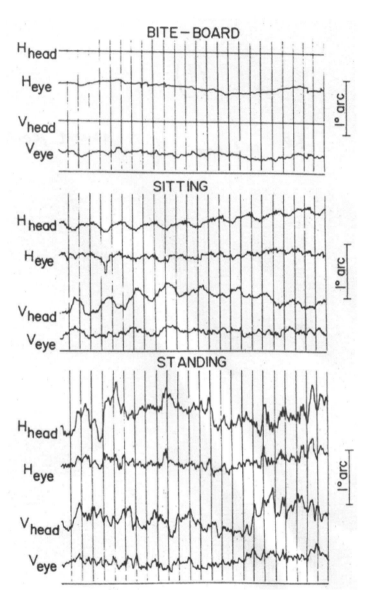

FIGURE 11.1. Simultaneous horizontal (H) and vertical (V) head and eye recordings of a subject when the head is supported by a bite bar compared with sitting and standing as still as possible without artificial support. Reprinted from *Vis. Res.*, 19: 675–683, 1979, Skavenski, A. A. et al., Quality of retinal image stabilisation during small natural and artificial body rotations in man, with permission from Elsevier Science.

11.2.1   Fixation Cells and Pathways

Primate studies on the parietal cortex and superior colliculus have uncovered a number of important features underpinning fixation control. First, parietal neurons discharge during fixation and not during pursuit and vice versa (Lynch et al., 1977). Second, the rostral superior colliculus has been implicated in maintaining fixation and preventing saccades until a trigger signal arrives from the higher centers. Certain cells (fixation cells) in this area are active during periods of steady fixation and attenuate their activity during saccades (Munoz and Wurtz, 1992). An increase in the discharge of the fixation cells delays the initiation of saccades. Moreover, saccades can be interrupted when increases in activity of the fixation cells occur mid-flight. In addition, when fixation cells are inhibited the saccadic latency decreases and the monkeys display difficulties maintaining fixation and suppressing unwanted saccades. There is also evidence to suggest that these fixation cells project to both the saccade-related cells in the caudal superior colliculus and the omni-pause cells in the brain stem. Thus, during fixation, it has been hypothesized that only the rostral fixation cells are active and suppress saccadic generation through direct inhibition of the caudal saccade-related cells and/or the omni-pause cells, and that a decrease in activity of the fixation neurons may constitute a neural substrate of fixation disengagement. Such local fixation activity is almost certainly orchestrated by cortical (e.g., frontal eye fields, posterior parietal areas) or cerebellar regions (e.g., fastigial nucleus) that have strong inputs into the superior colliculus.

The fixation system, which functions during primary gaze, has two distinct components:

1. The visual system's ability to detect retinal image drift and programme corrective eye movements.

2. The ability to attend to, or "engage," a particular target of interest.

Failure will bring about a disruption of steady fixation, resulting in two types of abnormal fixation-saccadic intrusions/oscillations and nystagmus. The essential difference between them lies in the initial movement that takes the line of sight off the object of regard. In the case of saccadic intrusions or saccadic oscillations it is an inappropriate fast movement (Figures 11.2a and b). On the other hand, with nystagmus, it is a slow drift or "slow phase," which moves the eyes off target (Figures 11.2c – f).

11.2.2   Saccadic Intrusions and Oscillations

Involuntary, sporadic fast eye movements that interfere with fixation are called saccadic intrusions (Dell'Osso and Daroff, 1999). They are seen in many guises and include square wave intrusions, macro-square wave intrusions, saccadic pulses, double saccadic pulses, and sporadic ocular bobbing.

Whilst lesions of the brain stem and cerebellum can cause saccadic intrusions, they can also be observed in otherwise healthy individuals. A number of re-

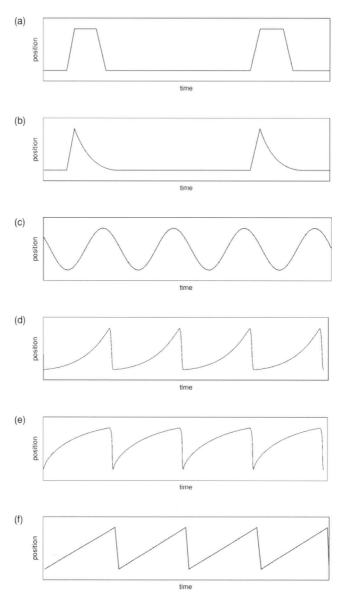

FIGURE 11.2. A schematic illustration of saccadic intrusions and nystagmus. Six possible slow phase and fast phase combinations are shown: (a) saccadic intrusions (fast phase followed by a fast phase); (b) saccadic pulses (fast phase followed by a slow phase); (c) pendular oscillations (slow phase followed by a slow phase); (d) an accelerating velocity exponential slow phase jerk nystagmus (slow phase followed by a fast phase); (e) a decelerating velocity exponential slow phase jerk nystagmus (slow phase followed by a fast phase); (f) a linear or constant velocity slow phase jerk nystagmus (slow phase followed by a fast phase).

ports suggest that conjugate horizontal square wave intrusions (sometimes called square wave jerks) are to be found in over 20% of the population (Herishanu and Sharpe, 1981; Shallo-Hoffmann, Peterson and Mühlendyck 1989; Abadi, Scallan and Clement, 2000). Typically they move in a direction away from fixation, have an inter-saccadic interval of 200 ms, are 0.5-1.5 deg in amplitude, and in common with many saccades, exhibit dynamic overshoots (Figure 11.3). The frequency of the saccadic intrusions are idiosyncratic for each observer, although fatigue and age tend to increase the probability of their presence (Abadi et al., 2000).

Saccadic pulses are also created by small, horizontal saccades in the direction away from fixation (Figure 11.2b). However, unlike square wave intrusions the saccade is followed almost immediately by a slow decreasing velocity return drift. The waveform can easily be confused with a manifest latent jerk nystagmus (Figure 11.2e). The difference being that the initiation of the saccadic pulse is a saccade made in a direction away from fixation, whereas the saccade in a manifest latent jerk nystagmus returns the eye back toward fixation. Saccadic pulse amplitudes can be as large as 5 deg, and their overall duration are generally around 500 ms. Saccadic pulses have been reported in multiple sclerosis and brainstem disease.

When saccadic intrusions become regular and sustained, they are termed saccadic oscillations. Invariably, saccadic oscillations reflect an underlying neurological disorder. Bursts of uniplanar back-to-back saccades are often referred to as ocular flutter. Both saccadic pulses and flutter have been reported recently in visually deprived infants (Gage et al., 2001). An apparent continuum has been reported by Abel and his colleagues (1984) who described the progression of saccadic intrusions into saccadic oscillations, and by Tychen and his team (1990) who reported that saccadic intrusions evolved into continuous, multiplanar, conjugate, back-to-back saccades of varying amplitude (i.e., opsoclonus). Very recently, Gage and her colleagues (2001) reported the progression of saccadic intrusions into a manifest latent nystagmus when they examined young aphakic and pseudophakic infants. Although burst cells, omni-pause cells, and fixation cells are likely to be involved in fixation instabilities such as saccadic intrusions and oscillations, it still remains unclear whether it is specific structures and/or their inputs that are at fault.

11.3   Infantile Nystagmus

The two most common types of benign nystagmus seen in infancy are congenital nystagmus and manifest latent nystagmus (Abel, 1990; Abadi et al., 1991; Dell'Osso and Daroff, 1975; Harris, 1997). In both cases the oscillations are typically conjugate, horizontal, and jerky. Differential diagnosis is made on the basis that the slow phases are typically of an increasing exponential velocity form in congenital nystagmus (Figure 11.2d) whereas in manifest latent nystagmus the slow phases are decreasing (Figure 11.2e) or linear (Figure 11.2f) (Dell'Osso and

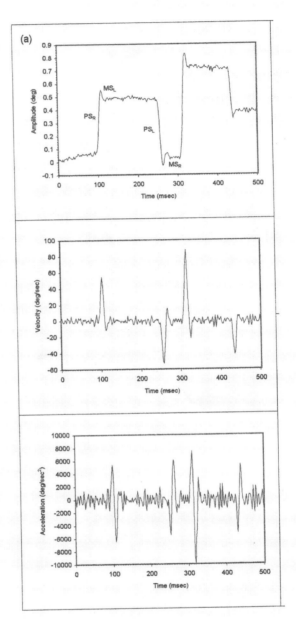

FIGURE 11.3. Subject with physiological saccadic intrusions (square wave jerks). Primary saccades to the right and left are indicated by PSR and PSL, and microsaccades to the right and left are indicated by MSR and MSL (a) Eye position, velocity, and acceleration traces. Reprinted from *Vis. Res.*, 40: 2813–2829, 2000, Abadi, R. V., The characteristics of dynamic overshoots in square-wave jerks, and in congenital and manifest nystagmus, with permission from Elsevier Science.

Daroff, 1975; Dell'Osso, 1985; Abadi and Dickinson, 1986; Abadi et al., 1991; Abadi and Scallan, 2001). In addition to its distinguishing slow phase, the fast phase of manifest latent nystagmus always beats toward the viewing eye. Both congenital and manifest latent nystagmus are associated with a variety of disorders, including albinism, optic nerve hypoplasia, and congenital cataracts. Congenital nystagmus may occur without ocular or central nervous system abnormalities (i.e., idiopathic congenital nystagmus).

11.3.1   Mechanisms Underlying Congenital Nystagmus

Over the years a number of mechanisms underlying congenital nystagmus have been proposed. These include abnormalities of the smooth pursuit, fixation, and optokinetic systems. To date, several distinct models have been constructed to account for congenital nystagmus. The first was provided by Dell'Osso (1967) who suggested that there was a high gain instability in the slow phase eye movement system. This was followed in 1984 by Optican and Zee's model in which the time constant of the neural integrator was lengthened by a velocity feedback signal. When the sign of the feedback signal is reversed, the small post-saccadic drift velocities are amplified by the unstable feedback loop and creates exponentially growing slow phases. Tusa and his colleagues (1992) extended this model by proposing that the fixation system has both normal and abnormal feedback loops. Thus, individuals who are unable to suppress their nystagmus either have only the abnormal feedback loop or cannot voluntarily manipulate the normal feedback loop. It is pertinent to note that central to both the 1984 and 1992 models is the need for neural mis-wiring. This seems somewhat untenable given the range of visual disorders associated with congenital nystagmus in the absence of chiasmal misdirection, the absence of an abnormal visual evoked response in idiopathic congenital nystagmusm and the finding of congenital nystagmus in achiasmic dogs (Dell'Osso and Williams, 1995) and humans (Dell'Osso, 1996). Fourth, Harris (1995) suggested that congenital nystagmus was due to excessive gain in an internal efference copy loop in the smooth pursuit system around a leaky neural integrator.

11.3.2   Are Congenital Nystagmus Waveforms Produced by Saccadic System Abnormalities?

In 2000, Broomhead and his colleagues offered a novel hypothesis to explain a range of fixation instabilities and, by using a displacement feedback model of normal saccadic eye movements, they were able to generate a variety of saccadic intrusions and nystagmus oscillations. Burst cells typically have an on-direction of eye movement, during which they fire and an off- direction, during which they are almost silent (Van Gisbergen, Robinson and Gielen, 1981). The amount of firing is a function of the motor error, which is the difference between the desired gaze direction and the current eye direction. A typical plot of burst cell firing against

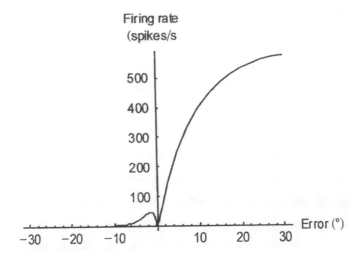

FIGURE 11.4. Plot of burst cell firing against motor error.

motor error has the form shown in Figure 11.4. Thus, the behaviour of a pair of mutually inhibitory right (r) and left (l) bursting neurons can be described by a fast/slow system of three differential equations (Broomhead, Clement, Muldoon, Whittle, Scallan and Abadi, 2000).

$$\begin{aligned} \epsilon \frac{dr}{dt} &= -r - \gamma r l^2 + f(m) \\ \epsilon \frac{dl}{dt} &= -l - \gamma l r^2 + f(-m) \\ \frac{dm}{dt} &= -r(-l) \end{aligned} \tag{11.1}$$

Where r and l are the right and left input to the muscle plant from the right and left burst cells,. ϵ is a small positive number, m is the motor error, and γ is a constant that determines the mutual inhibition between the burst cells.

Analysis of a simple model of this behaviour shows that there is a previously unsuspected instability inherent in the experimentally determined on and off behaviours of the burst cells.

A variety of waveforms can be produced by changing the parameters of the model. For example, if the range of motor errors over which the off component of burst cell operates is reduced, then microsaccadic oscillations occur.

11.3.3 A Dynamical Systems Approach to Understanding Intrusions and Oscillations

A dynamical system is one that can be modeled with equations that describe the way in which the state of the system evolves over time. A geometric picture of the behaviour of the system can be made by treating each of the variables, which specify the state of the system as co-ordinates in a vector space, referred to in this context as a phase or state space. The successive states of a system form a trajectory in phase space, which will eventually end up following a locus of points, referred to as an attractor. The attractor can be a fixed point (stable behaviour), a closed loop or limit cycle (periodic behaviour), a torus (quasiperiodic behaviour), or a portion of the phase space (chaotic behaviour) (Figure 11.5).

In 1997, Abadi and his colleagues (Abadi, Broomhead, Clement, Whittle and Worfolk, 1997) proposed that steady fixation of a normal oculomotor system corresponds to a fixed point of a deterministic control mechanism, while the eye movements that occur in congenital nystagmus are the result of loss of stability of this fixed point. Mathematically, the stability of a fixed point can be quantified by studying the evolution of nearby states. Close to a fixed point, the behaviour of a system is approximately linear, and can be characterized in terms of the eigenvalues and eigenvectors. Eigenvectors are special trajectories of the linear model that converge to, or diverge from, the fixed point at a rate determined by the eigenvalues. Such a characterization is important perceptually because in the neighborhood of the fixed point the eye is moving slowly with low acceleration and the image is close to the fovea. The low velocity portion, of the slow phase of nystagmus is referred to as the foveation period and has been shown to be correlated with the visual acuity of the subject (Abadi and Sandikcioglu, 1974).

The changes in the state of system can be recovered from experimental data by using the method of delays, in which a sliding window of n samples is moved through the data, generating a set of n-dimensional vectors. If the oculomotor system is in a state x and n consecutive measurements made on the system as it evolves from x are collected by the sliding window into a vector y, then the theory of the method of delays tells us that there is a transformation T such that $y = T(x)$. This transformation can be thought of as a change of co-ordinates taking us from the state space description of the oculomotor control system to the delay vector description we have constructed from the time series. In practice, it is enough to know that the dynamics of the oculomotor system and the dynamics of the delay vectors as the sliding window is moved through the data are closely related. Specifically, quantities that are independent of co-ordinates — for example, eigenvalues — can be computed using delay vectors, and theory tells us that they will be the same if they had been calculated using direct knowledge of the equations governing the behaviour of the oculomotor system.

Abadi and his colleagues (1997) found that in congenital nystagmus the stability of the fixed point can be characterized by three eigenvalues; a small positive eigenvalue, which describes the unstable drift away from the fixation direction, a large negative eigenvalue, which characterizes the stable corrective movement back to the point of fixation, and a neutral eigenvalue, which implies that rather than a single fixed point there is a line of fixed points (Figure 11.6). The dynamics of the nystagmus in the region of foveation near the fixed points have been found to be low dimensional and deterministic.

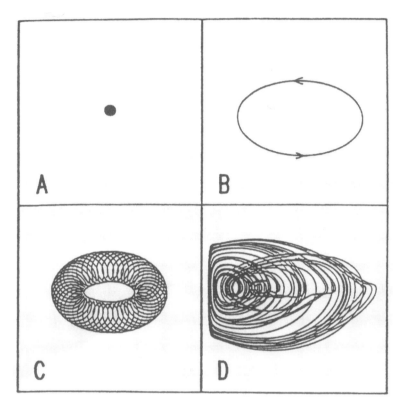

FIGURE 11.5. One way to gain insight into the dynamics of the oculomotor system is to consider all the possible initial conditions. Each possible initial condition can be represented as a point on a plane — the (x, v) — plane. Thus, the state of the oscillation at time t is the pair of values $(x(t), v(t))$, where x = position and v = velocity. It is possible to plot the state as time proceeds by simply plotting $(x(t), v(t))$ in the (x, v)-plane. The path taken is called the trajectory or attractor and the (x, v)-plane is called the phase plane. Each point represents a state and an orbit represents a chronological sequence of states. This figure illustrates four states of a system in the phase space. (a) A fixed point or static attractor represents a stable system. (b) A limit or closed cycle represents periodic behaviours such as seen in periodic pendular oscillations. (c) A torus represents a trajectory moving onto a two-dimensional surface. The behaviour of the system is quasi-periodic. (d) A strange attractor which signifies chaotic behaviour. Congenital nystagmus has similar attractors. (See Kaplan and Glass, 1995 for more details of non-linear dynamics and qualitative descriptions.)

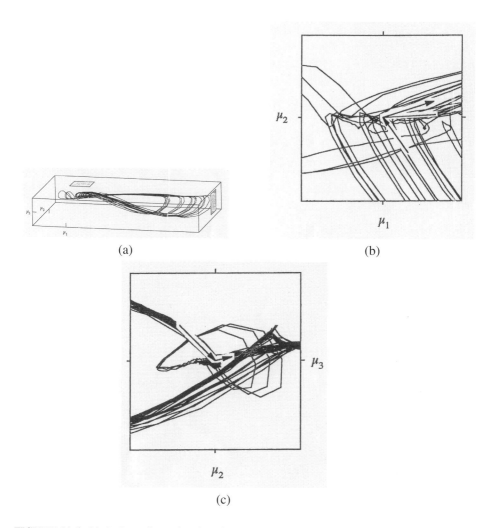

FIGURE 11.6. (a) A three-dimensional projection of the reconstructed phase space trajectory associated with 5 sec of data. Although 20 sec of data were used for the analysis,
the 5 sec portion is shown for the sake of clarity. The axes (μ_1, μ_2,μ_3) approximate to
the mean over the window of the eye position, velocity, and acceleration, respectively. The
tick marks correspond to the zero values of these means. The grey squares on the frame
surrounding the trajectories delineate the regions of the phase space which are shown in the
projections onto the μ_1, μ_2, and μ_3 planes plotted in (b) and (c) respectively. Both squares
have a side length of 3.0 units. Points within a radius of 0.8 units from the origin were used
to calculate the eigenvectors shown as black arrows in (b) and (c).

11.3.4   Manifest Latent Nystagmus

More often than not, the waveform of an individual with manifest latent nystagmus is made up of a decelerating slow phase followed by a quick phase (Figure 11.2e). Since manifest latent nystagmus occurs frequently in individuals who have congenital or uniocular visual loss, or who have experienced visual deprivation, it has been proposed that disturbances of egocentric localization may be in part responsible for these oscillations (Dell'Osso, Schmidt and Daroff, 1977; Abadi, 1980) together with the possibility that there is also an abnormality of extraocular proprioception (Ishikawa, 1979).

Recently, Abadi and Scallan (1999) have proposed a new hypothesis for the generation of particular nystagmus waveforms. Reporting on a unique case of horizontal manifest latent nystagmus that is converted to a congenital nystagmus on covering the subject's only seeing eye, they proposed that the patient had a low-gain neural integrator with an eccentric null. Early loss of the right eye (within the first week of birth) caused a recalibration of the straight-ahead position. More specifically, the neural integrator null was now no longer in the primary position but had moved to the right (here the neural integrator null is defined as that direction of gaze for which the neural integrator is not recruited to maintain eccentric gaze). On primary gaze the eye drifted toward the eccentric null position. Any effort to maintain fixation in the primary position caused leftward corrective saccades, resulting in a manifest latent nystagmus waveform (Figure 11.7). On covering the left eye (i.e., no visual fixation), the neural integrator null shifted to the primary position and in this way the beat direction was changed. Now, without any visual fixation, the gain of the neural integrator was driven up by the effort to see, and the underlying congenital nystagmus waveform became apparent. The reverse happened on uncovering.

The eccentric neural integrator null mechanism might also explain the presence of extended slow phases during relaxation. In this case the subject makes no special effort to maintain primary fixation, the eye drifts toward the neural integrator null position, and the saccadic components of the nystagmus drop out. On reaching the null the eye remains steady at this position. Thus, fast phases of the manifest latent nystagmus waveform reflect the influence of visual attention. The role of visual engagement and visual feedback will be considered in more detail in the following section on eccentric gaze holding.

11.4   Eccentric Gaze Holding

It is well established that the oculomotor neural integrator plays a crucial role during fixation away from the primary position, the eye position signal being created from the velocity command by integration with respect to time in the mathematical sense (Robinson, 1968; Cohen and Komatsuzaki, 1972; Robinson, 1975). Experimental studies have demonstrated that the time constant of the horizontal neural integrator is made up of a brain stem neural integrator with a short time

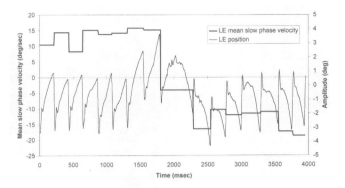

FIGURE 11.7. The effect of covering the left eye on the nystagmus waveform and the mean slow phase velocity. Prior to covering the left eye, the oscillation is a manifest latent nystagmus (i.e., a decelerating slow phase). On covering, the nystagmus becomes a congenital nystagmus (i.e., an accelerating slow phase). The transition phase is made up of a rightward slow eye movement followed by a leftward slow eye movement. The change in the mean slow-phase velocity is shown by the grey trace. Each horizontal line represents one slow phase of the nystagmus.

constant (~1.5 sec) and a cerebellar neural integrator that augments the brainstem integrator time constant to its normal value of 25 sec. Efficient holding of an eccentric eye position therefore requires a perfect neural integrator (Abel, Dell'Osso and Daroff, 1978; Eizenman, Cheng, Sharpe and Frecker, 1990). A variety of pathologies such as gaze palsies, cerebellar disease, as well as the side effects of sedatives, can modify the time constant of the integrator and bring about a gaze-evoked nystagmus (Leigh and Zee, 1999). Centripetal drift of the eyes away from the desired gaze position can also occur physiologically (Abel, Parker, Daroff and Dell'Osso, 1978; Eizenman et al., 1990; Abadi and Scallan, 2001). The presence of the correcting saccades is very dependent on visual feedback and attention. Eccentric gaze in the dark without the presence of a fixation target (i.e., attempted fixation of a remembered target position) invariably brings about a reduction in the slow-phase velocity (Figure 11.8). In spite of these changes, the desired gaze angle is largely maintained by slow eye movement control. Recently, Abadi and Scallan (2001) investigated how retinal image movement influenced eye position control during eccentric gaze by using a servo-controlled system to vary visual feedback. Generally, the larger the feedback gain (>0.5), the fewer were the number of centrifugal saccades and the lower the mean slow phase velocity of the end-point nystagmus. These experiments clearly highlight the importance of visual feedback and attentional mechanisms for sustained eccentric gaze holding.

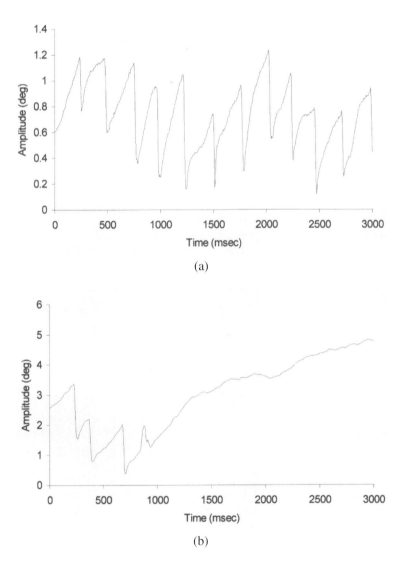

FIGURE 11.8. The effect of target presence on end-point oscillations during 40 deg left gaze. (a) Fixation of a seen target; (b) fixation of a remembered target. Reprinted from *Vis. Res.*, 41: 2895–2970, 2001, Abadi, R. V., and Scallan, C. J., Waveform characteristics of manifest latent nystagmus, with permission from Elsevier Science.

11.5   Concluding Remarks

There are many mechanisms that underpin stable fixation. Chief among them are the oculomotor control systems of fixation and eccentric gaze holding. Malfunction can bring about a spectrum of fixation instabilities ranging from uniplanar saccadic intrusions and/or oscillations through to a full-blown multiplanar nystagmus. In the past, conventional descriptions have relied heavily on the waveform characteristics gleaned from a time series, phase plane analysis, and scanpath plots. However, the recent use of non-linear dynamics to describe and understand the underlying behaviour of a system is proving most exciting. For example, the behaviour of the oculomotor system underlying nystagmus can be described by a limited number of cycles, called periodic orbits, such that identification of these orbits provides a method of characterizing the system. This approach has great implications for the control of the system's behaviour, since both periodic and chaotic behaviours (Figures 11.5b-d) can be controlled by making small oculomotor perturbations to move the state of the system from close to one periodic orbit to another.

References

Abadi, R. V. (1980), Pattern contrast thresholds in latent nystagmus. *Acta Ophthalmol.*, 58: 210-220.

Abadi, R. V., Broomhead, D. S., Clement, R. A., Whittle, J. P. and Worfolk, R. (1997). Dynamical systems analysis: a new method of analyzing congenital nystagmus waveforms. *Exp. Brain Res.*, 117: 335–361.

Abadi, R. V. and Dickinson, C. M. (1986). Waveform characteristics in congenital nystagmus. *Doc Ophthal.*, 64: 153–167.

Abadi, R. V., Dickinson, C. M., Pascal, E., Whittle, J. and Worfolk, R. (1991). Sensory and motor aspects of congenital nystagmus. In R. Schmid, and D. Zambarbieri (eds.), *Oculomotor Control and Congnitive Processes*, pp. 249–262, North Holland: Elsevier Science.

Abadi, R. V. and Sandikcioglu, M. (1974). Electro-oculographic responses in a case of bilateral idiopathic nystagmus. *Br. J. Phys. Optics*, 29: 73–85.

Abadi, R. V. and Scallan, C. J. (1999). Manifest latent and congenital nystagmus waveforms in the same subject. A need to reconsider the underlying mechanisms of nystagmus. *Neuro-ophthalmol.*, 21: 211–221.

Abadi, R. V. and Scallan, C. J. (2000). Waveform characteristics of manifest latent nystagmus. *Invest. Ophthalmol. Vis. Sci.*, 41: 3805–3817.

Abadi, R. V. and Scallan, C. J. (2001). Ocular oscillations on eccentric gaze. *Vis. Res.*, 41: 2895–2907.

Abadi, R. V., Scallan, C. J. and Clement, R. A. (2000). The charactersitics of dynamic overshoots in square-wave jerks, and in congenital and manifest latent nystagmus. *Vis. Res.*, 40: 2813–2829.

Abel, L. A. (1990). Ocular oscillations: congenital and acquired. In R. B. Daroff and A. Neetens (eds.), *Neurological Organization of Ocular Movements*, pp. 163–189. Amsterdam: Kluger and Ghedini.

Abel, L. A., Dell'Osso, L. F. and Daroff, R. B. (1978). Analog model for gaze-evoked nystagmus. *IEEE Transactions of Biomedical Engineering*, 25: 71–75.

Abel, L. A., Parker, L., Daroff, R. B. and Dell'Osso, L. F. (1978). End point nystagmus. *Invest. Ophthalmol. Vis. Sci.*, 17: 539–544.

Abel, L. A., Traccis, S., Dell'Osso, L. F. and Troost, B. T. (1984). Square wave oscillation: the relationship between saccadic intrusions and oscillations. *Neuro-ophthalmol.*, 4: 21–25.

Broomhead, D. S., Clement, R. A., Muldoon, M. R., Whittle, J. P., Scallan, C. and Abadi, R. V. (2000). Modelling of congenital nystagmus waveforms produced by saccadic system abnormalities. *Biol. Cybern.*, 82: 391–399.

Cohen, B. and Komatsuzakai, A. (1972). Eye movements induced by stimulation of the pontine reticular formation: evidence for integration in oculomotor pathways. *Exp. Neurol.*, 36: 101–117.

Dell'Osso, L. F. (1967). A model for the horizontal tracking system of a subject with nystagmus: visual and vestibular responses. *20th Annual Conference on Engineering and Medicine*. USA.

Dell'Osso, L. F. (1985). Congenital latent and manifest latent nystagmus: similarities, differences, and relation to strabismus. *Jpn. J. Ophthalmol.*, 29: 351–363.

Dell'Osso, L. F. (1996). See-saw nystagmus in dogs and humans: an international, across discipline, serendipitous collaboration. *Neurology*, 47: 1372–1374.

Dell'Osso, L. F. and Daroff, R. B. (1975). Congenital nystagmus waveforms and foveation strategy. *Doc Ophthalmol.*, 39: 155–182.

Dell'Osso, L. F. and Daroff, R. B. (1999). Nystagmus and saccadic intrusions and oscillations. In J. S. Glaser (Ed.), *Neuro-ophthalmology*, pp. 369–401, Philadelphia: Lippencott.

Dell'Osso, L. F., Schmidt, D. and Daroff, R. B. (1977). Latent, manifest latent and congenital nystagmus. *Arch. Ophthalmol.*, 97: 1877–1885.

Dell'Osso, L. F. and Williams, R. W. (1995). Ocular motor abnormalities in achiasmatic mutant Belgium sheepdogs: unyoked eye movements in a mammal. *Vis. Res.*, 35: 109–116.

Eizenman, M., Cheng, P., Sharpe, J. A. and Frecker, R. C. (1990). Endpoint nystagmus and ocular drift: an experimental and theoretical study. *Vis. Res.*, 30: 863–877.

Everling, S., Paré, M., Dorris, M. C. and Munoz, D. P. (1998). Comparison of the discharge characteristics of brain stem omni-pause neurons and superior colliculus neurons in monkeys: implications for control of fixation and saccadic behavior. *J. Neurophys.*, 79: 511–528.

Gage, J. E., Abadi, R. V., Lloyd, I. C. and Thompson, C. M. (2001). Fixation stability and infantile cataract. *Invest. Ophthalmol. Vis. Sci.*, 42: S164.

Harris, C. M. (1995). Problems modeling congenital nystagmus: towards a new model. In J. M. Findlay, R. Walker and R. W. Kentridge (eds.), *Eye Movement Research: Processes, Mechanisms, and Applications*, pp. 239–253. Amsterdam: Elsevier.

Harris, C. M. (1997). Nystagmus and eye movement disorders. In D. Taylor (ed.), *Paediatric Ophthalmology*. pp. 869–896. Oxford: Blackwell Press.

Herishanu, Y. O. and Sharpe, J. A. (1981). Normal square wave jerks. *Invest. Ophthalmol. Vis. Sci.*, 20: 268–272.

Ishikawa, S. (1979). Latent nystagmus and its etiology. In R. D. Reineke (ed.), *Third Meeting of the International Strabismological Association*, pp. 203-214. New York: Grune and Stratton.

Kaplan, D. and Glass, L. (1995). *Understanding Non-linear Dynamics*. New York: Springer-Verlag.

Leigh, R. J. and Zee, D. S. (1999). *The Neurology of Eye Movements*. Philadelphia: Davis.

Lynch, J. C., Mountcastle, V. B, Talbot, W. H. and Yin, T. C. (1977). Parietal lobe mechanisms for directed visual attention. *J. Neurophysiol.*, 40: 445–461.

Munoz, D. P. and Wurtz, R. H. (1992). Role of rostral superior colliculus in active visual fixation and execution of express saccades. *J. Neurophysiol.*, 67: 1000–1002.

Optican, L. M. and Zee, D. S. (1984). A hypothetical explanation of congenital nystagmus. *Biol. Cybern.*, 50: 119–134.

Robinson, D. A. (1968). Eye movement control in primates. *Science*, 161: 1219–1224.

Robinson, D. A. (1975). Oculomotor control signals. In G. Lennerstrand and P. Bach-y-Rita (eds.), *Basic Mechanisms of Ocular Motility and their Clinical Implications*, pp. 3370-374. Pergamon Press.

Shallo-Hoffmann, J., Peterson, J. and Mülendyck, H. (1989). How normal are "normal square wave jerks." *Invest. Ophthalmol. Vis. Sci.*, 30: 1009–1011.

Skavenski, A. A, Hansen, R. M., Steinman, R. M. and Winterson, B. J. (1979). Quality of retinal image stabilisation during small natural and artificial body rotations in man. *Vis. Res.*, 19: 675–683.

Tusa, R. J., Zee, D. S., Hain, T. C. and Simonsz, H. J. (1992). Voluntary control of congenital nystagmus. *Clin. Vis. Sci.*, 7: 195–210.

van Gisbergen, J. A. M., Robinson, D. A. and Gielens, S. (1981). A quantitative analysis of generation of saccadic eye movements by burst neurons. *J. Neurophys.*, 45: 417–441.

12

Plasticity of the Near Response

Clifton M. Schor

Voluntary binocular gaze shifts respond to both perceptual (spatiotopic) and retinal cues. When the gaze shifts are large, retinal disparity is too large to provide useful visual feedback and the primary cue for gaze is perceived spatial location. As the voluntary response to perceived location proceeds, retinal image disparity and blur are reduced into the effective stimulus operating range of a fine adjustment or foveal maintenance mechanism that utilizes visual feedback from retinotopic information. These two classes of stimuli are used to control the vergence response in a coarse to fine strategy, and their goal is to minimize horizontal, vertical, and cyclo components of retinal image disparity at the fovea.

Horizontal, vertical, and cyclo components of vergence respond to retinal image disparity; however, only horizontal vergence responds voluntarily to spatiotopic cues. Couplings with attentionally driven voluntary components of version and convergence guide involuntary vertical and torsional components of vergence. Couplings that link involuntary components of vergence with convergence are referred to collectively as the near response. These couplings exhibit plasticity in response to sensory demands placed on binocular vision. The adaptive near-response couplings are modeled as a combination of passive orbital mechanics and active gain control of the vertical ocular muscles that depends on convergence angle but is independent of gaze direction.

12.1   Introduction

The primary goal of binocular alignment is to optimize disparity stimuli for stereo-depth perception and binocular sensory fusion. This is accomplished for objects near the point of fixation by aligning their retinal images on corresponding retinal regions of the two eyes that evoke percepts in common visual directions (Hering, 1868). Images that are not aligned with corresponding retinal points are disparate. For the purpose of binocular eye alignment, disparity can be described with three degrees of freedom, including horizontal, vertical, and cyclo components. Binocular alignment minimizes all three components of disparity at the fovea with vergence control, and any residual disparities in the periphery are used to interpret relative depth and surface orientation. All three components of disparity are necessary to interpret space because by themselves, horizontal, vertical, and cyclo

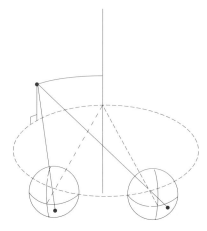

FIGURE 12.1. Optical geometry producing vertical disparity of points presented in tertiary directions at finite viewing distances.

disparities are ambiguous (Garding et al., 1995). The same horizontal disparity can correspond to many different depth intervals which depend on distance and direction of targets relative to the observer. Similarly, many different slant angles correspond to the same cyclo and shear disparities. These disparities are disambiguated with information about distance and direction of objects relative to the head. This information can be obtained from extraretinal information from the sense of eye position or from retinal information from vertical disparity.

Horizontal disparity is the primary metric for stereopsis. It is always subtended by objects lying nearer or farther than the fixation distance (i.e., the horopter). Horizontal disparities are quantified with respect to the longitudinal horopter, which is a surface in space describing all target locations that subtend zero horizontal disparity for a given angle of convergence. *Vertical disparity* is subtended by all near targets lying in tertiary directions above and below the visual plane. These disparities arise from the difference in distance of the target from the entrance pupils of the two eyes (Figure 12.1). The relative magnification or vertical size ratio of the two ocular images provides useful information for interpreting horizontal disparity as a relative depth cue. Vertical disparity can be used as a retinal cue to estimate the sagittal distance and horizontal eccentricity of targets relative to the head. Vertical disparity is useful in mapping horizontal disparity into linear depth percepts and to interpret horizontal slant relative to the head after correcting for horizontal direction of gaze (Gillam and Lowergren, 1983; Garding et al., 1995). Uniform cyclo disparities along both horizontal and vertical meridians result from cyclovergence errors. The uniform cyclodisparities produced by ocular cyclotorsion can be distinguished from non-uniform cyclodisparities produced by slanted surfaces which do not produce cyclodisparity between horizontal lines imaged on the two foveas (Howard and Kaneko, 1994).

12.2   A Coarse to Fine Strategy for Vergence

In order to interpret space correctly from retinal image disparity, it is important for the visual system to distinguish between disparity produced by optical geometry associated with target location relative to the head and errors in eye alignment. To accomplish this, the visual system relies upon very precise control of eye alignment. Two classes of stimuli evoke all three components of vergence. All three components of disparity stimulate vergence responses. Disparity is a closed-loop stimulus that provides feedback to guide small vergence response. Body referenced percepts of distance and direction are open-loop stimuli that initiate large vergence responses during voluntary gaze shifts (Schor, Alexander, Stevenson and Cormack, 1992). Under open-loop conditions, disparity is too large to provide an effective form of visual feedback. As the voluntary response to spatiotopic stimuli proceeds, eye alignment is refined and maintained with visual feedback from retinotopic or eye-referenced cues such as blur and disparity. The spatiotopic and retinotopic classes of stimuli are used to control the vergence response in a coarse to fine strategy (Schor, et al., 1992).

12.3   Cross-Coupling of Voluntary and Involuntary Motor Responses and the Near Response

While all three vergence components respond to retinal cues of horizontal, vertical, and cyclo disparity, only horizontal vergence responds voluntarily to spatiotopic cues. Vertical vergence and cyclo vergence are only under involuntary control, and they do not respond directly to perceived target location (Schor and McCandless, 1995; Bradshaw and Rogers, 1994; Stevenson, Lott and Yang, 1997). The voluntary and involuntary components of the vergence response are associated by several cross-couplings. Involuntary components of vergence are guided by the attentionally driven voluntary components of version and convergence that change direction and distance of gaze in response to spatiotopic cues. The coupled voluntary and involuntary components of the open-loop vergence response reduce all three components of retinal image disparity into the effective stimulus operating range of a fine adjustment or foveal maintenance mechanism under which all three components of vergence respond to their respective retinal disparities. Couplings that link involuntary components of vergence with convergence are referred to collectively as the near response (Allen and Carter, 1967).

12.4   What Geometric Properties of Stimuli for the Three Components of Vergence Make Coupling Possible?

Cross-links between voluntary and involuntary components of the near response are based upon predictable relationships between version and vergence components of eye position that reduce retinal image disparity to be zero. Horizontal, vertical, and cyclo disparities are determined by target location and the coordinate systems that control horizontal and vertical eye position. These predictable relationships allow the coupled involuntary responses to be scaled with version and horizontal vergence components of the voluntary response. Scale factors are used to describe the couplings quantitatively.

12.4.1   Horizontal Vergence Coupling

Coupling interactions have been described for all three components of vergence. The horizontal component of vergence, that is, under voluntary control, can be guided by accommodation under conditions where an occluder, such as the nose, blocks one eye's view of a target. The linkage between accommodation and convergence is possible because both respond to optical-geometric properties of a common viewing distance. The linkage between horizontal convergence (C) with the accommodation response (A) to blur is described by the accommodative vergence ratio (K_c).

$$C = K_c * A \qquad (12.1)$$

Ideally the K_c would be 1.0 for convergence expressed in meter angles and accommodation in diopters. Empirical measures demonstrate a normal $K_c = 0.66MA/D$ (Alpern and Ellen, 1956). Accommodation (A) can also follow convergence responses (C) to disparity with a different empirical scalar value (K_a) of 1.0 (Fincham and Walton, 1957).

$$A = K_a * C \qquad (12.2)$$

The scale value of less than 1.0 for Kc provides stability for the mutual interactions between accommodation and convergence (Schor, 1992).

12.4.2   Vertical Vergence Coupling

Binocular vertical alignment of foveal images is controlled by vertical vergence. In a Fick coordinate system that is often used to describe vertical disparity (Howard and Rogers, 1995), the amplitude of vertical vergence is scaled proportionally with the vertical image size ratios of the two ocular images of a target viewed in a tertiary direction (see Collewijn, 1994; Schor et al., 1994; Ygge and Zee, 1995). In this application, vertical vergence is quantified as the ratio of vertical positions of the left and right eyes (V_l/V_r) that is necessary for bifoveal fixation.

In symmetrical convergence, the ratio equals 1.0 and in asymmetric convergence it depends upon both convergence (C) and horizontal gaze eccentricity (H). Convergence is positive following a sign convention where left and down are positive values, and where convergence equals R-L eye position. In left-gaze, the ratio of left over right eye position is described by

$$V_l/V_r = K_v \tan^{-1}[\cos(H_l - C/2)] / \tan^{-1}[\cos(H_r + C/2)] \quad \text{(Fick coordinates)}$$
$$(12.3)$$

where H_l and H_r represent the azimuth of the left and right eye relative to primary position, and V_l and V_r represent the elevation of the left and right eye in Fick coordinates. Empirical measures demonstrate a normal scalar value of $K_v = 1.0$ (Schor et al., 1994). Note that when vertical vergence is described in a Helmholtz coordinate system,

$$V_l/V_r = K_v \quad \text{(Helmholtz coordinates)} \quad (12.4)$$

and K_v is independent of convergence or gaze angle (Schor et al., 1994). $K_v = 1.0$ for both the Fick or Helmholtz coordinate systems.

12.4.3 Cyclovergence Coupling

More than 150 years ago, Donders recognized that the torsional orientation of the eye in any given direction of gaze was independent of the path the eye took to reach that position (Donders 1848). The amount of torsion was described by Listing as though the eye had rotated from a primary position about an axis that was in a common plane with other axes describing ocular torsion in other directions of gaze. Helmholtz referred to this as Listing's plane (1910). Listing's law describes a coupling between cyclotorsion and versional eye position. Hering observed a binocular coupling between horizontal vergence, eye elevation, and cyclovergence (1868). He observed that in upward gaze, the eyes were intorted and in downward gaze, they were extorted relative to the orientations predicted by Listing's law (Nakayama, 1983). In a Fick coordinate system, in-cyclovergence forms an "A" pattern between the vertical meridians of the retinas and ex-cyclovergence forms a "V" pattern. Allen and Carter (1967) described the coupling between cyclovergence and eye elevation during convergence as part of the near response.

For over a century, the binocular coupling of cyclovergence was considered a violation or exception to Listing's law. However, recently the coupling between cyclovergence and eye elevation has been found to be consistent with Listing's law (Mok et al., 1992; van Rijn and van den Berg, 1993; Minken et al., 1994). During convergence, ocular torsion is still described as though the eyes had rotated about axes in two planes, one for each eye, from their respective primary positions. However, the orientation of the planes is different than the orientation of the classical Listing's plane that describes torsion when the visual axes are parallel and the eyes view distant targets. During near fixation, the orientation of

Listing' Non-Extended System (2 df)

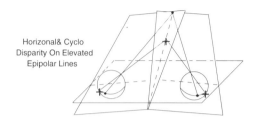

Horizonal& Cyclo
Disparity On Elevated
Epipolar Lines

Horizontal and Oblique Planes of Regard
Primary and Secondary Gaze
In The Midsagittal Plane

FIGURE 12.2. Torsional orientation of epipolar planes passing through the foveas with gaze horizontal and elevated gaze when Listing's law is not extended. The projections of the corresponding epipolar meridians (planes of regard) are extorted such that the horopter collapses to a single point where the visual axes intersect. Pure horizontal disparities (no vertical or cyclo disparity) only exist along the line at the intersection of the planes of regard.

Listing's planes for the right and left eye diverge from one another by approximately the amount that the eyes have converged (Tweed, 1997; Somani et al., 1998; Bruno and van den Berg, 1997). This has been referred to as a binocular extension of Listing's law (Mok et al.,1992; van Rijn and van den Berg, 1993; Tweed, 1997).

The change in orientation of Listing's planes with convergence makes it possible to have a horopter when gaze is elevated during symmetrical convergence. Figures 12.2 and 12.3 illustrate the torsional orientation of the two horizontal retinal meridians during two near fixation conditions. They illustrate variations of cyclotorsion for the classical Listing's coordinate system and the extended Listing's coordinate system. Both fixation conditions are illustrated for symmetrical convergence. Either the eyes have zero elevation or gaze is elevated in the midsagittal plane. The arcs across the two retinas represent sections of great circles that pass through the horizontal meridians of the retinas and the foveas. The arcs describe epipolar meridians of the eye with gaze elevated. The planes in the figures describe projections of the horizontal meridians of the retinas through the entrance pupils (planes of regard) when gaze is elevated. The locus of points where the two planes of regard intersect represent points in space that subtend pure horizontal disparity without any vertical or cyclo components. The circle in Figure 12.3 describes the geometric horopter where targets must lie to be imaged on corresponding points along the horizontal meridians of the retinas, and these points subtend zero disparity. Points shown in Figure 12.3 located in front or behind the horopter (e.g. X) in both eye's planes of regard subtend pure non-zero horizontal disparity without any vertical disparity components.

**Helmholtz and Listing's
Extended Systems (2 df)**

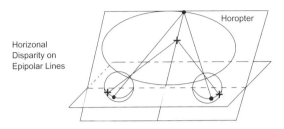

FIGURE 12.3. Coplanar orientation of epipolar lines and planes of regard passing through the foveas with horizontal and elevated gaze when Listing's law is extended. The longitudinal horopter exists in elevated gaze, and any point in the elevated visual plane subtends pure horizontal disparity on epipolar meridians.

During convergence, the only targets that can be viewed without retinal image cyclodisparity must lie in the horizontal plane that passes through the nodal points of the two eyes and intersects the two foveas (i.e., the visual plane). When the eyes are converged in the zero-elevation position, all points lying in the visual plane will subtend pure horizontal disparities because the visual plane contains both planes of regard (i.e., it intersects both horizontal retinal meridians). If the orientation of the Listing's planes did not change during convergence (Figure 12.2), then when gaze was elevated, the planes of regard would become extorted. Images of most objects lying in the inclined visual plane would be off the horopter and they would subtend complex combinations of horizontal, vertical, and incyclo disparities. The torted planes of regard would not be coplanar, and they would only intersect along a line in the midsagittal plane. Only points on the line of intersection would be imaged with pure horizontal disparities, and the fixation point would be the only point on the line of intersection that could be imaged on the horopter.

Figure 12.3 demonstrates how the oculomotor system remedies this situation and still obeys Listing's law during convergence. In a binocular extension of Listing's law (van Rijn and van den Berg, 1993; Tweed, 1997) cyclovergence intorts the eyes with gaze elevation and extorts the eyes with gaze depression to keep the horizontal meridians coplanar in symmetrical convergence (Allen and Carter, 1967; Mok et al., 1992; van Rijn and van den Berg, 1993; Tweed, 1997; Somani et al., 1998). When gaze is elevated, an incyclo rotation of the planes of regard makes them coplanar so that it is possible to image real targets on the longitudinal horopter. All targets in the inclined visual plane, whether they are nearer or farther from the fixation point, will be imaged on corresponding epipolar meridians of the eye with pure horizontal disparities. These gaze-dependent changes in cyclovergence are consistent with an outward or yaw rotation (divergence) of Listing's planes (Mok et al., 1992; Van Rijn and van den Berg, 1993; Tweed, 1997). Stereoscopic depth perception would be compromised if this torsional compensation did not occur (Schreiber et al., 2001).

During symmetrical convergence, the torsional alignment of the horizontal meridians of the retinas is controlled by cyclovergence (T) that is scaled proportionally by K_t with combinations of convergence (C) and vertical eye position (V). Following a right-hand rule sign convention, left, down, and clockwise ocular rotations are positive and convergence equals $R - L$ eye position. Eye position is described in Helmholtz coordinates so that foveal alignment is achieved by equal elevation of the two eyes. The cyclovergence necessary for coplanar alignment of the planes of regard is described by

$$T = K_t * 4(\tan^{-1}[\tan(C/4) * \tan(V/2)]) \quad \text{(Helmholtz, 1910)} \quad (12.5)$$

When expressed in radians

$$T = K_t * V * C/2 \quad \text{(Somani et al., 1998)} \quad (12.6)$$

When expressed as rotation vectors relative to the straight-ahead reference position (Haustein, 1989), the change in cyclotorsion of each eye with gaze elevation is used to calculate the yaw tilt orientation of the Listing's plane (Tweed, 1997). Figure 12.4 illustrates that changes of cyclovergence with gaze elevation correspond to a divergence or yaw tilt difference between the two eyes' Listing's planes (Bruno and van den Berg, 1997). Primary position is defined as the reference direction that is perpendicular to Listing's plane (Tweed et al., 1997), and accordingly the two eyes' primary positions diverge during convergence (Mok et al., 1992; Van Rijn and Van den Berg, 1993). The coupling between the yaw orientation of Listing's planes and convergence can be described by the scalar K_t as a ratio of the change in yaw tilt difference between the two eyes' Listing's planes ($\triangle Y$) over the change in convergence ($\triangle C$).

$$K_t = \triangle Y / \triangle C \quad (12.7)$$

Empirical measures demonstrate a normal scalar value for K_t of 0.5 to 0.8 (Mok et al., 1992; Mikhael et al., 1995; Minken and van Gisbergen, 1994; Somani et al., 1998) that is slightly less than the ideal of 1.0 for maintaining coplanar alignment of the horizontal meridians of the two retinas during gaze elevation in symmetrical convergence. The lower value is interpreted as a compromise between obtaining an economy of movement and an optimization of retinal image disparity for stereoscopic depth perception (Tweed, 1997).

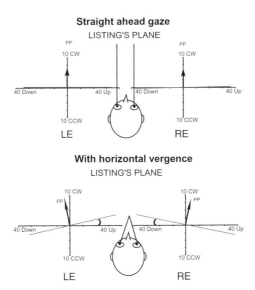

FIGURE 12.4. Changes in the yaw-tilt difference between Listing's planes with gaze at infinity (upper panel) and during convergence (lower panel). Primary position (PP) is orthogonal to Listing's plane. During distance viewing the primary positions are nearly parallel, and during near viewing they diverge by approximately the angle of convergence.

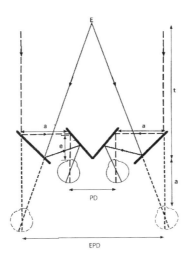

FIGURE 12.5. Telestereoscope increased the stimulus to convergence and decreased the stimulus to accommodation by using mirrors to widen the interpupillary distance optically.

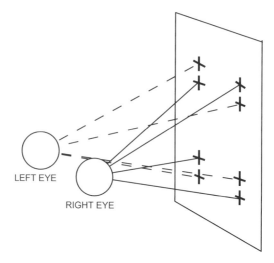

FIGURE 12.6. Vertical disparity produced by magnifiers placed over the abducting eye in asymmetric convergence.

12.5 To What Degree are These Couplings Fixed and Can They be Modified in Response to Sensory Demands Placed on Binocular Vision?

12.5.1 Horizontal Vergence

Plasticity has been demonstrated for all three components of vergence. The gain of the accommodative-vergence coupling can be increased or decreased by adapting to appropriate mismatches between the stimulus to accommodation and convergence (Judge and Miles, 1985; Eadie et al., 2000). Mismatches between accommodation and vergence stimuli were produced with a telestereoscope (Figure 12.5) that increased the stimulus to convergence and decreased the stimulus to accommodation by using mirrors to optically widen the interpupillary distance. After wearing this apparatus for 1 hr, the scalar (Kc) relating convergence and accommodation in equation (12.1) increased 75% from 0.66 to 1.16 (Judge and Miles, 1985). The ratio has also been lowered by reducing the stimulus for convergence while increasing the stimulus to accommodation in a binocular head mounted video display (Eadie et al., 2000).

12.5.2 Vertical Vergence

Vertical vergence can be adapted to either increase or decrease in tertiary gaze, and to vary with convergence in response to appropriate optical distortions (Schor and McCandless, 1997; Maxwell and Schor, 1994). Vertical disparity produced by optical geometry in tertiary gaze (Figure 12.1) can be exaggerated by placing a magnifier over the abducting eye in asymmetric convergence. Schor and Mc-

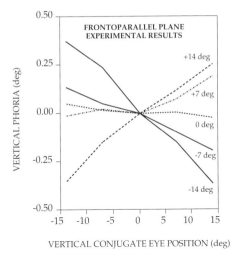

FIGURE 12.7. The ratio of vertical eye positions (left/right) measured under open-loop conditions increased by 2.6 percent after adapting the 6 percent magnifiers shown in Figure 12.6.

Candless (1997) used an apparatus that introduced a 6% magnifier over the right eye in right gaze and over the left eye in left gaze. Subjects alternated fixation for 1 hr between targets that were separated horizontally and vertically by 28 deg. The magnifiers produced vertical disparities whose sign depended on both horizontal and vertical eye position (Figure 12.6). The ratio of vertical eye positions (K_v) measured under open-loop conditions increased by 2.6% after adapting the 6% magnifiers (Figure 12.7). The result demonstrate that vertical vergence measured subjectively under open-loop conditions can be modified to vary with specific combinations of horizontal and vertical eye position. In another experiment McCandless and Schor (1997) varied vertical disparity by placing an 8% vertical magnifier over the right eye when viewing two targets separated vertically by 28 deg., and an 8% vertical magnifier over the left eye when the eyes viewed the same two targets through 15 diopter base-out prism that stimulated approximately 8.5 deg. of convergence (Figure 12.8a). They also varied the sign of vertical disparity (Figure 12.8b). The stimulus produced an aftereffect that was measured under open-loop conditions. The ratio of V_l/V_r for open-loop measures of vertical eye position varied with convergence. The ratio was less than 1.0 without convergence, and greater than 1.0 with convergence (Figure 12.9). In both experiments the vertical phoria changed by approximately half of the optically induced vertical disparity. The results demonstrate plasticity in the coupling between open-loop vertical vergence and specific combinations of vertical and horizontal eye position and horizontal convergence.

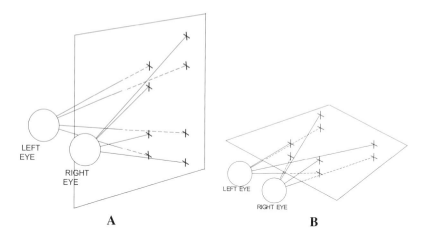

FIGURE 12.8. Vertical disparity produced by 8% magnifiers placed over the right eye when viewing a distant pair of targets and over the left eye when converging on a near pair of targets.

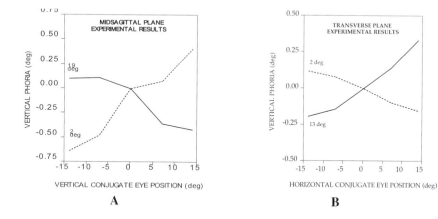

FIGURE 12.9. The ratio of vertical eye positions (left/right) measured under open-loop conditions increased by 3.5 percent after adapting the 8 percent magnifiers shown in Figure 12.8.

12.5.3   Cyclovergence

Variations of open-loop cyclovergence responses with gaze elevation during convergence also exhibit plasticity in response to environmental factors such as torsional disparities that vary with vertical eye position (Schor, Maxwell and Graf, 2001). Two adaptation procedures were conducted, one that exaggerated the normal pattern of distance-dependent cyclovergence variations with gaze elevation and another that reversed the pattern. In the exaggerated condition, 5 deg cyclodisparities were presented in the midsagittal plane with excyclodisparity in 10 deg upward gaze and in-cyclodisparity in 10 deg downward gaze during far fixation (zero convergence), and the opposite pattern of cyclodisparity with vertical eye position was presented in near fixation (10 deg convergence). In the reversed condition, the pattern of cyclodisparities was opposite of that in the exaggerated condition. The two patterns of cyclodisparity are shown schematically in Figures 12.10a and b. The disparate fusion patterns shown in Figure 12.10c consisted of a 53 deg. rectangular grid with three concentric circles superimposed on its center. The subject viewed the targets sequentially from one position to the next approximately at 10-sec intervals for a 2-hr period.

Three-dimensional eye position was recorded objectively using videooculography, or VOG (SMI, Germany). Open-loop cyclovergence (cyclophoria) responses to a non-fusible stimulus were measured objectively in Fick coordinates at the beginning and end of the 2-hr training period at 25 target locations in the fronto-parallel plane. The open-loop stimuli were presented at the two test distances (0 deg convergence and 10 deg convergence). The measures were transformed to rotation vectors (Haustein, 1989) and fit by least squared analysis to a plane. The fit parameters of the calculated planes were used to describe the yaw tilt angle of the resulting displacement planes relative to a straight-ahead reference direction. The yaw tilt corresponds to the horizontal slope about the vertical axis of the plane. The yaw orientations of displacement planes were doubled to obtain estimates of the orientation of Listing's plane (Tweed and Vilis, 1990).

Figure 12.11 illustrates an example of pre-adapted front and top-down views of displacement planes for a straight-ahead reference direction of the right and left eye. The planes are roughly fronto parallel and they do not diverged from one another. Changes in the horizontal yaw component of primary position were used to quantify the changes produced by the two training procedures. Figure 12.12 plots pre- and post-adapted measures of YTD for the reversed and exaggerated conditions. The light lines connect pre and post measures for individual subjects, and the solid line plots the mean result. Positive yaw-tilt differences (YTD) between primary positions represent an increase in excyclovergence in down gaze and a divergence of primary positions of the two eyes. Results following training that conformed to the torsional disparity pattern presented in the reversed training condition would be expected to have an increased positive or decreased negative YTD at the far distance and a decreased positive or increased negative YTD at the near convergence distance.

Figure 12.12 (top) illustrates that for the reversed condition, at the far distance

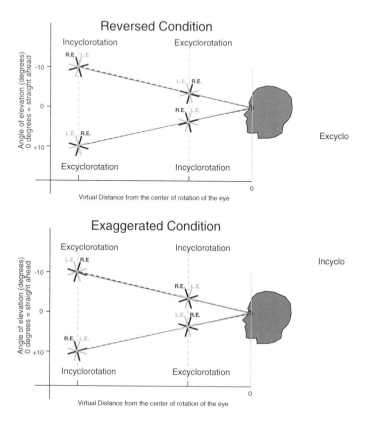

FIGURE 12.10. The patterns of torsional disparity used in the reversed (top) and exaggerated (bottom) conditions are shown schematically. Ten-deg changes in cyclodisparity are presented at a far and a 10 deg near convergence distance. At one distance incyclodisparity is presented in upward gaze and excyclodisparity in downward gaze and the reverse pattern is presented at the other convergence distance. The fusion pattern consisted of a 53 deg rectangular grid with three concentric circles superimposed on its center.

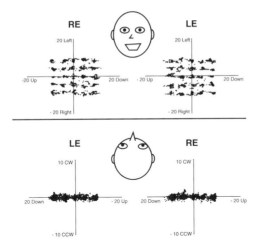

FIGURE 12.11. An example of pre-adapted front and top-down views of displacement planes for a straight-ahead reference direction of the right and left eye. Measurements were taken with the two eyes viewing an open-loop stimulus presented in 25 locations at the far test distance.

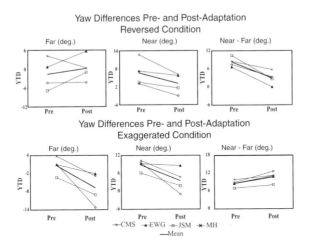

FIGURE 12.12. Reversed condition: Thin lines connect pre and post measures of the YTD between right and left eye primary positions for individual subjects, and the heavy line represents the mean. Results for the far (zero degrees convergence) and 10 deg convergence distance are plotted separately, and a third panel illustrates the difference between far and near measures. Upper and lower sets of panels illustrate results for the reversed and exaggerated conditions, respectively.

		Subject K-Value				
		CMS	JSM	EWG	MH	Average
Reversed	Pre	0.74	1.04	0.66	0.85	**0.82**
Condition	Post	0.33	0.26	0.00	0.57	**0.29**
Exaggerated	Pre	0.84	0.94	0.82	0.66	**0.82**
Condition	Post	1.23	1.09	1.07	0.76	**1.04**

TABLE 12.1. Pre-adapted and post-adapted gains.

(left panel) the post-adapted yaw angle was more divergent and at the near distance (middle panel) the post-adapted yaw angle was more convergent. The right panel illustrates the change in the primary position vergence between the far and near viewing distances. Following training, the near-far YTD increased in the convergent direction. Figure 12.12 (bottom) plots pre- and post-adapted measures of YTD for the exaggerated condition. Results following training that conformed to the torsional disparity pattern presented in the exaggerated condition would have the opposite pattern of changes of YTD of primary position. The left and middle panels illustrate that following training, there was a large increase in convergence of primary positions at the far viewing distance and a smaller increase in convergence at the near viewing distance. The right panel illustrates that following training, the near-far YTD increased in the divergence direction.

The changes in the yaw tilt difference between Listing's planes can be described as gain changes where the change in primary position vergence angle (yaw tilt difference) is divided by the change in horizontal vergence angle (equation 12.7). The pre-adapted and post-adapted gains are shown in Table 12.1. Ideally the perceptual adaptation would have a gain of 1.0; however, the pre-adapted gains averaged 0.82 for our four subjects. This value is similar to those reported in earlier studies (Mok et al., 1992, Mikhael et al., 1995; Minken and van Gisbergen, 1994; Somani et al., 1998). The reversed adaptation condition decreased these gains to an average value of 0.29, which is a 65% reduction. The exaggerated condition increased the gain to 1.04, which is a 27% elevation. Complete adaptation to the cyclo disparity stimuli would have changed the ratio by $+10.6$ for the exaggerated condition and -10.6 for the reversed condition; however, our largest after-effects were only one-tenth of this magnitude.

12.6   How Might These Changes in the Near Response Be Implemented?

12.6.1   Convergence and Accommodation

The coupling between accommodation and convergence can be modified by interactions between the various sub-components of these two motor systems. Both accommodation and convergence have dual modes of control (Schor and Kotulak,

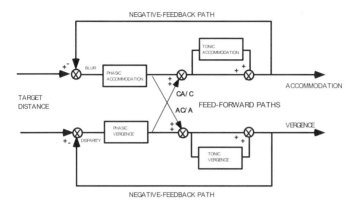

FIGURE 12.13. Block diagram of the cross-links between a dual control model of accommodation and convergence.

1986) (Figure 12.13). A rapid phasic component responds to both spatiotopic and retinal cues to distance, and an adaptable tonic component stores activity of the phasic systems (Schor et al., 1992). The phasic element enables rapid changes in viewing distance to obtain clear and single binocular vision. However, the phasic system lacks durability and it fatigues easily. It also has a very limited range of a few diopters or a few degrees that it can sustain comfortably. The slower tonic system extends the range of the fast system so that we can accommodate or converge large amplitudes and maintain these large responses to static or slow changing stimuli for long periods of time. Phasic and tonic mechanisms are arranged serially in a negative feedback loop that keeps their summed response from exceeding the stimulus amplitude. There is a trade-off between phasic and tonic responses such that as tonic adaptation increases, the stimulus to the phasic systems is reduced by negative feedback and its response is lowered.

The coupling between the two systems is stimulated principally by the phasic component, and adaptable tonic responses do not directly stimulate the crosslinks (Schor and Kotulak, 1986; Jaing, 1996; Hasebe et al., 2001). Consequently, the cross-link activity is greater in response to rapid (phasic) than gradual (tonic) dynamic responses. In addition, the tonic adapters respond to both direct and cross-linked phasic activity of accommodation and convergence (Schor and Kotulak, 1986). Factors that reduce tonic activity of one system (accommodation or convergence), such as fatigue following rapid alternating changes in viewing distance, cause cross-link activity of the fatigued system to be elevated and the cross-link activity of the non-fatigued system to be reduced (Schor and Tsuetaki, 1987). For example, fatigue of adaptable tonic accommodation causes the AC/A ratio to increase and the CA/C ratio to decrease. Thus, the magnitude of the cross-link interaction can be modulated by the activity of the adaptable tonic components of accommodation and convergence.

12.6.2   Cyclovergence and Eye Elevation During Convergence

How might the changes in orientation of Listing's planes be accomplished? Physiological studies indicate that simple gain changes of the obliques and perhaps the vertical recti might be involved. Mays et al. (1991) found that convergence-dependent changes of cyclovergence with gaze elevation were associated with a reduced discharge rate of trochlear motor neurons and an implied relaxation of the superior oblique muscle during convergence. The modulation of trochlear activity with convergence varied systematically with gaze elevation, and was largest in downward gaze. The fact that these authors observed no net increase in trochlear activity when the eyes incyclorotate with eye elevation during convergence indicates that the forces of other vertical ocular muscles were modulated during convergence to account for torsional adjustments in upward directions of gaze. Enright's measures of ocular translation also suggested that the superior oblique relaxes during convergence (Enright, 1992).

This hypothesis was tested by simulating 3-D eye position with Orbit^{TM}, a biomechanics model that simulates binocular eye position based upon the relationships of the six extraocular muscles, their tendons and supportive connective tissues including muscle sheaths or pulleys, innervation level and motor nucleus connection weights (innervation gain) according to equations given, in part, by Robinson (1975) and Miller and Robinson (1984). Orbit was designed to follow both Hering's and Listing's laws for distance viewing, but currently does not automatically implement the binocular extension of Listing's law. Simulations were conducted with 15% gain reductions to the obliques and 15% gain increases to the vertical recti. In this simulation the bilateral innervation to the medial rectus was increased and the innervation to the lateral rectus was decreased to produce 20 deg of convergence. Hering's law was simulated by finding the innervation to an assumed normal following eye that would produce the same gaze direction as that of the fixating eye. Orbit simulates binocular alignment when the two eyes are dissociated (i.e., vergence is open-loop), such that one eye fixates various target directions while the following eye is guided by Listing's and Hering's innervations. Parameters of either eye or both eyes may be modified, and torsion is allowed to deviate from Listing's law in both the fixating and following eye.

Eye positions were simulated during 20 deg of convergence while lateral gaze was varied over ±30 deg horizontally and vertically from the point of fixation. The simulated eye positions were converted from Fick coordinates to rotation vectors and fit to planes. Without any gain adjustments, the resulting orientation of Listing's plane at the near convergence distance was fronto-parallel. With gain adjustments, the primary positions diverged by 20 deg for a simulated convergence of 20 deg (Figure 12.14). These gain changes might be modified to describe the adaptive plasticity of K_t. The adaptation results of the exaggerated condition could be simulated with greater gain changes, and the results of the reversed condition could be simulated with smaller gain changes. The simulation demonstrates that simple convergence-related gain changes of the vertical ocular muscles are sufficient to transform the innervation pattern appropriate for torsion

Displacement Plane Rotations with 20° Convergence

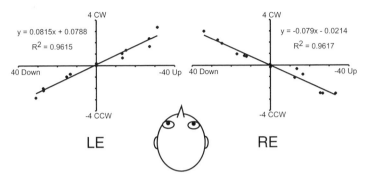

FIGURE 12.14. A top-down view of simulated displacement planes for a straight-ahead reference direction are shown the right and left eye during 20 deg of convergence. Planes were derived from Orbit simulations of three-dimensional eye position with modified gains of the obliques and vertical recti. The Orbit simulations are for a 15% decreased gain of the obliques and 15% increased gain of the vertical recti. Torsion, plotted in degrees, is simulated for vertical and horizontal changes in eye positions over a 60 deg horizontal and vertical range. Fifteen test points ranged from the near central fixation point in 30 deg horizontal increments and 15 degree vertical increments in a 3 × 5 rectangular matrix. YTD of the two displacement planes equals 9 deg, which corresponds to a YTD between primary positions of 18 deg and a K value equal to 0.9.

at far viewing distances into ones consistent with Listing's extended law at near viewing distances. These results strongly suggest that Listing's extended law responds to perceptual demands of binocular vision and that these modifications result from the combination of a central neural process and passive forces determined by biomechanical properties of the orbit.

12.6.3   Vertical Eye Alignment in Tertiary Gaze

Vertical eye alignment in tertiary gaze was preserved during the gain adjustments to the vertical recti and obliques that produced the binocular extension of Listing's law. Figure 12.15 plots the simulated open-loop vertical position of the following eye against closed-loop vertical position of the fixating eye during 20 deg of convergence while vertical and lateral gaze varied over a ±30 deg range of eye positions. Results from all horizontal test positions are combined into a single plot. Vertical eye position is specified in Helmholtz coordinates such that equal vertical position of the following eye and fixating eye (slope = 1.0) corresponds to binocular vertical eye alignment with the fixation target. The top plot shows an amplitude ratio of (following eye)/(fixating eye) of $K_v = 1.0$ indicating that changes in cyclophoria that were consistent with Listing's extended law did not disrupt vertical eye alignment in tertiary gaze. The bottom plot shows a similar amplitude ratio of the two eyes with normal (unaltered) gains to the vertical recti and obliques. In this simulation, the yoked innervation for vertical eye position

Simulated Vertical Eye Alignment
Helmholtz Coordinates

**(20 degrees convergence:
Vertical recti increased15%, Obliques reduced 15%)**

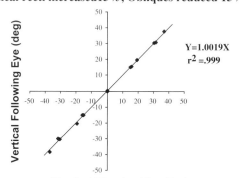

Y=1.0019X
$r^2 =.999$

Vertical Following Eye (deg)

Vertical Fixating Eye (deg)

(20 degrees convergence: No change to vertical recti or obliques)

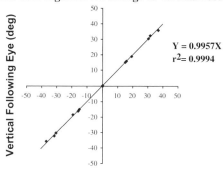

Y = 0.9957X
$r^2= 0.9994$

Vertical Following Eye (deg)

Vertical Fixating Eye (deg)

FIGURE 12.15. Simulations of open-loop vertical position of the following eye is plotted against closed-loop vertical position of the fixating eye for each of fifteen fixation test points. Simulated directions of gaze ranged from the near central fixation point in 30 deg horizontal increments and 15 deg vertical increments in a 3 × 5 rectangular matrix. The Orbit simulations are for 20 deg of convergence with the gain of the obliques decreased by 15% and vertical recti increased by 15% (top) and with a normal (unaltered) set of parameters (bottom).

to the following eye is the same as during distance fixation with parallel lines of sight, but orbital mechanics produces a different elevation of the two eyes when horizontal position of the following eye is modified by asymmetric convergence. As a consequence, the same innervation aligns the lines of sight with distant or near tertiary targets when the fixing eye is aimed in a direction common to both target distances. For normal eye alignment, it is not necessary to alter the innervation to vertical vergence to obtain binocular alignment of tertiary targets at near and far viewing distances (Schor et al., 1994; Ygge and Zee, 1995). The simulation suggests that binocular vertical eye alignment is primarily a consequence of Hering's law and the passive biomechanics of the oculomotor system (Miller and Demer, 1992; Enright, 1992; Porrill et al., 2000). The orbital pulleys cause the eyes to rotate vertically in a Helmholtz-like coordinate system (Porril et al., 2000).

Binocular control of cyclovergence and vertical vergence result from a calibrated neuromuscular interface. Binocular eye alignment is achieved by matching the innervation for horizontal, vertical, and cyclovergence to the physical constraints set by the extraocular muscles and orbital connective tissues. Orbital mechanics are organized to simplify the neural control needed to achieve precise cyclovergence and vertical vergence. Although cyclovergence varies with convergence and vertical eye position, the gain of the vertical muscles only needs to be modified with convergence, and orbital mechanics constrain cyclovergence with eye elevation. Similarly, in tertiary gaze, vertical vergence varies with both convergence and versional eye position to null vertical disparity. In natural viewing conditions, orbital mechanics constrains vertical vergence in tertiary gaze and achieves binocular alignment during convergence with the same innervation patterns used to align the eyes while viewing distant targets that do not subtend vertical disparities. In cases of pathology or optical distortion, it is possible to modify the open-loop innervation to vertical vergence to achieve binocular alignment.

The normal pattern of vertical eye alignment in tertiary gaze can be exaggerated by adapting to magnification differences between the two ocular images, and vertical phoria can be also adapted to misalign the eyes during convergence (Schor and McCandless, 1997; McCandless and Schor, 1997). These demonstrations illustrate that it is possible for neural mechanisms to modify the normal passive biomechanical coupling between vertical vergence, lateral gaze, and convergence. These examples could be the consequence of convergence-dependent gain alterations of the vertical extraocular muscles without regard to position of the eye in the orbit. These changes can be simulated with Orbit by altering the gains of the vertical ocular muscles. For example, if the gains of the left and right superior obliques are decreased 15% and the gains of the left and right superior recti are increased 15%, the normal changes of vertical phoria, described in Fick coordinates as a ratio of V_l/V_r, increase by 5%. Only one of our studies has demonstrated an adaptive response to a nonmonotonic change in vertical disparity with eye elevation that would require innervation to vary with information about both direction and distance of gaze (McCandless, Schor and Maxwell, 1996). Non-linear couplings have been modeled with lookup tables or an associa-

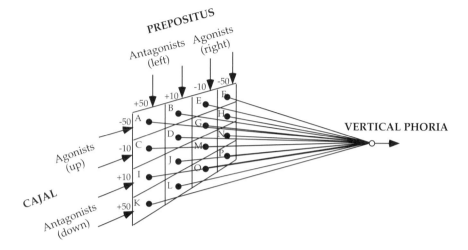

FIGURE 12.16. An association matrix illustrates a possible coupling of vertical vergence with various combinations of horizontal and vertical eye position that are coded by neurons in prepositus and the nucleus of Cajal respectively. Reprinted from *Network. Comput. Neural Syst.*, 8: 239-258, 1997, McCandless, J. W., and Schor, C. M., An association matrix model of context-specific vertical vergence adaptation, with permission of IOP Publishing Ltd.

tion matrix that, for example, can be used to derive vertical vergence innervation from various combinations of horizontal and vertical eye position coded by neurons in the prepositus nucleus and the nucleus of Cajal respectively (McCandless Schor and Maxwell, 1996; McCandless and Schor, 1997; Schor and McCandless, 1997) (Figure 12.16). Such a computation could occur in the cerebellum. This adaptive mechanism could calibrate vertical eye alignment and compensate for changes in orbital biomechanics.

12.7   Summary

The near response is composed of cross-coupled interactions between convergence and other distance-related oculomotor responses including accommodation, vertical vergence, and cyclovergence. These couplings serve to guide involuntary motor responses during voluntary shifts of distance and direction of gaze without feedback from retinal image disparity. They function to optimize the disparity stimulus for stereoscopic depth perception, and they can be modified by optically induced sensory demands placed on binocular vision. In natural viewing conditions, the coupling of accommodation and convergence is modulated by the activity of adaptable tonic components of both motor responses. The binocular extension of Listing's law could be achieved parsimoniously by a combination of passive orbital mechanics and an active gain control of the vertical ocular muscles that depended on convergence angle but were independent of gaze direction or elevation. The normal coupling of vertical vergence with convergence could be a by-product of passive orbital mechanics that would not require active convergence dependent neural gain control mechanism. Adapted changes of vertical vergence gain (K_v) in response to unequal ocular magnification (aniseikonia) and adapted changes in the orientation of Listing's planes (K_t) in response to torsional disparities could be achieved by a combination of passive orbital mechanics and active gain control of the vertical ocular muscles that depended on convergence angle but were independent of gaze direction. However, several adaptation studies suggest that it is possible to achieve non-linear changes in coupling of both vertical vergence and cyclovergence with gaze direction (Schor, Maxwell and Graf, 2001) that could be achieved with changes in neural control that depend upon both convergence and direction of gaze (McCandless and Schor, 1997).

References

Allen, M. J. and Carter, J. H. (1967). The torsion component of the near reflex. *Am. J. Optom.*, 44: 343–349.

Alpern, M. and Ellen, P. (1956). A quantitative analysis of the horizontal movements of the eyes in the experiments of Johannes Müller. I. Methods and results. *Am. J. Ophthal.*, 42: 289–303.

Bradshaw, M. F. and Rogers, B. J. (1994). Is cyclovergence state affected by the inclination of stereoscopic surfaces? *Invest. Ophthal. Vis. Sci.*, 35: 1316.

Bruno, P. and Van den Berg, A. V. (1997). Relative orientation of primary positions of the two eyes. *Vis. Res.*, 37: 935–947.

Collewijn, H. (1994). Vertical conjugacy: What coordinate system is appropriate? In Fuchs, A. F., Brandt, T., Buttner, U. and Zee, D. S. (eds), *Contemporary Ocular Motor and Vestibular Research: A Tribute to David A. Robinson*, pp. 296–303, Stuttgart: Thieme.

Donders, F. C. (1848). Beitrag zur Lehre von den Bewegungen des menschlichen Auges. *Hollandische Beitrage zu den anatomischen und physiologischen Wissenschaften*, 1: 105–145.

Eadie, A. S., Gray, L. S., Carlin, P. and Mon-Williams, M. (2000). Modeling adaptation effects in vergence and accommodation after exposure to a simulated virtual reality stimulus. *Ophthal. Phys. Optics*, 20: 242–251.

Enright, J. T. (1992). Unexpected role of the oblique muscles in the human vertical fusion reflex. *J. Physiol. (Lond.)*, 451: 279–293.

Fincham, E. F. and Walton, J. (1957). The reciprocal actions of accommodation and convertgence. *J. Physiol. (Lond)*, 137: 488=-508.

Garding, J., Porrill, J., Mayhew, J. E. W. and Frisby, J. P. (1995). Stereopsis, vertical disparity and relief transformations. *Vis. Res.*, 35: 703–722.

Gillam, B. and Lawergren, B. (1983). The induced effect, vertical disparity, and stereoscopic theory. *Percept. Psychophys.*, 34: 121–130.

Hasebe, S., Graf, E. W. and Schor, C. M. (2001). Fatigue reduces tonic accommodation. *Ophthal. Phys. Optics*, 21: 151–160.

Haustein, W. (1989). Considerations on Listing's law and the primary position by means of a matrix description of eye position control. *Biol. Cyb.*, 60: 411–420.

von Helmholtz, H. (1910). *Treatise on Physiological Optics*. Southall, J. P. C. (ed.), New York: Dover.

Hering, E. (1868). *The Theory of Binocular Vision*, B. Bridgeman and L. Stark (eds.). New York: Plenum.

Howard, I. P. and Kaneko, H. (1994). Relative shear disparities and the perception of surface inclination. Cyclotorsion vs. slant and disparity. *Vis. Res.*, 34: 2505–2517.

Howard, I. P. and Rogers, B. J. (1995). *Binocular Vision and Stereopsis*. Oxford: Oxford University Press.

Jaing, B. C. (1996). Accommodative vergence is driven by the phasic component of accommodative controller. *Vis. Res.*, 36: 97–102.

Judge, S. J. and Miles, F. A. (1985). Changes in the coupling between accommodation and vergence eye movements induced in human subjects by altering the effective interocular separation. *Percept.*, 14: 617–629.

Maxwell, J. and Schor, C. M. (1994). Mechanisms of vertical phoria adaptation revealed by time-course and two-dimensional spatiotopic maps. *Vis. Res.*, 34: 241–251.

Mays L. E., Zhang, Y., Thorsdtad, M. H. and Gamlin, P. D. R. (1991). Trochlear unit activity during ocular convergence. *J. Neurophysiol.*, 65: 1484–1491.

McCandless, J. W. and Schor, C. M. (1997). An association matrix model of context-specific vertical vergence adaptation. *Network: Comput. Neural Syst.*, 8: 239–258.

McCandless, J. W., Schor, C. M. and Maxwell, J. (1996). A cross coupling model of vertical vergence adaptation. *IEEE Trans. BioMed Engineer.*, 43: 24–34.

Mikhael, S., Nicolle, D. and Vilis, T. (1995). Rotation of Listing's plane by horizontal, vertical, and oblique prism-induced vergence. *Vis. Res.*, 35: 3243–3254.

Miller, J. M. and Demer, J. L. (1992). Biomechanical analysis of strabismus. *Binoc. Vis. and Eye Muscle Surgery Quart.*, 7: 233–248.

Miller, J. M. and Robinson, D. A. (1984). A model of the mechanics of binocular alignment. *Comput. Biomed. Res.*, 17: 436–470.

Minken, A. W. H. and van Gisbergen, J. A. M. (1994). A three-dimensional analysis of vergence movements at various levels of elevation. *Exp. Brain Res.*, 101: 331–345.

Mok, D., Cadera, A., Ro, W., Crawford, J. D. and Vilis, T. (1992). Rotation of Listing's plane during vergence. *Vis. Res.*, 32: 2055–2064.

Nakayama, K. (1983). Kinematics of normal and strabismic eyes. In C. M. Schor and K. Ciuffreda (eds) *Vergence Eye Movements: Basic and Clinical Aspects*, pp. 543–564, Boston: Butterworths.

Porrill, J., Warren, P. A. and Dean, P. (2000). A simple control law generates Listing's positions in a detailed model of the extraocular muscle system. *Vis. Res.*, 40: 3743–3758

Robinson, D. A. (1975). A quantitative analysis of extraocular muscle cooperation and squint. *Invest. Ophthal. Vis. Sci.*, 14: 801–825.

Schor, C. M. (1992). A ynamic model of cross-coupling between accommodation and convergence: simultation of step and frequency responses. *Optom. Vis. Sci.*, 69: 258–269.

Schor, C. M., Alexander, J., Cormack, L. and Stevenson, S. (1992). A negative feedback control model of proximal convergence and accommodation. *Ophthal. Physiol. Optics*, 12: 307–318.

Schor, C. M. and Kotulak, J. (1986). Dynamic interactions between accommodation and convergence are velocity sensitive. *Vis. Res.*, 26: 927–942.

Schor, C. M., Maxwell, J. and Graf, E. W. (2001). Plasticity of convergence-dependent variations of cyclovergence with vertical gaze. *Vis. Res.*, 41: 3353–3369.

Schor, C. M., Maxwell, J. and Stevenson, S. B. (1994). Isovergence surfaces: the conjugacy of vertical eye movements in tertiary positions of gaze. *Ophthal. Physiol. Optics*, 14: 279–286.

Schor, C. M. and McCandless, J. W. (1995). Distance cues for vertical vergence adaptation. *Optom. Vis. Sci.*, 72: 478–486.

Schor, C. M. and McCandless, J. W. (1997). Context-specific adaptation of vertical vergence to multiple stimuli. *Vis. Res.*, 37: 1929–1938.

Schor, C. M. and Tsuetaki, T. (1987). Fatigue of accommodation and vergence modifies their mutual interactions. *Invest. Ophthal. Vis. Sci.*, 28: 1250–1259.

Schreiber, K., Crawford, J. D., Fetter, M. and Tweed, D. (2001). The motor side of depth vision. *Nature*, 410: 819-822.

Somani, R. A. B., DeSouza, J. F. X., Tweed, D. and Vilis, T. (1998). Visual test of Listing's law during vergence. *Vis. Res*, 38: 911–923.

Stevenson, S., Lott, L. and Yang, J. (1997). The influence of subject instruction on horizontal and vertical vergence tracking. *Vis. Res.*, 37: 2891–2898.

Tweed, D. (1997). Visual-motor optimization in binocular control. *Vis. Res.*, 37: 1939–1951.

Tweed, D. and Vilis, T. (1990). Geometric relations of eye position and velocity vectors during saccades. *Vis. Res.*, 30: 1110–127.

van Rijn, L. J. and van den Berg, A. V. (1993). Binocular eye orientation during fixations: Listing's law extended to include eye vergence. *Vis. Res.*, 33: 691–708.

Ygge, J. and Zee, D. S. (1995). Control of vertical eye alignment in three-dimensional space. *Vis. Res.*, 35: 3169–3181.

13

Population Coding of Vergence Eye Movements in Cortical Area MST

A. Takemura, K. Kawano, C. Quaia, and F. A. Miles

13.1 Introduction

In recent years there has been considerable interest in the coding of information by the activity of populations of neurons, and particularly in the process whereby information becomes available at the level of the population that is not available at the level of the individual neurons. The information could relate to sensory events, such as the motion of a visual stimulus (Maunsell and Van Essen, 1983; Priebe, Churchland and Lisberger, 2001; Steinmetz, Motter, Duffy and Mountcastle, 1987) or the orientation of the head (Schor, Miller and Tomko, 1984), or to motor responses, such as the magnitude and direction of a saccadic eye movement (Anderson, Keller, Gandhi and Das, 1998; Büttner, Büttner-Ennever and Henn, 1977; Henn and Cohen, 1976; Lee, Rohrer and Sparks, 1988; Munoz and Wurtz, 1995; Optican, 1995; Quaia, Lefèvre and Optican, 1999; Schlag-Rey and Schlag, 1977; Sparks, Lee and Rohrer, 1990; Thier, Dicke, Haas and Barash, 2000; Van Gisbergen, Van Opstal and Tax, 1987; Van Opstal and Van Gisbergen, 1989; Wurtz, 1996) or the direction of a hand movement (Georgopoulos, Kalaska, Caminiti and Massey, 1982; Georgopoulos, Kettner and Schwartz, 1988; Georgopoulos, Schwartz and Kettner, 1986; Kalaska, Caminiti and Georgopoulos, 1983; Maynard, Hatsopoulos, Ojakangas, Acuna, Sanes, Normann and Donoghue, 1999; Moran and Schwartz, 1999a, b; Schwartz, 1993, 1994; Schwartz and Moran, 1999, 2000). The population coding here involved pooling the activity of neurons with rather broad, overlapping tuning functions to achieve a more accurate representation. In the present chapter we will review a recent study (Takemura, Inoue, Kawano, Quaia and Miles, 2001) in which we described the activity of neurons in the medial superior temporal (MST) area of cortex, which has traditionally been treated as a pure sensory area,[1] in relation to a simple sensory-motor

---

[1] An exception is the suggestion of Sakata, Shibutani and Kawano (1983) and Newsome, Wurtz and Komatsu (1988) that some MST neurons carry efference copy signals.

paradigm. We found that the discharges of individual neurons in MST each encode only some very limited aspect of the sensory stimulus (and/or possible associated motor response) but when pooled together provide a complete description of the motor response. At the level of the individual cells there was little or no hint of the information encoded at the level of the population of cells, just as there is generally no hint of the information conveyed by a word in its individual letters: the information is an emergent property of the population activity.

The sensory input in our study was a disparity step applied to a large stationary random-dot pattern, and the motor output was a vergence eye movement that was elicited at ultra-short latency. The neurons were recorded in cortical area MST because lesions of this area have been shown to cause deficits in these eye movements (Takemura, Inoue and Kawano, 2000; Takemura, Inoue, Kawano, Quaia and Miles, 1999). We will first review spatial aspects of the stimulus–response relationships and their associated single unit discharges in MST, concentrating initially on the spatial information carried by the individual cells and then on the spatial information encoded in the population activity. We will then review temporal aspects of the stimulus–response relationships and their associated single unit discharges in MST, again starting with the individual cells before considering the population activity. Finally, we will discuss some possible consequences of the coding of information in the activity of populations of neurons.

13.2   The Sensory-Motor Paradigm: Short-Latency, Disparity-Vergence Eye Movements

Brief (e.g., 200 msec) horizontal disparity steps applied to large correlated random-dot patterns (in which the two eyes see identical patterns of dots) elicit vergence eye movements at short latencies (Busettini, FitzGibbon and Miles, 2001; Busettini, Miles and Krauzlis, 1996): for an example, see the continuous traces in Figure 13.1a (Masson, Busettini and Miles, 1997). Disparity tuning curves, describing the dependence of the amplitude of the initial vergence responses on the amplitude of the disparity steps, resemble the derivative of a Gaussian, and indicate that appropriate servo-like behavior is seen only for small disparity steps. Thus, increases in the disparity input resulted in roughly proportional increases in the vergence output (in the compensatory direction) only for steps of less than a degree or so: see the closed symbols in Figure 13.1b. Similar disparity steps applied to anticorrelated random-dot patterns (in which the dots seen by the two eyes were of opposite contrast so that each black dot seen by one eye was matched to a white dot seen by the other eye) elicit similar vergence eye movements except that they have the opposite sign and hence are said to be "anticompensatory" (Masson et al., 1997): see the dotted traces in Figure 13.1a and the open symbols in Figure 13.1b. An interesting difference between the correlated and anticorrelated stimuli used in this study is that the former are associated with a change in per-

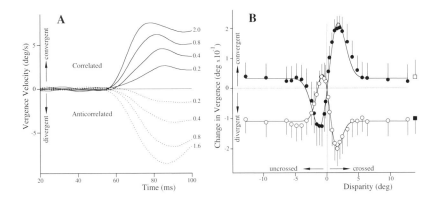

FIGURE 13.1. Vergence eye movements: dependence on the amplitude and direction of the disparity step. (a) mean vergence velocity over time in response to crossed disparity steps applied to correlated (continuous line) and anticorrelated (dotted line) random-dot patterns. Upward deflections denote increased vergence and the numbers on the traces indicate the magnitudes of the steps in degrees. (b) The mean change in vergence position in degrees (closed circles, correlated patterns; open circles, anticorrelated patterns), measured over the 33-ms period starting 60 ms after the disparity step, is plotted against the magnitude of the step in degrees: disparity tuning curves. Continuous lines are least-square, best-fit Gabor functions. Data from 1 monkey. Error bars, 1 SD. From Masson, Busettini and Miles (1997). *Nature*, 389: 283–286, with permission.

ceived depth, whereas the latter are not (Cogan, Lomakin and Rossi, 1993; Cumming and Parker, 1997; Masson et al., 1997). These findings with anticorrelated patterns have been used to argue that the short-latency vergence eye movements are generated independently of perception (Masson et al., 1997).

13.3   Neuronal Responses in MST: Spatial Coding by Individual Cells

Twenty percent of the neurons recorded in MST were sensitive to disparity steps applied to large correlated random-dot patterns, and disparity tuning curves describing the dependence of their initial (open-loop) discharges on the magnitude of the disparity step were constructed. Using objective criteria and the fuzzy c-means clustering algorithm (Bezdek, 1981), the neuronal disparity tuning curves were sorted into four groups based on their shapes: see Figure 13.2. These four groups had features in common with four of the classes of disparity-selective neurons that others have described in striate cortex (Poggio, Gonzalez and Krause, 1988), although groups 2, 3, and 4 appeared to be part of a continuum. About one-half of the cells responding to disparity steps applied to correlated patterns were also tested with disparity steps applied to anticorrelated random-dot patterns and all modulated significantly ($p < 0.005$, 1-way ANOVA).

A few of these neurons had disparity tuning curves whose shapes closely resembled the shapes of the tuning curve for vergence when the stimuli were either correlated or anticorrelated stimuli, but never for both types of stimuli. This was examined quantitatively by fitting the neurons' disparity tuning curves to the disparity tuning curve for vergence (gain and offset free parameters). The goodness of these fits was assessed by computing the fraction of the disparity-induced variation in the data accounted for by the fits (r^2). Figure 13.3 plots the r^2 values obtained with correlated stimuli against the r^2 values obtained with anticorrelated stimuli for each of the cells examined with both types of disparity stimuli (circular symbols). It is evident that there is considerable scatter in these data, the goodness of the fit with the correlated stimulus clearly having no relationship to the goodness of the fit with the anticorrelated stimulus. The upper right corner of the plot, which is where pure vergence encoding cells would reside, is devoid of any cells. Thus, the activity of the individual cells at best encoded only some aspect of the disparity stimulus and/or the associated vergence motor response and there were no pure vergence-encoding cells.

13.4   Neuronal Responses in MST: Spatial Coding by the Population of Cells

Most of the unit data were recorded from two monkeys (designated N and Q), and when the tuning curves of all of the cells obtained from either monkey were simply summed together (using a simple, as opposed to weighted, sum) they fitted the tuning curves for the vergence responses of that same monkey very well, always accounting for at least 93% of the disparity-induced variability.[2] This was true for the unit data obtained with both correlated and anticorrelated stimuli: see Figure 13.4, and the symbols N and Q plotted in Figure 13.3 (the latter representing the r^2 values for the summed activity of the two monkeys from which all of the data in this figure were recorded). Interestingly, the shapes of the disparity tuning curves for the vergence data obtained with correlated stimuli were rather similar for the two animals (compare Figs. 13.4a and b), but the disparity tuning curves for the vergence responses obtained with anticorrelated stimuli differed significantly (compare Figures 13.4c and d). The differences in the vergence data obtained with anticorrelated stimuli were such that the summed neural activity

---

[2]One manipulation of the data was critical to achieve the good fits: the sign of those tuning curves that had a negative slope in the important servo range, ± 1 deg, was inverted: this involved all cells in Group 1 and one in Group 2. This meant that all cells in all groups would make a positive net contribution to the population vergence signal because the sign of the vergence responses, which is entirely determined by our convention that increases in convergence are positive, was positive over the range in question. Of course, sign differences like this can be achieved by appropriate excitatory/inhibitory connections.

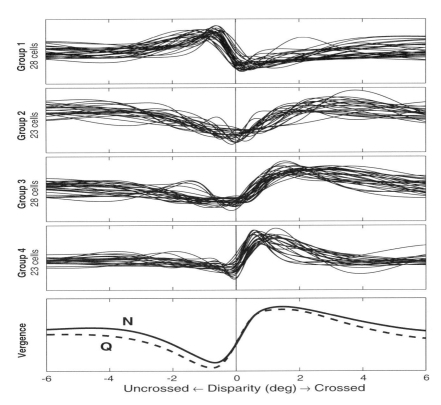

FIGURE 13.2. Disparity tuning curves for individual MST cells (correlated stimuli). Upper four graphs: mean change in discharge rate over the 60-ms period starting 40 ms after the disparity step is plotted against the magnitude of the disparity step; curves are normalized and arranged in four groups based on the outcome of the fuzzy c-means clustering algorithm of Bezdek (1981). Bottom: the disparity tuning curves for the vergence responses of the two monkeys that yielded most of the data (*N* and *Q*). Traces are spline interpolations. From Takemura, Inoue, Kawano, Quaia and Miles (2001). *J. Neurophysiol.*, 85: 2245–2266, with permission.

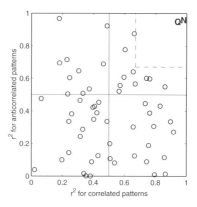

FIGURE 13.3. Coding of vergence by individual cells (correlated and anticorrelated stimuli). The disparity tuning curves of the individual cells were fitted to the disparity tuning curves of the associated vergence responses and r^2 values for the least-squares best fits were computed. This graph plots the r^2 values for the data obtained with correlated patterns against those obtained with anticorrelated patterns. No cell had r^2 values that exceeded 0.67 for both stimuli (indicated by the dashed lines). Also shown (indicated by their identifying letters) are the r^2 values for the fits between the summed activity and the vergence responses for the two monkeys, N and Q (from which all the unit data plotted here were obtained). From Takemura, Inoue, Kawano, Quaia and Miles (2001). *J. Neurophysiol.*, 85: 2245–2266, with permission.

from monkey N gave a very poor fit to the vergence data obtained from monkey Q ($r^2 = 0.29$) and vice versa ($r^2 = 0.35$). Thus, the summed neural activity reproduced the idiosyncratic differences between the vergence responses of the two monkeys, a finding that convinces us that the ensemble coding of vergence in MST has biological significance.

The question arises as to the relative contributions of the four groups of cells (identified by the fuzzy cluster analysis, based on the shapes of the disparity tuning curves obtained with correlated stimuli) to the population coding of vergence: It is possible that one or more groups actually make the fits between the aggregate activity and the vergence responses worse. Sufficient data to examine this question were available only for correlated stimuli and these suggested that all four groups of cells were necessary to get the good fit between the summed activity and vergence, especially for monkey N. Thus, excluding all of the cells in any one of the four groups always decreased the goodness of fit in monkey N. Further, randomly excluding equivalent numbers of cells indicated that the probability of achieving these results by chance was always <0.008: bootstrap statistic (Efron and Tibshirani, 1991). The data from monkey Q were less compelling and only exclusion of one of the four groups (Group 1) resulted in a fit that was significantly worse. Using a genetic algorithm, it was possible to get an estimate of the subsets of neurons whose tuning curves, when summed together, gave the best fit to the vergence tuning curves, and these subsets invariably included cells from all four groups. These findings indicate that the encoding of vergence depended on

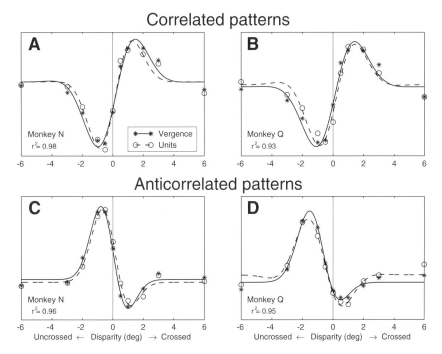

FIGURE 13.4. Spatial coding of vergence by populations of cells. (a, b) Correlated stimuli. (c, d) Anticorrelated stimuli. Disparity tuning data for vergence (stars) and for the
least-squares, best-fit summed activity (circles) for monkey N (a, c) and monkey Q (b,
d). Note that the individual unit curves with negative slopes around zero disparity were
inverted before summing. Curves are the least-square, best-fit Gabor functions for the vergence data (continuous line) and the summed activity data (dashed line). From Takemura,
Inoue, Kawano, Quaia and Miles (2001). *J. Neurophysiol.*, 85: 2245–2266, with permission.

contributions from across the entire spectrum of tuning curves found in disparity selective MST cells. Thus, the population code relies on the aggregate activity of a heterogeneous collection of cells.

13.5   Neuronal Responses in MST: Temporal Coding

Additional analyses of the spike trains elicited by disparity steps revealed considerable variation across cells in the latency, amplitude, and time course of the changes in discharge rate: see Figure 13.5, which shows the discharge frequency profiles of 20 cells recorded from monkey N in response to crossed disparity steps. When all of the spike trains elicited by a given disparity step (which were recorded from a given monkey) were summed together to give an average discharge profile for the whole population of cells, many were rather noisy (n = 20, 10 disparity steps and 2 monkeys), ruling out any possibility that they might match the profile of the associated vergence responses. However, other summed discharge profiles were much cleaner and matched the temporal profile of the vergence velocity response quite well (free parameters: gain, y-offset, and x-offset). An example is seen at the bottom of Figure 13.5: compare the dashed trace, depicting the summed neural activity, with the continuous trace, depicting the vergence velocity profile. In the example shown, the summed activity accounted for 93% of the disparity-induced variation in vergence. In fact, r^2 values were greater than 0.9 for 40% of the fits (8/20). In view of the noise problems inherent in spike trains and the fact that many discharge profiles of the individual cells showed a strong initial transient, we were surprised that so many summed-activity profiles approximated the vergence velocity profile, which generally showed a monotonic rise with no hint of an initial transient (as in Figure 13.5). Thus, the initial phasic components seen in the individual discharge profiles largely disappear in the summed-activity profiles, presumably due to the latency jitter: smoothing by temporal summation.

13.6   Population Coding

Our findings show that the magnitude, direction, and time course of the initial (open-loop) vergence velocity responses associated with disparity steps applied to large textured patterns are well correlated with the summed activity of the disparity-sensitive cells that we recorded in MST. The activity of the individual cells in MST correlated only poorly with the vergence responses, hence one can view the encoding of vergence velocity in MST as an emergent property of the population activity. Thus, it is tempting to conclude that the individual MST neurons carry sensory signals while the population of MST cells as a whole encodes the complete motor output, vergence velocity. Our latency data suggest that this activity in MST occurs early enough to play an active role in the genera-

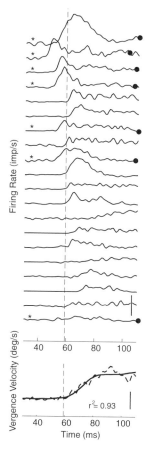

FIGURE 13.5. Time course of the neuronal responses (correlated stimuli). Upper traces show the changes in mean discharge rate over time in response to 2 deg crossed disparity steps for each of the 20 units that modulated most with this stimulus (ranked in descending order of their mean discharge rates over the period 30–110 ms after the step). The vertical dashed line is the estimated latency of the vergence response (59 msec). Seven of the units, indicated by asterisks, had response latencies <51 msec, and the dots on the traces indicate the estimated times at which the closing of the disparity feedback loop could first influence the discharges of these units. (The response latencies of the remaining units were too long for the disparity-feedback loop to close during the 110 msec time window shown.) The bottom traces show the changes over time in vergence velocity (continuous line) and in the least-squares, best-fit summed activity (dashed line, with a time lag of 18 msec). Calibration bars: 500 imps/sec, 5deg/sec. Data from monkey N. From Takemura, Inoue, Kawano, Quaia and Miles (2001). *J. Neurophysiol.*, 85: 2245–2266, with permission.

tion of even the earliest vergence eye movements, consistent with other findings that lesions of MST result in significant deficits in such eye movements (Takemura et al., 1999, 2000). We therefore hypothesize that the summed activity of the disparity-selective neurons in MST represents a motor signal and drives the vergence system. In this scheme, the motor signal is nothing more than the arithmetic sum of a set of sensory signals and the sensory-motor transformation could be accomplished by a simple summing junction. This would mean that the discharges in MST could be viewed as both sensory and motor depending on the level of scrutiny — the individual cells or the population — and would represent a clear example of parallel (rather than serial or hierarchical) signal processing. Interestingly, it has been argued that parallel processing is necessary to achieve the ultra-short latency that characterizes the motor responses in this study (Miles, 1998).

13.6.1   Subsequent Signal Processing

Various detailed schemes have been proposed for the decoding of the information embedded in the activity of populations of neurons: for recent reviews, see Abbott and Sejnowski (1999). Based on the effects of excluding any one of the four groups of cells from the population sum and on the outcome of the genetic algorithm, we concluded that the encoding of vergence motor responses in MST requires contributions from across the entire spectrum of disparity-selective cells that we recorded. This raises the possibility that a random selection of these cells would suffice to generate the vergence responses and that the projection from MST to the next stage in the processing of the vergence drive signal need not involve complex connectivity rules. That the aggregate activity of the disparity-selective cells in MST encodes vergence velocity is consistent with the commonly held view that MST is exclusively involved in the processing of motion signals: see Eifuku and Wurtz (1998, 1999) for recent references. There are a number of anatomical routes by which the vergence signals in MST might gain access to the vergence premotor centers (Takemura et al., 2001). Contrary to earlier notions, it is now clear that medial rectus motoneurons carry both position and velocity signals for the control of vergence (Gamlin and Mays, 1992), and both types of signals are evident in the midbrain neurons thought to carry the command signals for vergence eye movements (Judge and Cumming, 1986; Mays, Porter, Gamlin and Tello, 1986; Zhang, Gamlin and Mays, 1991). This means that the vergence velocity signal conveyed by the population activity in MST would need to be supplemented with a position signal in order to provide the complete command signal to the midbrain neurons generating vergence eye movements. This could be achieved by a single integration in the projection pathways, similar to that which has long been postulated to occur in the pathways from the semicircular canals to the oculomotor motoneurons (Galiana and Outerbridge, 1984; Skavenski and Robinson, 1973). It is interesting that the initial phasic component that is often seen in the discharge profiles of individual MST cells (see Figure 13.5) — and might be thought to represent a vergence acceleration signal, albeit rather crude

— is filtered out of the population response by the latency jitter and temporal summation.

13.6.2 Multiplexing

Population coding raises the possibility that these same MST cells can participate in other functions unrelated to vergence. This might involve subgroupings of the cell population that we have recorded operating through other output pathways to achieve some other function necessitating disparity information. Another possibility is that these same cells also carry signals unrelated to disparity and are members of other groupings/populations of cells that combine to achieve other purposes through their shared connections. In such a distributed network, the functionality depends critically on the pooling achieved by the shared connectivity. Clearly, such multiplexing would involve some delicate balancing of inputs to minimize cross-talk and thereby render the different functions of the individual cells orthogonal at the population level. Of course, failure to do this at one level of the system might be corrected by adding appropriate compensatory signals at subsequent levels. An example of this occurs in canal-ocular pathways during so-called suppression of the vestibulo-ocular reflex: see the chapter by Cullen, Roy, and Sylvestre in this volume.

References

Abbott, L. and Sejnowski, T. J. (1999). *Neural Codes and Distributed Representations: Foundations of Neural Computation*. Cambridge: MIT Press.

Anderson, R. W., Keller, E. L., Gandhi, N. J. and Das, S. (1998). Two-dimensional saccade-related population activity in superior colliculus in monkey. *J. Neurophysiol.*, 80: 798–817.

Bezdek, J. C. (1981). *Pattern Recognition with Fuzzy Objective Function Algorithms*. New York: Plenum.

Busettini, C., FitzGibbon, E. J. and Miles, F. A. (2001). Short-latency disparity vergence in humans. *J. Neurophysiol.*, 85: 1129–1152.

Busettini, C., Miles, F. A. and Krauzlis, R. J. (1996). Short-latency disparity vergence responses and their dependence on a prior saccadic eye movement. *J. Neurophysiol.*, 75: 1392–1410.

Büttner, U., Büttner-Ennever, J. A. and Henn, V. (1977). Vertical eye movement related unit activity in the rostral mesencephalic reticular formation of the alert monkey. *Brain Res.*, 130, 239–252.

Cogan, A. I., Lomakin, A. J. and Rossi, A. F. (1993). Depth in anticorrelated stereograms: Effects of spatial density and interocular delay. *Vis. Res.*, 33: 1959–1975.

Cumming, B. G. and Parker, A. J. (1997). Responses of primary visual cortical neurons to binocular disparity without depth perception. *Nature*, 389: 280–283.

Efron, B. and Tibshirani, R. (1991). Statistical data analysis in the computer age. *Science*, 253: 390–395.

Eifuku, S. and Wurtz, R. H. (1998). Response to motion in extrastriate area MSTl: Center-surround interactions. *J. Neurophysiol.*, 80: 282–296.

Eifuku, S. and Wurtz, R. H. (1999). Response to motion in extrastriate area MSTl: Disparity sensitivity. *J. Neurophysiol.*, 82: 2462–2475.

Galiana, H. L. and Outerbridge, J. S. (1984). A bilateral model for central neural pathways in vestibuloocular reflex. *J. Neurophysiol.*, 51: 210–241.

Gamlin, P. D. R. and Mays, L. E. (1992). Dynamic properties of medial rectus motoneurons during vergence eye movements. *J. Neurophysiol.*, 67: 64–74.

Georgopoulos, A. P., Kalaska, J. F., Caminiti, R. and Massey, J. T. (1982). On the relations between the direction of two-dimensional arm movements and cell discharge in primate motor cortex. *J. Neurosci.*, 2: 1527–1537.

Georgopoulos, A. P., Kettner, R. E. and Schwartz, A. B. (1988). Primate motor cortex and free arm movements to visual targets in three-dimensional space. II. Coding of the direction of movement by a neuronal population. *J. Neurosci.*, 8: 2928–2937.

Georgopoulos, A. P., Schwartz, A. B. and Kettner, R. E. (1986). Neuronal population coding of movement direction. *Science*, 233: 1357–1460.

Henn, V. and Cohen, B. (1976). Coding of information about rapid eye movements in the pontine reticular formation of alert monkeys. *Brain Res.*, 108: 307–325.

Judge, S. J. and Cumming, B. G. (1986). Neurons in the monkey midbrain with activity related to vergence eye movement and accommodation. *J. Neurophysiol.*, 55: 915–930.

Kalaska, J. F., Caminiti, R. and Georgopoulos, A. P. (1983). Cortical mechanisms related to the direction of two-dimensional arm movements: relations in parietal area 5 and comparison with motor cortex. *Exp. Brain Res.*, 51: 247–260.

Lee, C., Rohrer, W. H. and Sparks, D. L. (1988). Population coding of saccadic eye movements by neurons in the superior colliculus. *Nature*, 332: 357–360.

Masson, G. S., Busettini, C. and Miles, F. A. (1997). Vergence eye movements in response to binocular disparity without depth perception. *Nature*, 389: 283–286.

Maunsell, J. H. R. and Van Essen, D. C. (1983). Functional properties of neurons in middle temporal visual area of the macaque monkey. I. Selectivity for stimulus direction, speed, and orientation. *J. Neurophysiol.*, 49: 1127–1147.

Maynard, E. M., Hatsopoulos, N. G., Ojakangas, C. L., Acuna, B. D., Sanes, J. N., Normann, R. A. and Donoghue, J. P. (1999). Neuronal interactions improve cortical population coding of movement direction. *J. Neurosci.*, 19: 8083–8093.

Mays, L. E., Porter, J. D., Gamlin, P. D. R. and Tello, C. A. (1986). Neural control of vergence eye movements: neurons encoding vergence velocity. *J. Neurophysiol.*, 56: 1007–1021.

Miles, F. A. (1998). The neural processing of 3-D visual information: Evidence from eye movements. *Eur. J. Neurosci.*, 10: 811–822.

Moran, D. W. and Schwartz, A. B. (1999a). Motor cortical activity during drawing movements: population representation during spiral tracing. *J. Neurophysiol.*, 82: 2693–2704.

Moran, D. W. and Schwartz, A. B. (1999b). Motor cortical representation of speed and direction during reaching. *J. Neurophysiol.*, 82: 2676–2692.

Munoz, D. P. and Wurtz, R. H. (1995). Saccade-related activity in monkey superior colliculus. II. Spread of activity during saccades. *J. Neurophysiol.*, 73: 2334–2348.

Newsome, W. T., Wurtz, R. H. and Komatsu, H. (1988). Relation of cortical areas MT and MST to pursuit eye movements. II. Differentiation of retinal from extraretinal inputs. *J. Neurophysiol.*, 60: 604–620.

Optican, L. M. (1995). A field theory of saccade generation: temporal-to-spatial transform in the superior colliculus. *Vis. Res.*, 35: 3313–3320.

Poggio, G. F., Gonzalez, F. and Krause, F. (1988). Stereoscopic mechanisms in monkey visual cortex: binocular correlation and disparity selectivity. *J. Neurosci.*, 8: 4531–4550.

Priebe, N. J., Churchland, M. M. and Lisberger, S. G. (2001). Reconstruction of target speed for the guidance of pursuit eye movements. *J. Neurosci.*, 21: 3196–3206.

Quaia, C., Lefèvre, P. and Optican, L. M. (1999). Model of the control of saccades by superior colliculus and cerebellum. *J. Neurophysiol.*, 82: 999–1018.

Sakata, H., Shibutani, H. and Kawano, K. (1983). Functional properties of visual tracking neurons in posterior parietal association cortex of the monkey. *J. Neurophysiol.*, 49: 1364–1380.

Schlag-Rey, M. and Schlag, J. (1977). Visual and presaccadic neuronal activity in thalamic internal medullary lamina of cat: a study of targeting. *J. Neurophysiol.*, 40: 156–173.

Schor, R. H., Miller, A. D. and Tomko, D. L. (1984). Responses to head tilt in cat central vestibular neurons. I. Direction of maximum sensitivity. *J. Neurophysiol.*, 51: 136–146.

Schwartz, A.B. (1993). Motor cortical activity during drawing movements: population representation during sinusoid tracing. *J. Neurophysiol.*, 70: 28–36.

Schwartz, A. B. (1994). Direct cortical representation of drawing. *Science*, 265: 540–542.

Schwartz, A. B. and Moran, D. W. (1999). Motor cortical activity during drawing movements: population representation during lemniscate tracing. *J. Neurophysiol.*, 82: 2705–2718.

Schwartz, A. B. and Moran, D. W. (2000). Arm trajectory and representation of movement processing in motor cortical activity. *Euro. J. Neurosci.*, 12: 1851–1856.

Skavenski, A. and Robinson, D. A. (1973). Role of abducens motoneurons in the vestibulo-ocular reflex. *J. Neurophysiol.*, 36: 724–738.

Sparks, D. L., Lee, C. and Rohrer, W. H. (1990). Population coding of the direction, amplitude, and velocity of saccadic eye movements by neurons in the superior colliculus. *Cold Spring Harbor Symp. Quant. Biol.*, 55: 805–811.

Steinmetz, M. A., Motter, B. C., Duffy, C. J. and Mountcastle, V. B. (1987). Functional properties of parietal visual neurons: radial organization of directionalities within the visual field. *J. Neurosci.*, 7: 177–191.

Takemura, A., Inoue, Y. and Kawano, K. (2000). The role of MST neurons in short-latency visual tracking eye movements. *Soc. Neurosci. Abstr.*, 26: 1715.

Takemura, A., Inoue, Y., Kawano, K., Quaia, C. and Miles, F. A. (1999). Evidence that disparity-sensitive cells in medial superior temporal area contribute to short-latency vergence eye movements. *Soc. Neurosci. Abstr.*, 25: 1400.

Takemura, A., Inoue, Y., Kawano, K., Quaia, C. and Miles, F. A. (2001). Single unit activity in cortical areas MST associated with disparity-vergence eye movements: evidence for population coding. *J. Neurophysiol.*, 85: 2245–2266.

Thier, P., Dicke, P.W., Haas, R. and Barash, S. (2000). Encoding of movement time by populations of cerebellar Purkinje cells. *Nature*, 405: 72–76.

van Gisbergen, J. A. M., Van Opstal, A. J. and Tax, A. A. M. (1987). Collicular ensemble coding of saccades based on vector summation. *Neurosci.*, 21: 541–555.

Van Opstal, A. J. and van Gisbergen, J. A. M. (1989). A nonlinear model for collicular spatial interactions underlying the metrical properties of electrically elicited saccades. *Biol. Cybern.*, 60: 171–183.

Wurtz, R. H. (1996). Vision for the control of movement. The Friedenwald Lecture. *Invest. Ophthal. Vis. Sci.*, 37: 2131–2145.

Zhang, Y., Gamlin, P. D. and Mays, L. E. (1991). Antidromic identification of midbrain near response cells projecting to the oculomotor nucleus. *Exp. Brain Res.*, 84: 525–528.

14

Tendon End Organs Play an Important Role in Supplying Eye Position Information

Martin J. Steinbach

14.1 Preamble: Professor Howard Hires Soon-to-be-Professor Steinbach

In 1968, I was just finishing my Ph.D. when my advisor, Dick Held, introduced me to Ian Howard. It was at the Eastern Psychological Association meeting, *the* meeting for presenting visual perception results in those days. Ian, Dick told me, was recruiting new faculty for a new university (York) in Toronto, and that it was Ian's intention to build strength in vision research because the long-established University of Toronto had virtually no one in that area. Dick told me to listen carefully to what Ian had to say.

I listened very carefully. But I didn't understand a word Ian was saying, having never encountered a northern English accent before. I was pretty sure at one point we were talking about the eye, but Ian kept referring to a structure I had never heard of and I was loathe to betray ignorance. Where, I searched my memory, could a "fuvEEah" be?

After what sounded like an invitation to interview, I came to York to look around. My first impressions were discouraging because my taxi driver from the airport insisted on taking me to the Royal York Hotel, having never heard of York University. Worse, when we finally did get to the University, we entered what was a giant construction site where the helpful but misinformed guard told the driver that my destination, the Behavioural Sciences Building, "hadn't been built yet!"

Despite the inauspicious start, the interview went well, and, 33 years later, it looks like things worked out all right. I've never regretted taking up Ian's offer of a job, but it took me a long time to find the "fuvEEah."

14.2   Lessons on Visual Direction

Ian educated all of us on the nature of visual direction. He had, of course, just written, with Brian Templeton, the groundbreaking and universally admired book *Human Spatial Orientation* (Howard and Templeton, 1966). In those earliest days, we had weekly research meetings with Ian, Brian, Hiroshi Ono, Adrian Wilkinson, and myself. I came to appreciate the analytical brilliance and intellectual rigour of Ian, and his ability to reduce arguments and experiments to their absolute essence. He certainly made me think more about the elements involved in describing an object's place in space, and our ability to get to that object. Eye position information is of course a part of that loop. At that time, the gospel was that eye position could only be assessed from the outflowing, efferent, signal sent to the eye muscles, or their internal copy, the corollary discharge. My MIT teachers, Dick Held and Luke Teuber, had instilled that belief as part of the MIT catechism (see, e.g., Teuber, 1966), and, as a good novitiate, I had absorbed the lessons well. My dissertation was all about efference in the control of eye movements in hand-eye tracking (see, e.g., Steinbach and Held, 1968).

14.3   Testing the Outflow Theory of Eye Position Sense and Finding It Wanting

My first sabbatical was spent at the Smith Kettlewell Institute of Visual Sciences in San Francisco where the legendary strabismologist Dr. Arthur Jampolsky taught me about strabismus and its treatment by surgery. While observing Dr. Jampolsky operate, it occurred to me that the surgically treated patient was an ideal "preparation" for studying the outflow theory of eye position. Without afferent feedback, the surgical rotation of the eye by the surgeon should not be known to the patient's central nervous system until after a period of post-operative visual feedback and adaptation.

The surprising result was that some patients we tested after surgery clearly had information that the eye had been rotated to a new position in the orbit. We assessed this by an open-loop pointing task (Steinbach and Smith, 1981). But another group of patients didn't seem to have this information, and, by examining their histories, the difference became apparent: newly operated patients had eye muscle afference while patients who were having the same muscles operated on for a second or third time did not have this information. We speculated about tendon organs at the site of much of the surgeries being the source of the information, but it wasn't until Frances Richmond was drawn into collaboration that we were able to identify the presence of palisade endings, presumed proprioceptors, at the musculotendinous junction (Richmond et al., 1984).

Further studies gave us even more reason to believe that the palisade ending is playing a role in supplying eye position information. We compared two forms of muscle-weakening procedures done to correct strabismus, the marginal myotomy and the recession (see, e.g., von Noorden, 1996, pp. 93–101). The myotomy procedure must disturb the receptors because there is crushing and cutting at the musculotendinous junction. The recession procedure involves cutting just at the insertion into the sclera, and this is some distance away from these receptors. The myotomy was more deafferenting, again measured by open-loop pointing done before and after the surgery, than was the simple recession (Steinbach, Kirshner and Arstikaitis, 1987).

Dengis et al. (1998) looked at the results of pharmacological weakening of eye muscle, achieved by injecting microgram doses of botulinum toxin (botox), on changes in registered eye position. The botox procedure, developed by Scott (1980), corrects a strabismus by allowing the antagonist muscle to draw the strabismic eye into a straight (orthotropic) position by weakening the muscle that keeps the eye in the incorrect position. From animal studies, it was known that the botox injection has an immediate effect on afferent responses recorded from the trigeminal nerve (Manni et al., 1989). By measuring open loop pointing responses in patients, we found that the botox had no effect in the first hour following the injection. Over days and weeks, however, there were changes that could only have resulted from a proprioceptively derived signal about eye position. We interpeted these changes as resulting from the palisades, and not from the muscle spindles also found in human eye muscles and known to be proprioceptors in other skeletal muscle (e.g., Boyd and Gladden, 1985).

These studies of visual direction shifts due to proprioceptive changes in registered eye position have all been carried out in strabismic patients. There is a large and growing literature implicating inflow sources of eye position information, in both normal subjects and in patients with other eye muscle pathologies. For example, Lewis and Zee (1993) studied a patient with an abnormal synkinesis (moving the jaw caused a movement of the eye) and showed how the registration of eye position could have only come about by the contributions to eye position of the palisade ending. Dell'Osso et al. (1999), working with a breed of dog that has congenital nystagmus, showed that cutting the recti muscles at the insertion, and then reattaching them, dampened the nystagmus. This finding has led to a clinical trial for treatment of congenital nystagmus in humans by tenotomy of the recti muscles (Hertle et al., 2001). Early results are encouraging.

The literature on the roles proprioception may play in eye movement control is burgeoning. The reader is referred to Donaldson (2000), for an excellent, comprehensive, and up-to-date review.

14.4   Eye Muscles Are Special

It is clear that the muscles that move the eye are specialized in a number of ways, not only because of the presence of palisade endings. Büttner-Ennever et al. (2001) describe them as "among the most complicated muscles in the body." The muscles contain both twitch (singly innervated) and slow (multiply innervated) fibres, and the slow fibres are like those found in amphibia (reviewed in Porter et al., 1995). Both types of fibres are found in bands that run next to the globe — the global layer — and those that run next to the orbit — the orbital layer. Further, it appears that the orbital layers do not extend all the way to the insertion on the sclera, but rather end in "pulleys" of connective and elastic tissue that serve as the functional origin for the muscle (see Clark et al., 2000). The kinematics of eye movements (e.g., Listing's law) are claimed to be simplified by the existence of these pulleys (Demer et al., 2000). But how the ocular motor nuclei distribute efferent signals to the orbital and global fibres to control, respectively, the pulley positions and the eye rotations, remains to be discovered (Demer, 2001).

14.5   Palisade Endings Are Motor?!

The alleged role of the palisade endings in directly supplying afferent position information is also becoming more "interesting." These endings are found in the musculotendinous region and are exclusive to the multiply innervated global fibres (Richmond et al., 1984). I have argued that the palisades are the most important proprioceptors in human eye muscle (Steinbach, 2000). There now is some reason to doubt that the palisades may be entirely sensory in function. Büttner-Ennever et al. (2001) studied the motoneurons of twitch and non-twitch muscle fibres in the ocular motor nuclei. They injected retrograde tracers into monkey eye muscles in different locations. The tracer injected into the musculotendinous region labelled apparent motoneurons in peripheral regions of the ocular motor nuclei. Their "most parsimonious" conclusion: palisade endings are motor.

It is possible to reconcile the anatomical finding that palisades are motor with the behavioural findings that damage to the musculotendinous region leads to reduction of eye position information. The intramuscular spindle is a proprioceptor that has an efferent supply to the intrafusal muscle (the "gamma-efferent") which alters the spindle afferent response. This gain-control mechanism is important because the spindle is in parallel to the extrafusal muscle that changes length with contraction and relaxation (e.g., Guyton and Hall, 1996, pp. 687–691). Jendrassik's maneuver, where gamma-efferent loading of the spindle leads to an enhanced stretch reflex, demonstrates how the efferent supply can alter the afferent response (Walton, 1993, p. 53). If we assume that the palisade ending is part of an equivalent structure to the muscle spindle, then damage to that ending will still result in an altered afferent discharge, although the receptor and path for this afference remains to be discovered. Robinson (1991) speculated about the palisade being part of an inverted, inside-out, spindle, and Richmond et al. (1984, Fig. 2) showed a palisade fibre travelling some length along the multiinnervated global fibre. Perhaps these are all hints as to the true physiology of the palisade ending, the multiply-innervated global fibre, and their structure and function.

14.6   Coda

I am honoured and pleased to be a part of this Festschrift celebration and to be able to acknowledge my debt to Ian Howard as an employer, a teacher, and as a role model. He has been an inspiration in a number of ways and I count myself among the very lucky people who have him as a mentor and friend. Along with his wife, Toni, they have established a hospitable and warm "home" for me and for other lucky vision scientists who came to work at York University. It also can be noted that I now can understand virtually every word he says (fuvEEah = fovea!).

Acknowledgements

My research has been supported over the years by the Natural Sciences and Engineering Research Council (A7664), the Medical Research Council of Canada, the Jackman Foundation, the Sir Jules Thorn Charitable Trust, and Atkinson College. I thank Rosanne Steinbach and Linda Lillakas for their help with this manuscript, and Dr. J. A. Büttner-Ennever for providing a preprint of her *Journal of Comparative Neurology* paper.

References

Boyd, I. A. and Gladden, M. H. (1985). *The Muscle Spindle*. New York: Stockton Press.

Büttner-Ennever, J. A., Horn, A. K. E., Scherberger, H. and d'Ascanio, P. (2001). Mononeurons of twitch and non-twitch extraocular muscle fibres in the abducens, trochlear and oculomotor nuclei of monkeys. *J. Comp. Neurol.*, 438: 318–335.

Clark, R. A., Miller, J. M. and Demer, J. L. (2000). Three-dimensional location of human rectus pulleys by path inflections in secondary gaze positions. *Invest. Ophthal. Vis. Sci.*, 41: 3787–3797.

Dell'Osso, L. F., Hertle, R. W., Williams, R. W. and Jacobs, J. B. (1999). A new surgery for congenital nystagmus: effects of tenotomy on an achiasmatic canine and the role of extraocular proprioception. *J. Am. Assoc. Ped. Ophthal. Strab.* 3: 166–182.

Demer, J. L. (2001). Beyond origins and insertions: new concepts of extraocular muscles for the strabismus surgeon. *Binoc. Vis. Strab. Quart.* 16: 130–133

Demer, J. L., Oh, S. Y. and Poukens, V. (2000). Evidence for active control of rectus extraocular muscle pulleys. *Invest. Ophthal. Vis. Sci.*, 41: 1280–1290.

Dengis, C. A., Steinbach, M. J. and Kraft, S. P. (1998). Registered eye position: Short- and long-term effects of botulinum toxin injected into eye muscle. *Exp. Brain Res.*, 119: 475–482.

Donaldson, I. M. L. (2000). The functions of the proprioceptors of the eye muscles. *Phil. Tran. Roy. Soc. Lond. B.*, 355: 1685–1754.

Guyton, A. C. and Hall, J. E. (1996). Motor functions of the spinal cord; the cord reflexes. In *Textbook of Medical Physiology*, pp. 687-691. Philadelphia: W. B. Saunders Company.

Hertle, R. W., Dell'Osso, L. F., FitzGibbon, E. J., Thompson, D. J., Yan, D. and Mellow, S. D. (2001). Horizontal rectus tenotomy in the treatment of congenital nystagmus (CN): results of a study in ten adult patients (phase 1). *Invest. Ophthal. Vis. Sci.*, 42: S319.

Howard, I. P. and Templeton, W. B. (1966). *Human Spatial Orientation*, New York: Wiley.

Lewis, R. F. and Zee, D. S. (1993). Abnormal spatial localization with trigeminal-oculomotor synkinesis. *Brain*, 116: 1105–1118.

Manni, E., Bagolini, B., Pettorossi, V. E. and Errico, P. (1989). Effect of botulinum on extraocular muscle proprioception. *Doc. Ophthal.*, 72: 189–198.

Porter, J. D., Baker, R. S., Ragusa, R. J. and Brueckner, J. K. (1995). Extraocular muscles: basic and clinical aspects of structure and function. *Surv. Ophthalmol.*, 39: 451–484.

Richmond, F. J. R., Johnston, W. S. W., Baker, R. S. and Steinbach, M. J. (1984). Palisade endings in human extraocular muscle. *Invest. Ophthal. Vis. Sci.*, 25: 471–476.

Robinson, D. A. (1991). Overview. In R. H. S. Carpenter (ed.) *Vision and Visual Dysfunction*, Vol. 8, Eye Movements, pp. 320-331. Boca Raton: CRC press.

Scott, A. B. (1980). Botulinum toxin injection into extraocular muscles as an alternative to strabismus surgery. *Ophthalmol.*, 87: 1044–1049.

Steinbach, M. J. (2000) The palisade ending: An afferent source for eye position information in humans. In G. Lennerstrand and J. Ygge (eds.), *Advances on Strabismus Research: Basic and Clinical Aspects*, pp. 33-42. London: Portland Press.

Steinbach, M. J. and Held, R. (1968). Eye tracking of observer-generated target movements. *Science*, 161: 187–188.

Steinbach, M. J., Kirshner, E. L. and Arstikaitis, M. J. (1987). Recession vs. marginal myotomy surgery for strabismus: effects on spatial localization. *Invest. Ophthal. Vis. Sci.*, 28: 1870–1872.

Steinbach, M. J. and Smith, D. R. (1981). Spatial localization after strabismus surgery: evidence for inflow. *Science*, 213: 1407-1409.

Teuber, H. L. (1966). Alterations of perception after brain injury. In J. C. Eccles (ed.), *Brain and Conscious Experience.* New York: Springer-Verlag.

von Noorden, G. K. (1996). *Binocular Vision and Ocular Motility*, 5th ed., pp. 93-101. St. Louis: Mosby.

Walton, J. (1993). Disorders of function in the light of anatomy and physiology. In J. Walton (ed.) *Brain's Diseases of the Nervous System.* Oxford: Oxford University Press.

Part IV

Perception of Orientation and Self-Motion

15

Levels of Analysis of the Vestibulo-Ocular Reflex: A Postmodern Approach

Laurence R. Harris, Karl Beykirch, and Michael Fetter

15.1   Introduction

This chapter examines some of the ways in which a basic motor act, the vestibulo-ocular reflex, can be considered. It takes its inspiration in part from Dr Ian Howard. Ian's functional approach, with his emphasis on explanation-by-demonstration, has been hugely influential on the first author.

Which frame of reference is used by the reflexes that tend to maintain gaze during head movements is a question to which the answer has tended to be unnecessarily constrained by the anatomically rigid frame of the vestibular end organ embedded in the skull. Neural processes have no such constraint and indeed the frame they use cannot be based in the head since it is not possible to code head movement in a head-based co-ordinate system (Harris, 1997; Harris, Zikovitz and Kopinska, 1998). Here we model the eye movements evoked while rotating around various axes, as the output of a system consisting of three channels. The orientations of the channels needed to best model the eye movements evoked by rotation before and after an adaptation procedure have been determined. One was found close to the roll direction with the other two roughly in Listing's plane and approximately equidistant from all three canal pairs.

15.1.1   Channel Theory

The nervous system processes information. Information can be most efficiently processed if it is handled by a dedicated system or channel that carries information about only one thing. That way, the decoder that receives the output can make assumptions about the significance of activity in that channel. However, even for a dedicated channel, some of that activity will not be relevant to the information the system is transmitting, even though it might be potentially useful and interesting information about something else. If the information within the channel can be carried as a difference between multiple sub-channels that are all equally affected by at least some sources of noise, then correlated noise is cancelled out. This is channel theory. The idea of channels in neural systems emerged from communication theory (Shannon and Weaver, 1949; see Regan, 2000) and has since had a long and useful history in modelling neural information processing (Campbell and Tegeder, 1991; Harris, 1997). It is closely connected to the modular approach to modelling the brain (e.g., Zeki, 2001) and both channel and modules are the descendants of Müller's Principle of Specific Nerve Energies (Müller, 1840). Müller's principle states that modality of information is coded by which nerves or brain areas are active, leaving the decoder to specialize in decoding an already-known parameter.

Channels have been explicitly used to model many aspects of sensory information processing (Blum, 1991), but not overtly to describe "lower-level" information processing such as that involved in the essentially unconscious processes underlying so-called reflex eye movement control. Here we apply the concept of channels to model the coding the pattern of eye movements evoked by passive head rotation in the dark.

15.1.2   Compensating for Instability

As we move around the world our movement threatens the precarious stability of our retinal image and our posture. Various compensatory responses allow us to minimize perceptual and physical instability and thus help us to continue to function as well as possible during the movement. Potential threats to ocular stability stimulate visual, vestibular, and other proprioceptive systems, which together, in the context of the task in hand, drive compensatory movements of the eye, head, and body that tend to neutralize the threat. Each of the contributing systems, defined in terms of its sensory input and motor output, is capable of operating on its own, and each system has traditionally been investigated alone with the implication that each can be viewed as an essentially independent sub-system.

The vestibulo-ocular reflex (VOR) in such a view is the sub-system that measures head movement with the vestibular organs and generates a compensatory oculomotor response. The VOR has been regarded as a classic example of a simple reflex (Lorente de Nó, 1933; Szentágothai, 1950): the head moves to the left, the eyes move reflexively to the right, thus cancelling the retinal disturbance that would otherwise have occurred. More recently an amazing flexibility has become apparent in the VOR system and it has come to be regarded as a context-dependent motor response with an extraordinary adaptive ability. The VOR is able to adjust to a remarkable array of demands that are placed on it by virtue of the position of the two eyes in the head and the geometry of their relationship to objects in the outside world, and it can be recalibrated in response to both long- and short-term demands of the environment, especially visual demands.

15.1.3   A Postmodern Approach

Considering the vestibulo-ocular reflex as dependent on the ongoing context reflects the postmodern movement that, originating in architecture, now pervades many aspects of knowledge. Postmodernism rejects the modernist approach that celebrated the autonomous individual: a VOR capable of working perfectly in glorious isolation. In a postmodern world actions and thoughts can only be interpreted as part of the total environment in which they occur.

Early models of the VOR were based on the three-neurone arc concept (Lorente de Nó, 1933) that implied that the horizontal VOR (lateral eye movement evoked in response to yaw head rotation; see Figure 15.1) was processed as an independent mechanism within the VOR system. Since yaw movement stimulates primarily the horizontal semicircular canal pair (Curthoys, Markham and Blanks, 1975), this approach implied that other canal pairs also had their own independent connections to the appropriate eye muscles. The idea of a set of three subsystems defined by the semicircular canals within the skull underlying the generation of the VOR around any axis received support from the spatial tuning of sensory responses in various parts of the brain known to be involved in the neural interpretation of the VOR (Leonard, Simpson and Graf, 1988; Simpson, Graf and Leonard, 1981; Graf, Simpson and Leonard, 1988; Oyster, Takahashi and Collewijn, 1972). Furthermore, the direction of pull of the oculomotor muscles themselves also seem to be arranged in planes roughly aligning with the orientation of the semicircular canal planes (Graf and Simpson, 1981).

As the rather sophisticated properties of the VOR emerged, regarding its neural substrate as simple, independent connections between the canals and the oculomotor muscles became less and less tenable. Amongst the VOR's features are the flexibility of its gain (the speed of the eye movement associated with a given speed of head movement) (Malcolm and Jones, 1970) and direction (Gonshor and Jones, 1976) in response to changes in visual demands (see Berthoz and Melvill Jones, 1985, for a review). Even the coupling between canal stimulation and which muscles are activated are flexible. The changes in the VOR in response to changes in the coupling required during natural changes in the course of some animal's lives (Graf and Baker, 1983, 1990) or imposed in the laboratory (Peng, Baker and Peterson, 1994) have shown how vestibularly transduced movement around a given axis can be coupled to eye movements around various axes depending on circumstances. The VOR also depends on the eye's instantaneous position in the orbit (Fetter, Hain and Zee, 1986; Fetter, Misslisch, Sievering and Tweed, 1995; Misslisch, Tweed, Fetter, Dichgans and Vilis, 1997; Misslisch and Hess, 2000), vergence (Mok, Ro, Cadera, Crawford and Vilis, 1992), the distance of regard (Biguer and Prablanc, 1981; Paige, Telford, Seidman and Barnes, 1998), and the orientation of the rotation axis relative to the head and relative to gravity (Hess and Angelaki, 1997). Although the VOR has been thought of as a purely angular phenomenon, nearly all rotations of the head cause a physical translation of the eyes because the centre of rotation of the eyes are not on the axis of most head rotations. The VOR is sensitive to this (Biguer and Prablanc, 1981; Viirre and

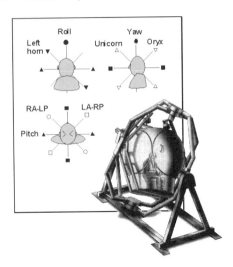

FIGURE 15.1. The axes used in this study seen in above, behind, and side views. The orientation of the vertical canals are shown diagrammatically in the above view. The insert is a picture of the Tübingen 3-D Chair. The symbols and labels by each axis identify them throughout this chapter.

Demer, 1995; Viirre, Milner, Tweed and Vilis, 1986) and is even responsive to extending the distance between the eyes and the axis of head rotation (Snyder and King, 1992).

Thus, we have a postmodern reflex to consider: one that is sensitive to its context. We show that a simple three-channel model can form a core onto which this flexible family of modifications can be appended.

15.2   The Three-Dimensional Performance of the Vestibulo-Ocular Reflex

We measured the VOR evoked in response to passive whole-body rotation about various axes in the dark to quantify its performance, especially its alignment with the stimulating axis. Deviations have been reported around pitch, roll, and yaw axes (Biguer and Prablanc, 1981; Viirre and Demer, 1995; Viirre, Milner, Tweed and Vilis, 1986) but have been less well investigated in intermediate axes (but see Solomon, Straumann and Zee, 1997). We then altered the response around one axis and looked again at the variation of response amongst axes. Our aim was to test a channel-based model inspired by the success of channel-based models in describing other aspects of sensory processing. A detailed description of the methods can be found on the enclosed CD-ROM and in Harris, Beykirch and Fetter, (2001).

The axes we used are shown in Figure 15.1 and correspond to the cardinal axes (roll, pitch, and yaw) and half-way in between. The axes between roll and yaw we refer to as unicorn (forward tipping) and oryx (backward tipping). The axes between roll and yaw correspond approximately to the planes of the vertical semicircular pairs: right anterior-left posterior (RALP) and left anterior-right posterior (LARP). The axes between the yaw and pitch we refer to as horns, but only the left horn was available in the equipment's configuration. We used sum-of-sines stimuli, comprising four frequencies from 0.032 to 0.258 Hz with an amplitude of ± 20 deg for each component. The sum-of-sines stimulus ensured that the motion was unpredictable. Three-dimensional eye position signals were recorded using a coil system.

15.2.1   VOR Evoked by Rotation About Axes in the Fronto-Parallel Plane

When rotation was about axes in the fronto-parallel plane (yaw, pitch, and left horn axes), compensatory eye movements were evoked that accurately aligned with the stimulating axis. The mean deviation from accurate alignment was only 2.0 ± 0.4 deg with the largest deviation being for rotation about the left horn axis (4.6 deg). The exact values are given in Table 1 on the enclosed CD.

15.2.2   VOR Evoked by Rotation About Axes in the Horizontal Plane

The arrangement of the slow-phase velocity of the VOR evoked by rotation around axes lying in the horizontal plane (roll, LARP, pitch, RALP; see Figure 15.1) are illustrated in Figure 15.2 (right). Sine waves at the stimulus frequency were fitted to each of the three components (yaw, pitch, and roll) of the slow phase of the response. The amplitude of each best-fit sine wave was then divided by the stimulus velocity and treated as a components of a three-dimensional vector (see CD-ROM for details).

Rotation around the LARP axis evoked a response that was significantly larger in the left eye than the right eye, and for RALP the response was larger in the right eye than the left. These differences were primarily due to differences in the pitch component of the response. When the head rolled counterclockwise (right ear up), evoking a clockwise slow-phase eye roll component, it also evoked an upward component in the left eye and a downward component the right eye, that is, a vertical divergence. When rotation was around the RALP axis, the vertical component was larger in the right eye than in the left, thus again producing a vertical divergence. Similarly, when rotation was about the LARP axis, the left eye's vertical component was larger than in the right, thus also leading to a vertical divergence.

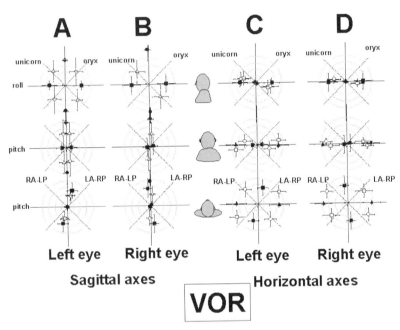

FIGURE 15.2. The orientation and gain of the slow phase of the vestibulo-ocular reflex for each eye elicited by rotation about axes in the sagittal plane (a,b) or horizontal plane (c,d) as seen from the side (top row), back (middle row), and above (bottom row). Each symbol (see Figure 15.1 for key) represents the tip of a vector obtained by fitting a sine wave to each of the roll, pitch, and yaw components of the desaccaded slow phase of the responses. The distance from the centre represents the magnitude of the response normalized to the stimulus speed. The circles are at 0.1 intervals. Also shown are the standard deviations in each dimension.

15.2.3   VOR Evoked by Rotation About Axes in the Sagittal Plane

The response to rotation around the roll, yaw, unicorn, and oryx axes in the sagittal plane of the head are shown on the left of Figure 15.2. The clockwise roll of the eyes evoked by the roll component of the roll, unicorn, and oryx rotations was associated with an upward motion of the left eye and a downward rotation of the right eye (Figures 15.2 a, b). The evoked vertical component had the effect of shifting the response vectors into a plane tilted from the sagittal by approximately 12 deg (Figures 15.2a, b, bottom row).

15.2.4   Uneven Gains Theory

Some deviations of the direction of VOR from its evoking stimuli have been noted previously (Crawford and Vilis, 1991) and have been suggested to arise because the different components of the response are generated by independent roll, pitch, and yaw systems with different gains (Crawford and Vilis, 1991; Fetter, Zee, Tweed and Koenig, 1994; Solomon, Straumann and Zee, 1997). The uneven gains theory explains the misalignments as following from the different gains of the response evoked by rotation around the cardinal axes of roll, pitch, and yaw. Since pitch has a higher gain than roll (pitch 50%; roll 30% in this study), the pitch component will dominate the response to a rotation with both pitch and roll components, pulling the overall response toward the pitch direction (Tweed, Sievering, Misslisch, Fetter, Zee and Koenig, 1994).

  However, the uneven gains theory always predicts deviations toward the higher gain cardinal axis. In fact responses to rotation about the unicorn and oryx axes (in between roll and yaw) were accurate (Figure 15.2a, b, top row) and not pulled toward the higher gain yaw axis. Thus, an uneven gains model based on the amplitude of yaw, roll, and pitch components measured separately is not adequate.

15.2.5   The VOR Compensates for Rotation and Translation of the Eyes Associated with Head Rotation

Interestingly, the "extra" components of the response that create the misalignments reported here — for example, the mysterious up/down components in response to rotation around axes in the sagittal plane — generally turn out to be in the appropriate direction to compensate for the translation of the eyes that necessarily accompanies head rotation. For example, because clockwise roll of the head shifts the left eye up and the right eye down, an asymmetrical pitch movement is appropriate (see also Jauregui-Renaud, Faldon, Gresty and Bronstein, 2001). Because the RALP and LARP rotation axes pass quite close to the left and right eyes, respectively (see illustration on the CD-ROM), rotation about these axes causes the more distant eye (left for LARP and right for RALP) to be translated vertically by a considerably larger amount than the eye that is closer to the axis. Clockwise slow-phase eye movement evoked by LARP stimulation was associated with more upward movement of the left eye than of the right, and the clockwise eye movement evoked by RALP stimulation was associated with more downward movement of the right eye than the left (see Figure 15.2). These puzzling directions are thus potentially useful. Attempting to compensate for translation when the two eyes are translating differently necessarily leads to ocular divergence. If the eyes diverge, they cannot maintain their gaze on an individual target. Thus, these findings lead to a similar conclusion as that of Groen, Bos and de Graaf (1999) who suggested that the goal of the compensatory eye movement system is more to stabilize the eye in space than to minimize retinal slip. A similar conclusion was reached by Harris and Mente (1996) who found the goal of adaptation of the VOR to be directed not toward minimizing retinal slip but to compensating for the best estimate of the movement of the head in space.

The fact that the gain and spatial characteristics of the roll response are not altered when the roll axis is aligned with gravity (Jauregui-Renaud, Faldon, Clarke, Bronstein and Gresty, 1996; Seidman and Leigh, 1989) and that the translation-related components reported here are found at low frequencies suggests that they do not result from detecting the translation directly. In fact, there is no sensory information available anyway to indicate the translation of the eyes due to head rotation in the dark. Although the evidence that these "extra" components are to compensate for translation is circumstantial, nevertheless, to call the deviations of the VOR for the stimulating axis a "misalignment" might be to malign a possible function. This argument is further developed in Harris, Beykirch and Fetter (2001). Our data are compatible with the notion that the vestibulo-ocular reflex represents an initial core of a compensation not only for angular rotation but also for the translation components of the eye associated with these rotations. Can such an apparently intelligent system be modelled by a simple three-channel system such as we initially proposed?

15.3   Modelling the VOR as a Simple Three-Channel System

We took the VOR evoked in each subject about each axis and fitted the whole set with a model that consisted of three channels. Each channel can be regarded as a vector with a gain. There is no intrinsic reference frame for the VOR — neither the structure of canals nor the Cartesian co-ordinate system of yaw, roll, and pitch have any special claims — and so the orientations of the channels as well as their gains were left as free parameters to be optimized by a programme (Sigmaplot 6) that chose values for the orientation and gain of the channels and compared the model's oculomotor performance with the data. The model's output was obtained by projecting each stimulus onto the three channels and then multiplying each projection by the channel's gain. The response was the vector sum of three channel's activities. This process is shown diagrammatically in Figure 15.3. The orientation and gains of the channels were systematically varied and the comparison of the simulation output with the recorded data repeated until a best fit was

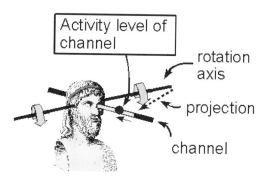

FIGURE 15.3. How the channels were estimated. Three channels (only one shown) were assigned random orientations and gains. The response of each of these hypothetical channels was then calculated by projecting the rotation axis onto the channel (gray bar). The projection was then multiplied by a gain, setting the channel's "activity level." The set of all three "activity levels" was then used as the vector co-ordinates determining the magnitude and direction of the overall response.

found.

The orientation and gains of the three channels that best fit the data are shown in Figure 15.4c. The output of this model in response to rotation around each of the eight axes used in this study are shown in Figure 15.4a, d. The model's output (open circles) reproduces all the major features of the pre-adapt data.

15.3.1   Minimum Gain Axes

From the distribution of responses to the tested axes, one can extrapolate the direction of the smallest response. This is an interesting direction in our endeavour to model the system with three channels because one channel must align with this "minimum gain axis." If the VOR is generated by the output of three channels, then no response can be less than the vector sum of the three outputs. That is, no one channel can have a gain less than the minimum response. Therefore if one channel has a gain significantly lower than the others (which it does), the location of the minimum response must correspond to when the stimulus is aligned with this channel. Although locating the minimum gain axis was not a method explicitly used to locate the channels, indeed its location (which can be seen from the distribution of gains in the horizontal plane shown in Figure 15.5), about 10 deg to the left of straight ahead in the left eye, does correspond to the location of one of the channels abstracted from the overall fit to all the data (see the above view of our proposed channels in the lower panel of Figure 15.4c). But it turns out that the minimum gain axis also fits into the other part of our story, the role of the VOR in compensating simultaneously for the translations, as well as the rotation, associated with head movement.

As described above, when the head rotates it causes both eyes to translate. The

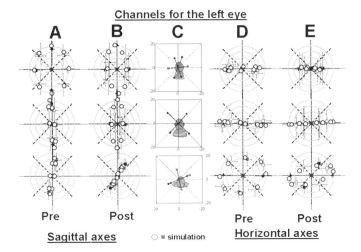

FIGURE 15.4. Simulating the VOR for the left eye. The orientation and relative length of our proposed channels are shown in the centre column (c) from side (top panel), back (middle panel), and above (bottom panel) views. For clarity, the results are divided into those obtained from rotation around axes in the sagittal plane (a, b) and the horizontal plane (d, e). The average data are shown with standard deviations as symbols (conventions as Figure 15.2). Superimposed on the data symbols are the output of the model for the pre- and post-adaptation conditions (large open circles). All the major features of the data are reproduced by the model both before and after adaptation.

only exception is when the axis of rotation of the head passes through the centre of rotation of the eye. Most natural rotations are about the atlanto-occipital joint which is substantially behind the eyes. Thus, the only natural head rotation axis that passes through an eye is around the line joining the atlanto-occipital joint (the approximate centre of head rotation) with the eye's centre of rotation. The minimum gain axis for each eye lines up closely with this line (see illustration on the CD-ROM). That is, the minimum gain axis for each eye is closely associated with the only naturally occurring head rotation axis not associated with any translation of that eye.

15.4   Testing the Model

To test the channel model with the gains and orientations obtained above, we adapted both the subjects and the model and compared the adapted model's predictions with the responses of our adapted subjects.

Our adaptation procedure was to use subject-stationary vision to lower the gain of the vestibulo-ocular reflex around the right-anterior left-posterior axis. Subjects sat in the apparatus and viewed text stuck on the wall of the 1 m radius sphere. They were instructed to read this text out loud during the adaptation pro-

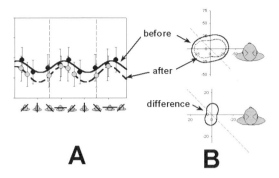

FIGURE 15.5. The effect of reducing the gain around the RALP axis. Comparison between before and after responses in the left eye evoked by rotations in the horizontal plane. The data are plotted in linear (a) and polar in above view (b) formats. Sine waves are plotted through the data to visualize the difference between the before and after responses. The difference between these two sine waves is plotted in polar co-ordinates in the lower panel of b. The major differences between before and after are found around the roll axis.

cedure. The sphere on which the letters were attached moved with the subject, thus requiring the VOR to be suppressed. Many experiments have indicated that this procedure results in an adaptive reduction of the gain of the VOR which outlasts the duration of stimulus and persists even when the VOR's performance is measured in darkness (see Berthoz and Melvill Jones, 1985, for a review). Subjects viewed the subject-stationary stimulus during 30 min of sinusoidal rotation around the right-anterior, left-posterior (RALP) axis at a frequency of 0.129 Hz with a peak velocity of 80 deg/s and therefore a maximum displacement of ±100 deg. Only the left eye's responses were measured after adaptation.

15.4.1   VOR After Adaptation Around the RALP Axis

Figure 15.4b and e shows the pattern of responses for axes in the sagittal and horizontal planes following adaptation. The roll, pitch, and yaw components of each of the average responses are also given in Table 1 on the CD-ROM. The VOR responses evoked by rotation about axes in the sagittal plane normally have a small pitch component which is not required to compensate for such rotations (see Section 15.2.3 and Figure 15.4b, bottom). This pitch component became substantially larger after adaptation. As a result, after adaptation the responses evoked by stimulation axes in the sagittal plane formed a plane tilted by 40 deg (Figure 15.4b, bottom). The response of the left eye to rotations in the horizontal plane were also substantially altered by our adaptation procedure (compare Figure 15.4d and e, bottom panels especially). These changes are examined in Figure 15.5 which plots a sine wave through the "before" and "after" responses. The difference is largely found around the roll axis where the response is much reduced and strongly deviated by the emerging vertical component after adaptation around the RALP axis.

15.4.2   Adapting the Model

Next the model was "adapted." To do this, first the hypothetical activity of each of the channels expected in response to the adapting stimulus was calculated using the gains and orientations of the channels obtained from the best fit to the pre-adapted data. For the RALP stimulation this produced activity in the ratio of 4 : 3 : 7.5 in the three channels The gains of each channel were then adjusted by an amount proportional to this activity to produce an "adapted" model (gains going from 0.51: 0.42: 0.34 to 0.33: 0.37: 0.12). The responses of the "adapted" model were then simulated by projecting each head rotation onto the "adapted" channels. The output of the adapted model (Figure 15.4b and e, open circles) showed an excellent fit to the adapted data, reproducing all the major features of the data described above.

This simple three-channel model thus predicts the normal 3-D response to rotation around various axes. The model's "deviations" also match the measured deviations in response both before and after adaptation. These "deviations" in turn match the needs of an ocular stabilizing system to deal with the movement of eyes displaced from the head rotation axis.

15.5   Significance of the Orientations of the Channels

The proposed channels (Figure 15.4c) do not correspond to canal or roll, pitch, and yaw co-ordinates. They are far from the planes of individual canals, with one close to roll and the others forming an X approximately in Listing's plane which can be expected to be tilted outward since the eyes were likely verged in the dark at a distance roughly corresponding to the screen (Mok, Ro, Cadera, Crawford and Vilis, 1992). A separate set of channels is needed for each eye because of this outward twist.

The orientations of the channels that best fit our data, while initially surprising, are in fact compatible with the known neural elements involved in the control of the VOR, but which have not been put together in this way before. One channel aligns fairly closely with an axis around which rotation is specially treated by the brain, namely, roll movement. Roll movement is processed separately (Crawford, Cadera and Vilis, 1991) and the response to roll rotation has some unique features (Anderson, 1981; Seidman, Paige and Leigh, 1997) suggesting that it is indeed a separate sub-system of the VOR mechanism. The other proposed channels lie roughly in Listing's plane, again not corresponding to any canal plane. In fact these two proposed channels lie close to the intersection of all the planes that bisect each possible pairing of canal planes. The channels are thus close as possible to being orthogonal to all the canal planes and are thus optimally configured to use information from all of them.

Listing's plane has a central role in the coding of eye movements, especially for saccades (Smith and Crawford, 2001). It has been implicated in the neural substrate of the VOR (Crawford and Vilis, 1991), but here we are making it explicitly the site of two of the VOR's channels or neural co-ordinates. The locations of the channels are thus constrained by lying in Listing's plane, using information from all canals, being roughly orthogonal to one another, and maintaining body symmetry.

Since the VOR is a phylogenetically ancient eye movement control system found even in animals with no saccadic system (Walls, 1962), the fact that the VOR uses a plane close to Listing's plane, defined in terms of optimal use of information from the canals, may be connected to the evolutionary origin of Listing's plane itself.

15.6   The VOR as a Postmodern Reflex With a Simple Mechanism

In summary, we can say that the VOR is a postmodern reflex. A central tenant of postmodernism is that nothing can escape its time — everything is context dependent. In this sense, can the concept of a simple, reflexive, input–output vestibulo-ocular system be maintained? The data presented in this chapter suggest that, despite the sophistication and flexibility of the compensatory oculomotor response to head rotation, the VOR can usefully be regarded as driven by a relatively simple core mechanism. Context-specific responses can then be built onto this base to make it a postmodern reflex.

Acknowledgements

Supported by the Deutsche Forschungsgemeinschaft (MF and KB), the Natural Sciences and Engineering Research Council of Canada (NSERC) (LRH), and the Centre for Research in Earth and Space Technology of Ontario, Canada (CRESTech) (LRH).

References

Anderson, J. H. (1981). Ocular torsion in the cat after lesions of the interstitial nucleus of Cajal. *Ann. New York Acad. Sci.*, 374: 865–871.

Berthoz, A. and Melvill Jones, G. (1985). *Adaptive Mechanisms in Gaze Control.* New York: Elsevier.

Biguer, B. and Prablanc, C. (1981). Modulation of the vestibulo-ocular reflex in eye-head orientation as a function of target distance in man. In A. F. Fuchs and W. Becker (eds.) *Progress in Oculomotor Research*, pp. 525-530. Amsterdam: Elsevier North Holland.

Blum, B. (1991). *Channels in the Visual Nervous System: Neurophysiology, Psychophysics and Models.* London: Freund.

Campbell, F. W. and Tegeder, R. W. (1991). A survey of channels and challenges, of information and meaning. In B. Blum (ed.) *Channels in the Visual Nervous System: Neurophysiology, Psychophysics and Models*, pp. 1-10. London: Freund.

Crawford, J. D., Cadera, W. and Vilis, T. (1991). Generation of torsional and vertical eye position signals by the interstitial nucleus of Cajal. *Science*, 252: 1551–1553.

Crawford, J. D. and Vilis, T. (1991). Axes of eye rotation and Listing's law during rotations of the head. *J. Neurophysiol.*, 65: 407–423.

Curthoys, I. S., Markham, C. H. and Blanks, R. H. I. (1975). *The Orientation of Middle and Inner Ear Structures in Cat and Man*. Los Angeles: UCLA Brain Information Service.

Fetter, M., Hain, T. C. and Zee, D. S. (1986). Influence of eye and head position on the vestibulo-ocular reflex. *Exp. Brain Res.*, 64: 208–216.

Fetter, M., Misslisch, H., Sievering, D. and Tweed, D. (1995). Effects of full-field visual input on the three-dimensional properties of the human vestibuloocular reflex. *J. Vestib. Res.-Equilib. Orientat.*, 5: 201–209.

Fetter, M., Zee, D. S., Tweed, D. and Koenig, E. (1994). Head position-dependent adjustment of the 3-dimensional human vestibuloocular reflex. *Acta Oto-Laryngologica*, 114: 473–478.

Gonshor, A. and Melvill Jones, G. M. (1976). Extreme vestibulo-ocular adaptation induced by prolonged optical reversal of vision. *J. Physiol. (Lond.)*, 256: 381–414.

Graf, W. and Baker, R. (1983). Adaptive changes of the vestibulo-ocular reflex in flatfish are achieved by reorganization of central nervous pathways. *Science*, 221: 777–779.

Graf, W. and Baker, R. (1990). Neuronal adaptation accompanying metamorphosis in the flatfish. *J. Neurobiol.*, 21: 1136–1152.

Graf, W. and Simpson, J. I. (1981). The relations between the semicircular canals, the optic axis and the extraocular muscles in lateral-eyed and frontal eyed animals. In A. Fuchs and W. Becker (Eds.) *Progress in Oculomotor Research*, pp. 411-420. New York: Elsevier.

Graf, W., Simpson, J. I. and Leonard, C. S. (1988). Spatial-organization of visual messages of the rabbit's cerebellar flocculus. 2. Complex and simple spike responses of Purkinje-cells. *J. Neurophysiol.*, 60: 2091–2121.

Groen, E., Bos, J. E. and de Graaf, B. (1999). Contribution of the otoliths to the human torsional vestibulo-ocular reflex. *J. Vestib. Res.-Equilib. Orientat.*, 9: 27–36.

Harris, L. R. (1997). The coding of self motion. In L. R. Harris and M. Jenkin (eds.) *Computational and Psychophysical Mechanisms of Visual Coding*, pp. 157-183. Cambridge: Cambridge University Press.

Harris, L. R., Beykirch, K. and Fetter, M. (2001). The visual consequences of deviations in the orientation of the axis of rotation of the human vestibulo-ocular reflex. *Vis. Res.*, 41: 3271–3281.

Harris, L. R. and Mente, P. (1996). When vision shifts the vestibulo-ocular reflex, what defines the goal? *J. Vestib. Res.-Equilib. Orientat.*, 6: S91.

Harris, L. R., Zikovitz, D. C. and Kopinska, A. (1998). Frames of reference with examples from driving and auditory localization. In L. R. Harris and M. Jenkin (Ed.) *Vision and Action*, pp. 66-81. Cambridge: Cambridge University Press.

Hess, B. J. M. and Angelaki, D. E. (1997). Kinematic principles of primate rotational vestibuloocular reflex. 2. Gravity-dependent modulation of primary eye position. *J. Neurophysiol.*, 78: 2203–2216.

Jauregui-Renaud, K., Faldon, M., Clarke, A., Bronstein, A. M. and Gresty, M. A. (1996). Skew deviation of the eyes in normal human-subjects induced by semicircular canal stimulation. *Neurosci. Let.*, 205: 135–137.

Jauregui-Renaud, K., Faldon, M., Gresty, M. A. and Bronstein, A. M. (2001) Horizontal ocular vergence and the three-dimensional response to whole-body roll. *Exp. Brain Res.*, 136: 79–92.

Leonard, C. S., Simpson, J. I. and Graf, W. (1988). Spatial-organization of visual messages of the rabbit's cerebellar flocculus. 1. Typology of inferior olive neurons of the dorsal cap of Kooy. *J. Neurophysiol.*, 60: 2073–2090.

Lorente de Nó, R. (1933) Vestibulo-ocular reflex arc. *Ann. Neurol. Psychiatry*, 30: 245–291.

Malcolm, R. and Melvill Jones, G. M. (1970). A quantitative study of vestibular adaptation in humans. *Acta Oto-Laryngologica*, 70: 126–135.

Misslisch, H. and Hess, B. J. M. (2000). Three-dimensional vestibuloocular reflex of the monkey: optimal retinal image stabilization versus Listing's law. *J. Neurophysiol.* 83: 3264–3276.

Misslisch, H., Tweed, D., Fetter, M., Dichgans, J. and Vilis, T. (1997). Interaction of smooth pursuit and vestibulo-ocular reflex in three dimensions. In M. Fetter, T. Haslwanter, H. Misslisch and D. Tweed (eds.), *Three-dimensional Kinematics of Eye, Head, and Limb Movements*, pp. 191–196. Amsterdam: Harwood.

Mok, D., Ro, A., Cadera, W., Crawford, J. D. and Vilis, T. (1992). Rotation of Listing's plane during vergence. *Vis. Res.*, 32: 2055–2064.

Müller, J. (1840). *Handbuch der Physiologie des Menschen.* Vol. II. Coblentz: Holscher.

Oyster, C. W., Takahashi, E. and Collewijn, H. (1972). Direction-selective retinal ganglion cells and control of optokinetic nystagmus in the rabbit. *Vis. Res.*, 12: 183-193.

Paige, G. D., Telford, L., Seidman, S. H. and Barnes, G. R. (1998). Human vestibuloocular reflex and its interactions with vision and fixation distance during linear and angular head movement. *J. Neurosci.*, 80: 2391–2404.

Peng, G. C. Y., Baker, J. F. and Peterson, B. W. (1994) Dynamics of directional plasticity in the human vertical vestibulo-ocular reflex. *J. Vest. Res.*, 4: 453–460.

Regan, D. M. (2000). *Human Perception of Objects.* Sunderland, MA: Sinauer.

Seidman, S. H. and Leigh, R. J. (1989). The human torsional vestibulo-ocular reflex during rotation about an earth-vertical axis. *Brain Res.*, 504: 264–268

Seidman, S. H., Paige, G. D. and Leigh, R. J. (1997). The VOR during head roll: distinctive properties related to visual demands. In M. Fetter, T. Haslwanter, H. Misslisch and D. Tweed (eds.) *Three-Dimensional Kinematics of Eye, Head, and Limb Movements*, pp. 171–176. Amsterdam: Harwood.

Shannon, C. E. and Weaver, W. (1949). *The Mathematical Theory of Communication.* Urbana, IL: University of Illinois Press.

Simpson, J. I., Graf, W. and Leonard, C. (1981). The coordinate system of visual climbing fibers to the flocculus. In A. F. Fuchs and W. Becker (eds) *Progress in Oculomotor Research*, pp. 475-484. North Holland: Elsevier.

Smith, M. A. and Crawford, J. D. (2001). Self-organizing task modules and explicit coordinate systems in a neural network model for 3-D saccades. *J. Comput. Neurosci.*, 10: 127–150.

Snyder, L. H. and King, W. M. (1992). Effect of viewing distance and location of the axis of head rotation on the monkey's vestibulo-ocular reflex. I. Eye movement responses. *J. Neurophysiol.* 67: 861–874.

Solomon, D., Straumann, D. and Zee, D. S. (1997). Three dimensional eye movements during vertical axis rotation: effects of visual suppression, orbital eye position and head position. In M. Fetter, T. Haslwanter, H. Misslisch, and D. Tweed (eds.) *Three-Dimensional Kinematics of Eye, Head, and Limb Movements*, pp. 197-208. Amsterdam: Harwood.

Szentágothai, J. (1950). The elementary vestibulo-ocular reflex arc. *J. Neurophysiol.*, 13: 395–407.

Tweed, D., Sievering, D., Misslisch, H., Fetter, M., Zee, D. and Koenig, E. (1994). Rotational kinematics of the human vestibuloocular reflex. 1. Gain matrices. *J. Neurophysiol.*, 72: 2467–2479.

Viirre, E. S. and Demer, J. L. (1995). Effect of target proximity on human vertical vestibuloocular reflex (VOR) during combined linear and angular-acceleration. *Invest. Ophthal. Vis. Sci.*, 36: S685.

Viirre, E. S., Milner, K., Tweed, D. and Vilis, T. (1986). A reexamination of the gain of the vestibuloocular reflex. *J. Neurophysiol.*, 56: 439–450.

Walls, G. L. (1962). The evolutionary history of eye movements. *Vis. Res.* 2: 69–79.

Zeki, S. M. (2001). Localization and globalization in conscious vision. *Ann. Rev. Neurosci.*, 24: 57–86.

16

Signal Processing in Vestibular Nuclei: Dissociating Sensory, Motor, and Cognitive Influences

Kathleen E. Cullen, Jefferson E. Roy, and Pierre A. Sylvestre

The vestibular sensory apparatus and associated vestibular nuclei are known to play an essential role in generating ocular and head stabilization reflexes and in controlling posture during our daily activities. In addition, vestibular sensory information is necessary for the performance of tasks that require accurate spatial orientation such as determining heading direction during self-motion and/or navigating through space in the absence of visual cues. Prior studies in head-restrained animals have shown that the afferent fibres of the vestibular nerve, as well as neurons within the vestibular nuclei to which they project, encode angular head-in-space velocity during passive whole-body rotation. However, to date, few studies have characterized the vestibular system during more natural behaviours, for example during self-generated head movements. It is well known that, in addition to direct inputs from vestibular afferents, the vestibular nuclei receive substantial projections from cortical, cerebellar, and other brainstem structures. Thus, given this diversity of inputs, it is natural to ask whether vestibular information is processed differentially by neurons in the vestibular nuclei in a manner that depends on the current behavioural goal.

In this study, we focused on the signal processing carried out by two classes of neurons in the monkey vestibular nuclei that receive direct input from the vestibular nerve: (i) position-vestibular-pause (PVP) neurons that mediate the vestibulo-ocular reflex (VOR), and (ii) vestibular-only (VO) neurons that mediate the vestibulo-collic reflex (VCR). The eye and head premotor commands generated by these vestibular reflexes, respectively, can be counterproductive during certain voluntary behaviours. The VOR functions to stabilize the visual axis in space by producing a compensatory eye movement of equal and opposite amplitude to the movement of the head. Thus, the eye movement response produced by the VOR does not lead to appropriate eye movements when gaze is redirected using a combination of eye and head movements; an intact VOR would generate an eye movement command in the direction opposite to the intended shift in gaze. Similarly, the VCR functions to stabilize the head in space, via activation of the neck musculature, during head motion. Thus, the stabilization response produced by the VCR would be counterproductive during active head movements.

We found that neither the VOR nor VCR are hardwired reflexes, but rather reflexes that are modulated in a behaviourally dependent manner. The head velocity signals carried by VOR interneurons (PVP neurons) are reduced when the goal is to redirect gaze in space. The head velocity signals carried by VCR interneurons (VO neurons) are reduced when the goal is to *move the head relative to the body*. To characterize the mechanisms that underlie this differential processing of vestibular inputs, PVP and VO neurons were tested during passive whole-body rotation, passive rotation of the head-on-body, active head-on-body movements, as well as during a task in which a monkey actively "drove" both its head and body together in space. We show that neither the activation of neck proprioceptive information nor the fact that the monkey has knowledge of its own motion influences the processing of vestibular information. Rather, we propose that VOR and VCR pathways use efference copies of gaze/eye and neck movement commands, respectively, for the differential processing of vestibular information.

We conclude that the vestibular nuclei do not reliably encode head-in-space velocity during the active head movements made during gaze shifts and gaze pursuit. We discuss the implications of this differential processing with respect to higher order vestibular functions such as the computation of spatial orientation and the perception of self-motion.

16.1   Introduction

The vestibular system is associated classically with detecting the motion of the head-in- space to generate the reflexes that are crucial for our daily activities, such as stabilizing the visual axis (gaze) and maintaining head and body posture. Angular head velocity is detected by vestibular hair cells that are located within the semicircular canals of the inner ear labyrinth. The afferent fibres of the vestibular nerve (VIIIth cranial nerve) project from the labyrinth directly to the vestibular nuclei of the brainstem. Bilateral loss of labyrinth function causes (i) unwanted motion of the visual world on the retina with head movements (i.e., lack of gaze stability), (ii) an inability to keep the head and/or body erect during common activities, and (iii) an inability to perform tasks that require accurate spatial orientation such as determining the direction of body motion or navigating through space in the absence of visual cues.

Clinically, vestibular function is commonly tested by passively rotating the patient in a chair and measuring the resultant eye movement response (vestibulo-ocular reflex: VOR). In this context, the VOR is considered to be a stereotyped reflex that effectively *stabilizes* gaze by moving the eye in the opposite direction to the applied head motion. However, the eye movement response produced by the VOR can be counterproductive during more natural behaviours. For example, primates and humans commonly use a combination of eye and head movements to redirect gaze from one target to another. During this orienting behaviour, the

VOR would produce an eye movement command in the direction opposite to that of the intended shift in gaze.

The vestibular system also plays a critical role in controlling head and body posture, and in computing spatial orientation. For example, head velocity-related information from the vestibular system is used to stabilize the head relative to the body via the vestibulo-collic reflex (VCR). As for the VOR, the stabilization response produced by the VCR can be counterproductive during voluntary behaviours. The VCR would oppose the intended head motion during voluntary eye-head shifts and tracking. Thus, it would be logical to modulate the VCR pathways in a behaviourally dependent manner. Given that most of our knowledge of vestibulo-spinal mechanisms is based on studies that applied passive head rotations to anesthetized and decerebrate animals, it is important to compare the processing of vestibular signals by these reflex pathways during active and passive head rotations. The activity of vestibular nuclei neurons has been well characterized in head-restrained monkeys during passive whole-body rotations. It has been demonstrated that several distinct classes of neurons exist within the vestibular nuclei and neighbouring nucleus prepositus hypoglossi (Chubb et al., 1984; Cullen and McCrea, 1993; Cullen et al., 1991; Fuchs and Kimm, 1975; Keller and Daniels, 1975; Lisberger and Miles, 1980; McFarland and Fuchs, 1992; Miles, 1974; Scudder and Fuchs, 1992; Tomlinson and Robinson, 1984). Of these, two classes of neurons receive strong direct projections from the vestibular nerve and are thought to play an important role in generating the vestibular reflexes described above: (i) position-vestibular-pause (PVP) neurons mediate the rotational VOR (Cullen and McCrea, 1993; McCrea et al., 1987; Scudder and Fuchs, 1992), and (ii) vestibular-only (VO) neurons mediate the VCR (Boyle, 1993; Boyle et al., 1996; McCrea et al., 1999). Over the normal range of head movements that we generate during our daily activities, the vestibular afferents, as well as the vestibular nuclei neurons to which they project, are generally thought to encode angular head velocity. Furthermore, the vestibular nuclei receive substantial projections from cortical, cerebellar, and other brainstem structures. Thus, given this diversity of inputs, the question arises: Are the responses of vestibular nuclei neuron to head velocity modified by these additional inputs during naturally occurring behaviours?

Here we have investigated, in the alert rhesus monkey, whether PVP and VO neurons differentially process vestibular information during self-generated versus passively applied head rotations. We have found that the head velocity sensitivity of PVP neurons was attenuated when the animal's behavioural goal was to redirect its gaze. In contrast, the head velocity sensitivity of VO neurons was attenuated only when the animal voluntarily moved its head relative to its body. While our findings are consistent with the function of these neurons in mediating the VOR and VCR, respectively, they also raise important questions with respect to other functions of the vestibular system such as our perception of self-motion in space. For example, the attenuation of head velocity signals in the vestibular nuclei could lead to severe disorientation during gaze shifts if the vestibular nuclei were the only route by which vestibular information reached structures involved in the further processing of sensory information. We propose that parallel vestibular afferent projections to the cerebellum are used in combination with the selectively modulated activity of vestibular nuclei neurons to generate an internal representation of current spatial orientation.

16.2   Methods

16.2.1   Surgical Procedures

Three monkeys (*macaca mulatta*) were prepared for chronic extracellular recording. The surgical preparation and extracellular recording techniques utilized have been recently described (Sylvestre and Cullen, 1999). Briefly, a stainless steel recording chamber and a stainless steel post, used to restrain the animal's head, were attached to the animal's skull using dental acrylic and cortical screws in a sterile surgical procedure. During the same procedure, a 19 mm in diameter eye coil (three loops of Teflon-coated stainless steel wire) was implanted in the right eye behind the conjunctiva to allow for the measurement of eye position using the magnetic search coil technique (Fuchs and Robinson, 1966). Animals were given two weeks to recover from the surgery before any experiments were performed. The experimental protocols were approved by the McGill University Animal Care Committee and complied with the guidelines of the Canadian Council on Animal Care.

16.2.2   Experimental Paradigms

During experimental sessions, the monkeys were seated in a primate chair that was fixed to the suprastructure of a vestibular turntable. Gaze and head positions were measured using the magnetic search coil technique (Fuchs and Robinson, 1966), and extracellular single-unit activity was recorded using enamel insulated tungsten microelectrodes (7–10 MΩ impedance, Frederick-Haer) as has been described elsewhere (Sylvestre and Cullen, 1999). The torque produced by the monkey against the head restraint was measured using a reaction torque transducer (Sensotec).

   The activity of each neuron was recorded initially in alert monkeys whose heads were restrained. The animals were given a fruit juice reward for tracking a laser target that was projected on a cylindrical screen. The horizontal and vertical positions of the laser were controlled using a pair of computer-controlled galvanometers. Neuronal responses were recorded during eye movements made to targets that were: (i) stepped between different horizontal positions over a range of ±30 deg, and (ii) moved sinusoidally (0.5 Hz, 80 deg/s peak velocity) in the horizontal plane. Neuronal sensitivities to head velocity were tested by passively rotating monkeys about an earth vertical axis (0.5 Hz, 80 deg/s peak velocity) in the dark (VORd) and while they cancelled their vestibulo-ocular reflex by fixating a target that moved with the vestibular turntable (VORc). The motion of the target and turntable, as well as the on-line data displays and data acquisition, were controlled by a UNIX-based real-time data acquisition system (REX; Hayes et al., 1982).

After each neuron had been fully characterized in the head-restrained condition, we slowly and carefully released the monkey's head. The monkey was then able to voluntarily rotate its head through the natural range of motion in the yaw (horizontal), pitch (vertical), and roll (torsional) axes. Neuronal activity was recorded during combined eye-head gaze shifts (15 to 65 degrees in amplitude) and combined eye-head gaze pursuit of a sinusoidal target. In order to confirm that isolation of the same neuron was maintained after the head-restrained to head-unrestrained transition, resting discharge rates were compared. In addition, the VORd and VORc paradigms were repeated for the majority of neurons following head release; for all the neurons included in this report, the neuronal modulation was found to be comparable to that observed during the initial head-restrained characterization. Because monkeys frequently generated voluntary head-on-body movements during VORd and VORc in the head-unrestrained condition, we also utilized these paradigms to simultaneously evaluate the neuronal sensitivities to the passive and to the active component of head-in-space motion.

To activate neck proprioceptive inputs passively, two different paradigm were used. First, the monkey's head was rotated manually on its stationary body. Second, the monkey's head was held stationary relative to the earth while its body was rotated passively. In addition, neuronal responses were tested in a "driving paradigm," in which head-restrained monkeys were trained to operate a steering wheel that controlled the rotation of the turntable on which they were seated. As the monkey drove, it controlled both the initiation and the velocity of the turntable rotation.

16.2.3   Analysis of Neuron Discharges

A Gaussian function was convolved with the spike train (standard deviation of 5 ms for saccades and gaze shifts and 10 ms for the remainder of the paradigms) to generate the spike density profile of the neuron (Cullen et al., 1996). Gaze and head position signals were filtered digitally at 125 Hz, and eye position was calculated from the difference between filtered gaze and head position signals. Position signals were differentiated digitally to produce velocity signals. Saccade and gaze shift onsets and offsets were determined using a 20 deg/s gaze velocity criterion.

Analysis of neuronal eye and head movement sensitivities were performed using custom algorithms (Matlab, Mathworks, Inc.). A least-squared regression analysis was used to determine each unit's eye position sensitivity and resting discharge (bias) during periods of steady fixation, and its eye position and velocity sensitivities during intervals of saccade-free smooth pursuit. A comparable analysis was used to determine each unit's eye position sensitivity (spikes/s/deg), phase shift relative to head velocity (deg), resting discharge (bias, spikes/s), and head velocity sensitivity (spikes/s/deg/s) during head-restrained VORd and VORc. We then determined if the results of our head-restrained whole-body rotation analysis could be used to predict the activity of neurons during (i) active head-on-body motion, (ii) passive head-on-body and body-under-head rotations, and (iii) combined active head and body motion. Details of this analysis have been described elsewhere (Roy and Cullen, 1998). Statistical significance was determined using a paired Student's t test.

This chapter will focus on the neural encoding of active versus passive head motion by two specific classes of neurons in the vestibular nuclei that receive direct input from the vestibular nerve: (i) position-vestibular-pause (PVP) neurons that mediate the vestibulo-ocular reflex (VOR), and (ii) vestibular-only (VO) neurons that mediate the vestibulo-collic reflex (VCR).

16.3   VOR Pathways: Active Versus Passive Head Motion

16.3.1   The Direct VOR Pathway

PVP neurons are thought to constitute most of the intermediate leg of the direct VOR pathway; they receive a strong monosynaptic connection from the ipsilateral semicircular canal afferents and, in turn, project directly to the extraocular motoneurons (Cullen et al., 1991; Cullen and McCrea, 1993; McCrea et al., 1987; Scudder and Fuchs, 1992). With respect to the horizontal VOR, the vast majority of PVP neurons send an excitatory projection to the motoneurons of the contralateral abducens nucleus (ABN), while a minority sends inhibitory projections to the motoneurons of the ipsilateral ABN (Figure 16.1a). PVP neurons derive their name from the signals they carry during head-restrained paradigms. They are sensitive to contralateral eye position, ipsilateral head velocity, and stop firing (i.e., pause) during ipsilaterally directed saccades and vestibular quick phases. The activity of a typical PVP neuron is illustrated in Figure 16.1b. During passive whole-body rotation in the dark (Figure 16.1b, left panel), this neuron increased its discharge in relation to ipsilateral head rotation during slow phase nystagmus, and paused during ipsilaterally directed vestibular quick phases (downward arrows).

16.3.2   The VOR During Gaze Redirection: VOR Cancellation and Gaze Pursuit

In order to dissociate neuronal sensitivities of PVP neurons to vestibular stimulation from their eye-movement related responses, vestibular physiologists have utilized a paradigm in which the monkey cancels its VOR by tracking a target that moves with the head. The example neuron was typical in that its head velocity signal was attenuated, on average, by 30% during the cancellation paradigm (Figure 16.1b, middle panel). It is important to note that during the VOR cancellation paradigm, the behavioural goal is different from that during the VOR paradigm: during VOR cancellation, the animal's goal is to redirect rather than stabilize its

A. The Direct VOR Pathways

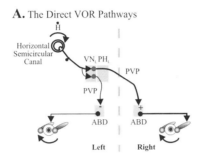

B. Position-vestibular-pause neuron

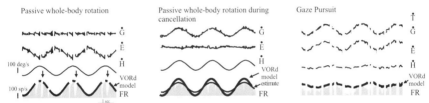

FIGURE 16.1. (a) Schematic diagram of the direct VOR pathways during rightward head rotation. Note the predominant excitatory projection from PVP neurons in the vestibular nucleus (VN) / nucleus prepositus hypoglossi (PH) complex to the contralateral abducens nucleus (ABD), and the weaker inhibitory projection to the ipsilateral abducens nucleus. (b) Left panel: Discharge of a typical PVP neuron during passive whole-body rotation. The thick solid trace superimposed on the firing rate is a model fit estimated from the data. Middle panel: Activity of the same PVP neuron during VOR cancellation. The thick solid trace superimposed on the firing rate is a prediction based on the model estimated during passive whole-body rotation (left panel). The VORc model fit (thin trace) was estimated from the data. Right panel: Activity of the example PVP neuron during eye-head gaze pursuit. The model fit is a prediction based on the model estimated during passive whole-body rotation (left panel). Abbreviations: $\dot{G}$, $\dot{E}$, $\dot{H}$, $\dot{T}$ gaze, eye, head and target velocity, respectively; FR, firing rate.

gaze in space. Accordingly, the attenuation of neuronal modulation is useful since the eye movement command generated by the direct VOR pathways would be in the opposite direction to the intended change in gaze. However, the VOR cancellation paradigm is artificial in that the head is restrained and head motion is externally applied. When the head is not restrained, primates commonly use a combination of head and eye motion to pursue a moving target of interest (i.e., gaze pursuit). We recorded from the same population of PVP neurons while monkeys generated voluntary head movements to pursue a sinusoidally moving target, and found that neuronal head velocity sensitivities were dramatically reduced (average: 56% attenuation; Figure 16.1b, right panel). Taken together, these results suggest that the gain of the direct VOR pathways is attenuated during active as well as passive head rotations if the behavioural goal is to redirect gaze in the direction of the head movement.

16.3.3   The VOR During Gaze Redirection: Gaze Shifts

To redirect their visual axis rapidly to a new stationary target in space, humans and monkeys will naturally generate coordinated eye-head movements that have been termed gaze shifts (humans: André-Deshays et al., 1988; Barnes, 1979; Guitton and Volle, 1987; Pélisson et al., 1988; Zangemeister and Stark, 1982a, b; and monkeys: Bizzi et al., 1971; Dichgans et al., 1973; Lanman et al., 1978; Morasso et al., 1973; Tomlinson and Bahra, 1986a, b; Tomlinson, 1990). As during gaze pursuit, an intact VOR would be counterproductive during gaze shifts: it would generate an eye movement signal in the direction opposite to that of the intended change in gaze.

To determine whether the modulation of the direct VOR pathways is attenuated during voluntary gaze shifts, we recorded the activity of PVP neurons. The head-velocity related discharges of PVP neurons were consistently attenuated (Figure 16.2a, intervals denoted by filled arrows) relative to those observed during passive whole-body rotation. Note, a model prediction based on the neuron's response during passive whole-body rotation is superimposed on the neuron's response during gaze shifts to facilitate a direct comparison (VORd model, heavy trace). Furthermore, the amount of neuronal attenuation increased as a function of gaze shift amplitude (Figure 16.2b). This neurophysiological trend mirrored the results of prior behavioural studies which have shown that the gain of the VOR decreases as a function of increasing gaze shift amplitude (Pélisson et al., 1988; Tabak et al., 1996; Tomlinson, 1990). Accordingly, we have concluded that the amplitude-dependent reduction of the head velocity signal carried by direct VOR pathways is responsible for the amplitude-dependent decrease in behavioural VOR gain observed during gaze shifts (Roy and Cullen, 1998). In contrast to the attenuation observed during gaze shifts, the head velocity sensitivity of PVP neurons recovered immediately once gaze was stable, although the monkey's head was still moving (Figure 16.2b, intervals denoted by open arrows). Thus, the VOR is not a hard-wired reflex, but rather a reflex that is modulated in a manner that depends on current gaze strategy: it is attenuated when the be-

A. Gaze shifts

B. Population

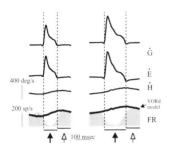
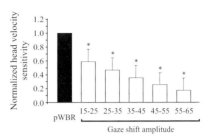

FIGURE 16.2. (a) Activity of a typical PVP neuron during eye-head gaze shifts. The VORd model fit is a prediction based on the model estimated during passive whole-body rotation. The filled arrows indicated the gaze shift intervals utilized for analysis, and the open arrows indicated the postgaze shift intervals. (b) The head velocity sensitivity of our sample of PVP neurons was always attenuated significantly during gaze shifts versus passive whole-body rotations. The degree of the attenuation was proportional to the amplitude of the gaze shift. Note that the head velocity sensitivities estimated during gaze shifts were normalized for each neuron to the value estimated during passive whole-body rotation.

havioural goal is to redirect gaze, and it is fully functional when the behavioural goal is to stabilize gaze.

16.4   Vestibulo-Spinal Pathways: Active Versus Passive Head Motion

16.4.1   Head-Restrained Activity and Projections of Vestibular-Only Neurons

A distinct population of vestibular nuclei neurons, called vestibular-only (VO) neurons, are known to receive direct monosynaptic projections from vestibular nerve afferents and are not sensitive to eye movements (Cullen and McCrea, 1993; Scudder and Fuchs, 1992). Like PVP neurons, which also receive direct inputs from the vestibular afferents, the activity of VO neurons increases for ipsilaterally directed head rotations. However, unlike PVP neurons, VO neurons do not play a role in mediating the VOR. Instead, VO neurons project to the cervical spinal cord, and are thought to mediate the VCR pathway (Figure 16.3a; Boyle, 1993; Boyle et al., 1996; McCrea et al., 1999). In addition, VO neurons project to the nodulus and uvula subdivisions of the cerebellum (Voogd et al., 1996; Wearne et al., 1998; Wylie et al., 1994). Recent lesion experiments have implicated these cerebellar structures in the control and coordination of head and body posture (Reisine and Raphan, 1992; Wearne et al., 1998; Yokota et al., 1992).

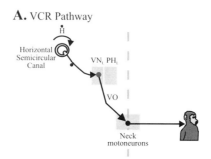

A. VCR Pathway

B. Vestibular-only neuron

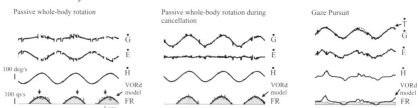

FIGURE 16.3. (a) Schematic diagram of the direct VCR pathway during rightward head rotation. Note that VO neurons project bilaterally to spinal motoneurons. (b) Left panel: Discharge of a typical VO neuron during passive whole-body rotation. The thick solid trace superimposed on the firing rate is a VORd model fit estimated from the data. Middle panel: Activity of the same VO neuron during VOR cancellation. The thick solid trace superimposed on the firing rate is a prediction based on the model estimated during passive whole-body rotation (see left panel). Right panel: Activity of the example VO neuron during eye-head gaze pursuit. The model fit is a prediction based on the model estimated during passive whole-body rotation (left panel).

16.4.2   Vestibular-Only Neurons: Active Gaze Pursuit and Gaze Shifts

To test whether the modulation of VO neuron activity itself plays an integral role in the control of head/body posture during natural self-generated behaviours, we recorded VO neuron responses during passive versus active head motion. An example VO neuron is illustrated in Figure 16.3 during VOR in the dark (Figure 16.3b, left panel.), VOR cancellation (Figure 16.3b: middle panel) and voluntary head-unrestrained gaze pursuit (Figure 16.3b, right panel). This VO neuron was typical in that it carried an identical head velocity signal during passive whole-body rotation in the dark and during VOR cancellation (Figure 16.3b, compare left and middle panels). This characteristic of VO neurons differentiated them from PVP neurons. However, during gaze pursuit, the head velocity signal of VO neurons was greatly attenuated with respect to that carried during passive whole-body rotation in the dark (VORd model; Figure 16.3b, compare left and right panels). For the example neuron, the head velocity signal actually changed direction. Across our sample of VO neurons, the head velocity modulation was attenuated on average by 60% during gaze pursuit. This significant decrease was much greater than that observed in PVP neurons.

The head velocity-related modulation of VO neurons was also recorded during gaze shifts (Figure 16.4a), and was found to be dramatically attenuated compared to that evoked by passive whole-body rotations (VORd model, Figure 16.4a; Roy and Cullen, 2001a; McCrea et al., 1999). The extent of the attenuation was similar across all gaze shift amplitudes (Figure 16.4B). This attenuation was also comparable during all active head movements regardless of whether the animal was stabilizing its gaze or redirecting its gaze to a new point in space (Cullen and Roy, 1998; Roy and Cullen, 2001a). For example, the head movement sensitivity of VO neurons was attenuated not only during combined eye-head gaze shifts (Figure 16.4a, filled arrows), but also immediately after gaze shifts, when gaze was stable in space but the head was still moving (Figure 16.4a, open arrows). Thus, the head velocity-related responses of VO neurons were significantly reduced for active head-on-body movements during redirection (i.e., gaze shifts and pursuit) as well as during gaze stabilization. In contrast, the head-velocity related activity of VOR interneurons (i.e., PVP neurons) was attenuated *only* while gaze was being redirected in space (Figure 16.2a, open versus filled arrows). On a neuron by neuron basis, both VO and PVP neurons showed attenuation levels that were comparable during gaze shifts and gaze pursuit with matching head movements (Figure 16.5a and b, respectively). However, for larger gaze shifts with faster head movements, the attenuation of PVP neurons was greater than during gaze pursuit, while it remained comparable for VO neurons. Hence, during active head movements, there are differences between the manner in which VO and PVP neurons process vestibular information.

A. Gaze shifts

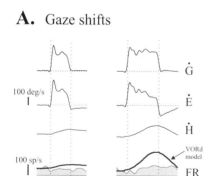

B. Population

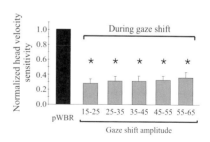

FIGURE 16.4. (a) Activity of a typical VO neuron during eye-head gaze shifts. The VORd model is a prediction based on the head velocity sensitivity estimated during passive whole-body rotation. (b) The head velocity sensitivity of our sample of VO neurons was always significantly attenuated during gaze shifts versus passive whole-body rotation. In contrast to PVP neurons, the degree of the attenuation was constant across all gaze shift amplitudes.

A. VO neurons

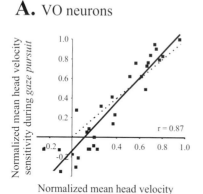

B. PVP neurons

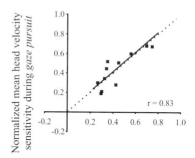

FIGURE 16.5. (a) VO neurons, and (b) PVP neurons, were similarly attenuated during combined eye-head gaze shifts and gaze pursuit. Each data point corresponds to a neuron's normalized head velocity sensitivity during the two conditions. Note that the head velocities during the gaze shifts selected for this analysis were matched to those generated during gaze pursuit. The dotted line has a unity slope and a zero y-intercept. The solid curves are regression lines through the data, and the correlation coefficients are indicated.

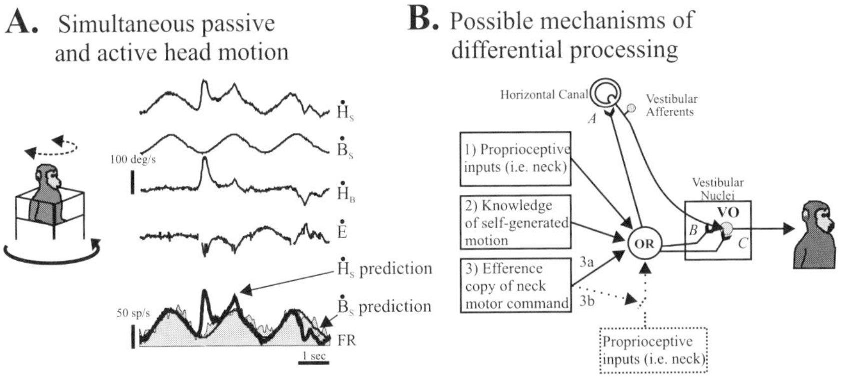

A. Simultaneous passive and active head motion

B. Possible mechanisms of differential processing

FIGURE 16.6. (a) Discharge patterns of a VO neuron during passive whole-body rotation in the head-unrestrained condition. The dotted arrow in the cartoon indicates that the animal was free to generate voluntary head movements. The $\dot{H}_S$ prediction is based on the head-in-space velocity (i.e., active + passive head velocity), and the $\dot{B}_S$ prediction is based on the body-in-space velocity (i.e., the passive whole-body rotation). (b) Possible mechanisms for the suppression of the head velocity sensitivity of VO neurons during active head-on-body movements. Note that the mechanisms may apply at the level of the primary vestibular afferents (site A), and/or pre- or post-synaptically (sites B and C, respectively) at the level of VO neurons in the vestibular nucleus. *Abbreviations:* $\dot{H}_S$, $\dot{H}_B$, $\dot{B}_S$, head-in-space, head-on-body and body-in-space velocities, respectively.

16.4.3   Vestibular-Only Neurons: Differential Encoding of Active Versus Passive Head Motion

One possible explanation for the suppression of head-velocity signals on VO neurons during active head movements is that during self-generated head motion, the vestibular afferent input to the VO neurons is cancelled out in its entirety. To investigate this possibility, we recorded from VO neurons during passive whole-body rotation while the animal was head-unrestrained. Data from an example neuron is illustrated in Figure 16.6a. The top two traces illustrate head velocity in space and the applied chair velocity, respectively, and the third trace illustrates the difference of these two traces, namely head velocity relative to the body (i.e., the active component of head motion). The light and heavy traces superimposed on the firing rate represent the predicted discharge of the neuron based on the head velocity signal in response to the body velocity and head velocity in space, respectively. It is clear that the neuron reliably encoded the passive head velocity signal that was generated by the chair rotation, indicating that the vestibular afferent input to the VO neurons was not simply gated out. Remarkably, the neuron did not respond to the active movements made by the monkey during this paradigm; the observed attenuation was specific to the self-generated component of head-in-space motion.

16.4.4   Vestibular-Only Neurons: Mechanisms of Attenuation

The vestibular nuclei receive inputs from multiple sources, as well as direct projections from the vestibular nerve. Inputs from neck proprioceptors (Anastasopoulos and Mergner, 1982; Boyle and Pompeiano, 1980), cortical structures (reviewed in Fukishima, 1997; Wilson et al., 1999) and cerebellar structures (reviewed in Voogd et al., 1996) converge on the vestibular nuclei. Accordingly, several possible mechanisms could be used to attenuate the responses of VO neurons to voluntary head-on-body motion made during gaze shifts and gaze pursuit at the level of the vestibular nuclei (Figure 16.6b). For example, a VO neuron's response to active head-on-body motion could be attenuated by (i) inhibitory inputs from neck proprioceptors, (ii) a signal of cortical origin representing the monkey's knowledge of its self-generated motion, and/or (iii) an efference copy of the motor behaviour (i.e., a neck movement) that is generated. We investigated the relative influence of each of these inputs.

16.4.5   Neck Proprioceptive Inputs

Activation of neck muscle spindle afferents can alter the activity of vestibular nuclei neurons in decerebrate animals (Anastasopoulos and Mergner, 1982; Boyle and Pompeiano, 1980) via a disynaptic pathway mediated by the central cervical nucleus (Sato et al., 1997). To investigate whether neck proprioception influences the discharges of VO neurons in alert animals, we carried out two tests. First, we passively rotated the monkey's head on its neck while recording neuronal activity. Passive head rotations were applied which (i) mimicked those generated during voluntary eye-head gaze shifts (frequency content > 2 Hz), and (ii) were in the frequency range of the rotations used in prior studies of decerebrate animals (0.05–0.5Hz). In all cases, we found that the passive activation of neck proprioceptors did not influence the head velocity sensitivity of VO neurons. A typical example is shown in Figure 16.7a, where the neuron's response was predicted well by its sensitivity to passive whole-body rotation (VORd model, heavy trace). In a second test, the monkey's body was rotated relative to its stationary head. During rotation of the body under the head, the example neuron (Figure 16.7b) was typical in that its discharge was not related to movement of the body (i.e., neck). Thus, in alert rhesus monkeys, the passive activation of neck proprioceptors had a negligible influence on VO neuron activity either in the presence (Figure 16.7a) or absence (Figure 16.7b) of vestibular stimulation.

16.4.6   The Role of Monkey's Knowledge of Its Self-Generated Motion

Cortical areas that have been implicated in spatial orientation, navigation, gaze and posture control also project to the vestibular nuclei (reviewed in Fukushima, 1997). Accordingly, we have investigated whether the monkey's cognitive percept of its self-generated head motion plays a role in modulating VO neuron responses

A. Passive head-on-body rotation

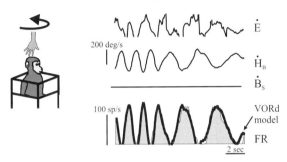

B. Passive body-under-head rotation

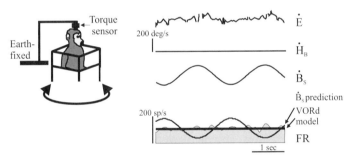

FIGURE 16.7. (a) Activity of a VO neuron during passive rotations of the head-on-body. The model fit is a prediction based on the head velocity sensitivity estimated during passive whole-body rotation. (b) Activity of the same neuron during passive rotations of the body under a stationary head. The VORd model is a prediction based on the neuron's activity estimated during passive whole-body rotations. The $\dot{B}_S$ prediction is based on the body-in-space velocity (which is equivalent to the neck proprioceptors activation). The torque generated by the animal against the head restraint was measured (inset), and only segments where the animal did not resist the passive movements were analyzed.

(Figure 16.6b). We devised a novel paradigm in which head-restrained monkeys drove their own head and body motion together in space by rotating a steering wheel connected to the controller of the vestibular turntable (Figure 16.8a, inset). Following training, monkeys consistently drove the chair to align their head and body position with a moving laser target. Because monkeys were head-restrained, the self-generated head-in-space movements did not involve the neck musculature but rather, in this case, the shoulder and limb musculature (to generate an arm movement to turn the steering wheel). All neurons tested responded robustly to the head motion actively generated during this task; no significant attenuation of their head velocity-related modulation was observed. Thus, a signal reflecting the monkey's knowledge of its own self-generated motion is *not used* to attenuate the modulation of VO neurons.

16.4.7   The Influence of Neck Motor Commands

To determine whether an efference copy of the motor command to the neck musculature has an inhibitory influence on VO neurons during active gaze pursuit and gaze shifts, we first recorded from neurons during a task in which the head was restrained and the neck musculature was activated. During head-restrained saccades, neck activation (measured via EMG and/or a torque sensor) is strongly coupled with eye movements (human: André-Deshays et al., 1991, monkey: Bizzi et al., 1971; Lestienne et al., 1984). We observed that monkeys generate small neck torque for saccades to laser targets (0.1 Nm), but can generate up to a 50 fold increase in torque for saccades to food targets (5 Nm). Thus, we recorded from VO neurons and measured neck torque during large (40–60 degs) head-restrained saccades made to food targets. In this task, the vestibular sensory apparatus is not activated, and the net activation of neck proprioceptors should be reduced relative to tasks where the head is free to move (Richmond and Abrahams, 1979). Accordingly, if neck efference copy signals directly influenced VO neuron activity, we predicted that the effect specific to this input should be evident: we would expect to see a substantial decrease in VO neuron firing rates during ipsilaterally directed saccades that would be dynamically coupled to the generation of significant neck torque. However, for all of the neurons tested, neuronal firing rates were unaffected during this task. Neuronal firing rates remained constant regardless of whether or not the monkey was generating neck torque (Figure 16.8b; mean firing rate $= 79 \pm 16.9$ sp/s vs. 79 ± 117.5, for low and high torque, respectively; $P > 0.3$; Roy and Cullen, 2001b). Similar results were obtained in a paradigm where the monkeys' bodies were rotated passively while their heads were held earth-stationary. In this task, neck proprioceptors were activated passively while the vestibular sensory apparatus was not. We presented food targets to entice the monkeys to generate neck motor commands either with or against the passive neck rotation. For the population of VO neurons, the firing rates were not affected when the monkeys generated large voluntary neck motor commands (mean firing rate $= 74 \pm 135.1$ for low torque ($\sim$0.2 Nm), and 72 ± 132.2 for high torque ($\sim$2.5 Nm)).

A. Active head/body motion

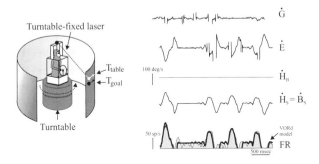

B. Is an efference copy responsible?

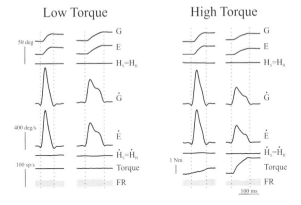

FIGURE 16.8. (a) Discharge patterns of a VO neuron when the monkey voluntarily drove the vestibular turntable to reorient to a target (see inset). The model fit is a prediction based on the head velocity sensitivity estimated during passive whole-body rotation. (b) Comparison of a VO neuron activity during head-restrained saccades that were accompanied by low torque levels (*left panel*) versus saccades to food targets accompanied by high torque levels (*right panel*). Abbreviations: G, E, H_S and H_B: gaze, eye, head-in-space and head-on-body position, respectively.

In summary, the attenuation of VO neuron modulation responses during active head-on-body motion (i.e., gaze shifts and gaze pursuit) is not mediated by neck proprioceptive inputs, by cognitive inputs, or by a direct copy of the neck motor command (Figure 16.6b, #1-3a). We propose that during active head movements, an efference copy of the neck motor command gates-in inhibitory signals from neck proprioceptors to reduce the head velocity sensitivities of VO neurons (Figure 16.6, #3b).

16.5   VOR Pathways: Testing Our Initial Hypothesis

We concluded Section 16.3 above by stating the following hypothesis: The VOR is not a hardwired reflex, but rather a pathway that is modulated in a manner that depends on the current gaze strategy. More specifically, the VOR is attenuated when the behavioural goal is to redirect gaze, and it is fully functional when the behavioural goal is to stabilize gaze. To rigorously test our proposal that the vestibular signals carried by PVP neurons depends only on the animal's current gaze strategy, we recorded from PVP neurons during the same tasks that we used above in Section 16.4 in our analysis of VO neurons. Specifically, we investigated whether (i) inputs from neck proprioceptors, (ii) a signal of cortical origin representing the monkey's knowledge of self-generated motion, and/or (iii) an efference copy of the motor behaviour (e.g., a neck movement), might affect neuronal activity. First, PVP neurons were tested during the same paradigms illustrated for VO neurons in Figure 16.7 (i.e. passive head-on-body rotation and passive body-under-head rotation). Similar to our results with VO neurons, we found no evidence that passive activation of neck proprioceptive inputs significantly influenced PVP neuron discharges. In addition, PVP neurons were tested during the driving paradigm illustrated in Figure 16.8a. Consistent with our hypothesis, PVP neuron discharges were attenuated only when the monkey redirected its gaze to the target light; when gaze was stable, neuronal modulation in response to head-in-space motion was identical for self-generated and passive whole-body rotations. Finally, we investigated whether neck motor command-related signals might influence PVP neuron discharges. PVP neurons were recorded while head-restrained animals generated significant neck torque during saccades made to food targets. Again, similar to our results with VO neurons, we found no evidence that neck motor command signals specifically influenced PVP neuron discharges.

16.6   Discussion and Conclusions

Our results demonstrate that the primate vestibular system can distinguish vestibular signals that arise from active self-motion of the head on the body, early in processing, at the level of the vestibular nuclei. This finding supports the proposal of von Holst and Mittelstaedt (1950) who suggested that afferent signals arising from an animal's own behaviour could be distinguished from afferent signals generated by external sources. They proposed that a copy of the motor command (i.e., a motor efference) is combined with the afferent signal to selectively remove the component caused by the motor behaviour. A similar mechanism has been reported in the electric fish, where an efference copy of the command to activate the electric organ converges centrally with electroreceptor afferent information, thereby reducing the response to self-generated electric fields (Bell, 1981; Zipser and Bennett, 1976).

Two important questions that arise from our studies of active head-on-body motion are: Do the vestibular nuclei lose track of vestibular information during active gaze shifts and gaze pursuit? And if they do, are we effectively operating as if we have a significant bilateral loss of labyrinth function during these self-generated behaviours? We addressed these points by utilizing a paradigm in which the monkey was able to generate voluntary head movements on its body (Figure 16.6a, dashed arrow in cartoon) while undergoing passive whole-body rotation (Figure 16.6a, solid arrow in cartoon). In this paradigm, head-in-space velocity is the sum of passive whole-body velocity and the *voluntarily* generated head-on-body velocity. Recall that VO neuron responses to the component of head-in-space motion arising from the monkey's voluntary head-on-body movements were relatively weak or negligible (Figure 16.6a). Yet, remarkably, neurons continued to respond robustly to the component of head-in-space motion produced by the passive rotation of the body. A summary of the population response is shown in Figure 16.9a. In contrast, when we tested PVP neurons during the same paradigm, their responses to both active and passive components of head-in-space motion were significantly reduced during combined eye-head gaze shifts. A summary of the population responses are shown in Figure 16.9b. Thus, during combined eye-head gaze shifts neither cell group reliably encoded the monkey's head-in-space motion (compare black and white bars, Figure 16.9). A similar trend was observed during gaze pursuit (not shown). Previous work in head-restrained animals has shown that passive horizontal head rotations are predominantly encoded at the level of the vestibular nuclei in the modulation of these two classes of neurons. However, since we have shown that neither cell type faithfully encodes head in space velocity under all conditions (Figure 16.10, pathway A), it appears that the vestibular nuclei actually do lose track of vestibular information that results from voluntary head motion during active gaze shifts and gaze pursuit.

The combined results of our studies of VO and PVP neuron discharges during active head-on-body motion also lead to a third important question: Does the brain have access to reliable vestibular sensory information during active gaze shifts and gaze pursuit via some route independent of the vestibular nuclei? Vestibular afferents project strongly to cerebellar regions involved in vestibular and eye movement control, namely, the nodulus/uvula, the flocullus, and the fastigial nucleus (reviewed in Voogd et al., 1996), as well as diffusely to other regions of the vestibulo-cerebellar vermis (Kotchabkakdi and Walberg, 1978). A reliable estimate of head-in-space motion may be encoded by these pathways during voluntary gaze shifts and gaze pursuit (Figure 16.10, pathway B). The projection to the

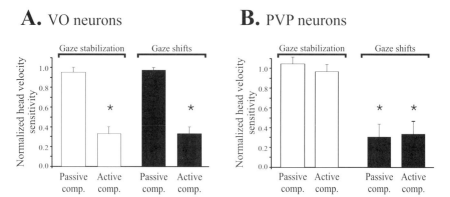

FIGURE 16.9. (a) Normalized head velocity sensitivity of our sample of VO neurons as a function of the animal's behavioural goal during head-unrestrained whole-body rotation. The sensitivities of the neuron to the passive and the active components of the head movement were estimated during gaze stabilization (open columns) and gaze shifts (solid columns). For VO neurons, only the responses to voluntary head-on-body movements were attenuated, independently of whether or not gaze was being stabilized or redirected. (b) In contrast to VO neurons, the head velocity sensitivity of our sample of PVP neurons was attenuated whenever the animal redirected its gaze. Whether the head movement was generated actively or applied passively did not affect the neuronal discharges. Note that both neuron types did not accurately encode head-in-space velocity during gaze shifts.

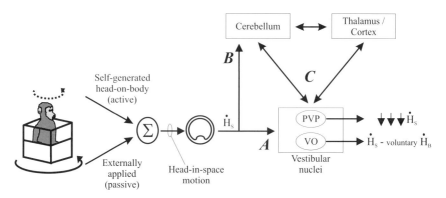

FIGURE 16.10. Proposed mechanism for the higher level processing of sensory information during gaze shifts. Head-in-space velocity information, which is the sum of actively and passively generated head movements, is relayed directly to both the vestibular nuclei (pathway A) and the cerebellum (pathway B) by primary vestibular afferents. During gaze shifts, the head-velocity sensitivity of PVP neurons is markedly reduced, and VO neurons only accurately encode the passive components of head-in-space ($\dot{H}_S - \dot{H}_B$). A likely route by which intact vestibular information (i.e., signal $= \dot{H}_S$) could reach higher centers is via the cerebellum. The cerebellum projects to the thalamus as well as to many regions of the cortex. These regions in turn project back to the cerebellum and vestibular nuclei (pathway C).

nodulus/uvula is of particular interest because this information might be used to produce an estimate of current orientation in space during self-generated motion. Previous studies have implicated the nodulus/uvula in the transformation of head-centered motion information into an inertial (gravity-centered) coordinates frame (Wearne et al., 1998; Angelaki and Hess, 1995). Vestibular information could also be relayed to the cortex via cerebellar-thalamic pathways. Indeed, vestibular related responses have been recorded in numerous cortical regions (e.g., area 7, area 3aV, parieto-insular vestibular cortex (PIVC), and premotor cortex) that are involved in spatial representation, navigation, and gaze control, and many of these areas project back directly to the vestibular nuclei (reviewed in Fukushima, 1997). For example, it has been proposed that neurons in the PIVC, which receive vestibular inputs via the thalamus as well as somatosensory and visual information, are involved in the perception of head motion (Grüsser et al., 1990). We suggest that the central nervous system computes an internal estimate of self-motion via the interconnections between the vestibular nuclei, the cerebellum and cortical structures (Figure 16.10, pathway C).

References

Anastasopoulos, D. and Mergner, T. (1982). Canal-neck interaction in vestibular nuclear neurons of the cats. *Exp. Brain Res.*, 46: 269–280.

André-Deshays, C., Berthoz, A. and Revel, M. (1988). Eye-head coupling in humans. I. Simultaneous recording of isolated motor units in dorsal neck muscles and horizontal eye movements. *Exp. Brain Res.*, 69: 399–406.

Angelaki, D. E. and Hess, B. J. M. (1995). Inertial representation of angular motion in the vestibular system of rhesus monkeys. II. Otolith-controlled transformation that depends on an intact cerebellar nodulus. *J. Neurophysiol.*, 73: 1729–1751.

Barnes, G. R. (1979). Vestibulo-ocular function during co-ordinated head and eye movements to acquire visual targets. *J. Physiol. (Lond.)* 287: 127–147.

Bell, C. (1981). An efference copy which is modified by reafferent input. *Science*, 214: 450–453.

Bizzi, E., Kalil, R. E. and Tagliasco, V. (1971). Eye-head coordination in monkeys: evidence for centrally patterned organization. *Science*, 173: 452–454.

Boyle, R. (1993). Activity of medial vestibulospinal tract cells during rotation and ocular movement in the alert squirrel monkey. *J. Neurophysiol.*, 70: 2176-2180.

Boyle, R., Belton, T. and McCrea, R. A. (1996). Responses of identified vestibulospinal neurons to voluntary eye and head movements in the squirrel monkey. *Ann. N. Y. Acad. Sci.*, 781: 244–263.

Boyle, R. and Pompeiano, O. (1980). Response of vestibulospinal neurons to sinusoidal rotation of neck. *J. Neurophysiol.*, 44: 633–649.

Chubb, M. C., Fuchs, A. F. and Scudder, C. A. (1984). Neuron activity in monkey vestibular nuclei during vertical vestibular stimulation and eye movements. *J. Neurophysiol.*, 52: 724–742.

Cullen, K. E., Belton, T. and McCrea, R. A. (1991). A non-visual mechanism for voluntary cancellation of the vestibulo-ocular reflex. *Exp.- Brain Res.*, 83: 237252.

Cullen, K. E. and McCrea, R. A. (1993). Firing behavior of brainstem neurons during voluntary cancellation of the horizontal vestibulo-ocular reflex. I. Secondary vestibular neurons. *J. Neurophysiol.*, 70: 828–843.

Cullen, K. E., Rey, C. G., Guitton, D. and Galiana, H. L. (1996). The use of system identification techniques in the analysis of oculomotor burst neuron spike train dynamics. *J. Comp. Neurosci.*, 3: 347–367.

Dichgans, J., Bizzi, E., Morasso, P. and Tagliasco, V. (1973). Mechanisms underlying recovery of eye-head coordination following bilateral labyrinthectomy in monkeys. *Exp. Brain Res.*, 18: 548–562.

Fuchs, A. F. and Kimm, J. (1975). Unit activity in vestibular nucleus of the alert monkey during horizontal angular acceleration and eye movement. *J. Neurophysiol.*, 38: 1140–1161.

Fuchs, A. F. and Robinson, D. A. (1966), A method for measuring horizontal and vertical eye movements in the monkey. *J. Physiol. (Lond.)*, 191: 609–631.

Fukushima, K. (1997). Corticovestibular interactions; anatomy, electrophysiology, and functional considerations. *Exp. Brain Res.*, 117: 1–16.

Grüsser, O.-J., Pause, M. and Schreiter, U. (1990). Localization and responses of neurones in the parieto-insular vestibular cortex of awake monkeys (*Macaca Fascicularis*). *J. Physiol. (Lond.)*, 430: 537–557.

Guitton, D. and Volle, M. (1987). Gaze control in humans: Eye-head coordination during orienting movements to targets within and beyond the oculomotor range. *J. Neurophysiol.*, 58: 427–459.

Hayes, A. V., Richmond, B. J. and Optican, L. M. (1982). A UNIX-based multiple process system for real-time data acquisition and control. *Proc. WESCON Conf.*, 2: 1–10.

Keller, E. L. and Daniels, P. (1975). Oculomotor related interaction of vestibular and visual stimulation in vestibular nucleus cells in the alert monkey. *Exp. Neurol.*, 46: 187–198.

Kotchabkakdi, G. G. H. and Walberg, F. (1978). Primary vestibular afferent projections to the cerebellum as demonstrated by retrograde axonal transport of horseradish peroxidase, *Brain Res.*, 142: 142–146.

Lanman, J., Bizzi, E. and Allum, J. (1978). The coordination of eye and head movement during smooth pursuit. *Brain Res.*, 153: 39–53.

Lestienne, F., Vidal, P. P. and Berthoz, A. (1984). Gaze changing behaviour in head restrained monkey. *Exp. Brain Res.*, 53: 349–356.

Lisberger, S. G. and Miles, F. A. (1980). Role of the primate vestibular nucleus in long-term adaptive plasticity of the vestibulo-ocular reflex. *J. Neurophysiol.*, 43: 1725–1745.

McCrea, R. A., Strassman, A. and Highstein, S. M. (1987). Anatomical and physiological characteristics of vestibular neurons mediating the horizontal vestibulo-ocular reflexes in the squirrel monkey. *J. Comp. Neurol.*, 264: 547–570.

McCrea, R. A., Gdowski, G. T., Boyle, R. and Belton, T. (1999). Firing Behavior of vestibular neurons during active and passive head movements: vestibulo-spinal and other non-eye movement related neurons. *J. Neurophysiol.*, 82: 416–428.

McFarland, J. L. and Fuchs, A. F. (1992). Discharge patterns of nucleus prepositus hypoglossi and adjacent vestibular nucleus during horizontal eye movement in behaving macaques. *J. Neurophysiol.*, 41: 319–332.

Miles, F. A. (1974). Single unit firing patterns in the vestibular nuclei related to voluntary eye movements and passive body rotation in conscious monkeys. *Brain Res.*, 71: 215–224.

Morasso, P., Bizzi, E. and Dichgans, J. (1973). Adjustment of saccade characteristics during head movements. *Exp. Brain Res.*, 16: 492–500.

Pélisson, D., Prablanc, C. and Urquizar, C. (1988). Vestibulooocular reflex inhibition and gaze saccade control characteristics during eye-head orientation in humans. *J. Neurophysiol.*, 59: 997–1013.

Reisine, H. and Raphan, T. (1992). Neural basis for eye velocity generation in the vestibular nuclei of alert monkeys during off-vertical axis rotation. *Exp. Brain Res.*, 92: 209-226.

Richmond, F. J. R. and Abrahams, V. C. (1979). Physiological properties of muscle spindles in dorsal neck muscles of the cat. *J. Neurophysiol.*, 42: 604–617.

Roy, J. E. and Cullen K. E. (1998). A neural correlate for vestibulo-ocular reflex suppression during voluntary eye-head gaze shifts. *Nature Neurosci.*, 1: 404–410.

Roy, J. E. and Cullen, K. E. (2001a). Selective processing of vestibular reafference during self-generated head motion. *J. Neurosci.*, 21: 2131–2142.

Roy, J. E. and Cullen, K. E. (2001b). Is an efference copy of the neck motor command responsible for the attenuation of VO neuron responses during self-generated head-on-body motion? Neural Control of Movement Meeting (NCM), Sevilla, Spain.

Sato, H., Ohkawa, T., Uchino, Y. and Wilson, V. J. (1997). Excitatory connections between neurons of the central cervical nucleus and vestibular neurons in the cat. *Exp. Brain Res.*, 115: 381–386.

Scudder, C. A. and Fuchs, A. F. (1992). Physiological and behavioral indentification of vestibular nucleus neurons mediating the horizontal vestibuloocular reflex in trained Rhesus Monkeys. *J. Neurophysiol.*, 68: 244–264.

Sylvestre, P. A. and Cullen, K. E. (1999). A quantitative analysis of abducens neuron discharges during saccadic and slow eye movements. *J. Neurophysiol.*, 82: 2612–2632.

Tabak, S., Smeets, J. B. J. and Collewijn, H. (1996). Modulation of the human vestibulooocular reflex during saccades: probing by high-frequency oscillation and torque pulses of the head. *J. Neurophysiol.*, 76: 3249–3263.

Tomlinson, R. D. (1990). Combined eye-head gaze shifts in the primate. III. Contributions to the accuracy of gaze saccades. *J. Neurophysiol.*, 64: 1873–1891.

Tomlinson, R. D. and Bahra, P. S. (1986a). Combined eye-head gaze shifts in the primate. I. Metrics. *J. Neurophysiol.*, 56: 542–1557.

Tomlinson, R. D. and Bahra, P. S. (1986b). Combined eye-head gaze shifts in the primate. II. Interactions between saccades and the vestibuloocular reflex. *J. Neurophysiol.*, 56: 1558–1570.

Tomlinson, R. D. and Robinson D. A. (1984). Signals in vestibular nucleus mediating vertical eye movements in the monkey. *J. Neurophysiol.*, 51: 1121–1136.

von Holst, E. and Mittelstaedt, H. (1950). Das Reafferenzprinzip. *Naturwissenschaften*, 37: 464–476.

Voogd, J., Gerrits, M. and Ruigrok, J. H. (1996). Organization of the vestibulocerebellum. *Ann. N. Y. Acad. Sci.*, 781: 553–579.

Wearne, S., Raphan, T. and Cohen, B. (1998). Control of spatial orientation of the angular vestibuloocular reflex by the nodulus and uvula. *J. Neurophysiol.*, 79: 2690–2715.

Wilson, V. J., Yamagata, Y., Yates, B. J., Schor, R. H. and Nonaka, S. (1983). Response of vestibular neurons to head rotations in vertical planes. III. Response of vestibulocollic neurons to vestibular and neck stimulation. *J. Neurophysiol.*, 64: 1695–1703.

Wilson, V. J., Zarzecki, P., Schor, R. H., Isu, N., Rose, P. K., Sato, H., Thomson, D. B. and Umezaki, T. (1999). Cortical influences on the vestibular nuclei of the cat. *Exp. Brain Res.*, 125: 1–13.

Wylie, D. R., De Zeeuw, C. I., Digiorgi, P. L. and Simpson, J. I. (1994). Projections of individual Purkinje cells of identified zones in the ventral nodulus to the vestibular and cerebellar nuclei in the rabbit. *J. Comp. Neurol.*, 349: 448–463.

Yokota, J. I., Reisine, H. and Cohen, B. (1992). Nystagmus induced by electrical stimulation of the vestibular and prepositus hypoglossi nuclei in the monkey: evidence for site of induction of velocity storage. *Exp. Brain Res.*, 92: 123–138.

Zangemeister, W. H. and Stark, L. (1982a). Gaze latency: variable interactions of head and eye latency. *Exp. Neurol.*, 75: 389–406.

Zangemeister, W. H. and Stark, L. (1982b). Types of gaze movement: variable interactions of eye and head movements. *Exp. Neurol.*, 77: 563–577.

Zipser, B. and Bennett, V. L. (1976). Interaction of electrosensory and electromotor signals in lateral line lobe of a mormyrid fish. *J. Neurophysiol.*, 39: 713–721.

Neural Encoding of Gaze Dependencies During Translation

Dora E. Angelaki and J. David Dickman

To maintain binocular visual acuity during head movements, compensatory eye movements in the translational vestibulo-ocular reflex (TVOR) exhibit a systematic dependence on gaze parameters, including vergence angle and eye position. To investigate if and how these dependencies are reflected in neural activities, the firing rates of eye movement–sensitive neurons in the rostral vestibular nuclei were examined during translation (0.5 – 5 Hz). Motion was delivered along different heading directions and animals were required to fixate head-fixed or earth-stationary targets at different distances and eccentricities. All cells exhibited changes in sensitivity with vergence angle during lateral translation, and these changes were appropriate to drive the respective eye movements. Furthermore, neurons also exhibited a dependence on gaze and heading directions, as expected from the equivalent dependence of eye velocity in the TVOR. These results argue against dynamic co-contraction as a major mechanism to explain the gaze dependencies in the TVOR. Interestingly, the firing rates of cells that carry both sensory head movement and motorlike signals during rotation were more strongly related to the motor output than to the vestibular sensory signal during translation. In fact, the main secondary neuron in the disynaptic rotational VOR (RVOR) pathways (position-vestibular-pause cell) that exhibits a robust modulation during RVOR suppression did not modulate during TVOR suppression. In contrast, a different class of potentially premotor neurons, the so-called eye-head cells, exhibits responses that are not mere replicas of the oculomotor behavior but might represent substrates for the sensorimotor transformations in the TVOR.

17.1   Introduction

Working in close synergy with lower-frequency visual tracking mechanisms, the vestibulo-ocular reflexes (VORs) transform information about head movements sensed by the vestibular organs into ocular deviations appropriate to maintain visual stability (Busettini et al., 1996a, b; 1997; Miles, 1993, 1998). While the rotational VOR (RVOR) has been well characterized at both behavioural and neurophysiological levels, the neural mechanisms and computations underlying the generation of compensatory binocular eye movements during translation (TVOR) are less well understood. An important characteristic of the TVOR is that ocular rotations must compensate for head translations. As a result of this geometrical scaling, maintenance of binocular fixation on near targets during translational disturbances requires the generation of eye movements whose amplitude and direction depend on target distance and eccentricity. Specifically, the amplitude of the elicited eye movement increases as a function of the inverse of viewing distance (Angelaki and McHenry, 1999; Busettini et al., 1994; Gianna et al., 1997; McHenry and Angelaki, 2000; Paige and Tomko, 1991a, b; Schwarz and Miles, 1991; Schwarz et al., 1989; Telford et al., 1997). In addition, the amplitude and direction of the elicited eye movements depend on eye position. This gaze dependence is stronger during fore-aft motion, where eye movement direction reverses for gaze to the right and to the left (Angelaki and McHenry, 1999; McHenry and Angelaki, 2000; Paige and Tomko, 1991b; Seidman et al., 1999). A more general form of gaze dependence is obeyed during movements along any heading direction (Angelaki and Hess, 2001; Tomko and Paige, 1992).

How and where in the premotor circuitry, otolith signals are dynamically transformed into the appropriate motor drives to generate the highly elaborate repertoire of eye movements necessary to maintain binocular gaze on near targets is currently unknown. One line of thought assumes that these transformations are implemented neurally, whereby premotor signals are scaled by a signal related to the inverse of target distance (e.g., vergence angle), as well as a signal proportional to a sinusoidal function of monocular eye position. Whereas a simple scaling would account for the vergence dependence, the gaze dependence is more complex and a function of heading direction (e.g., Angelaki and Hess, 2001; Tomko and Paige, 1992). Specifically, the velocity of each eye, $\dot{\theta}$, during translation with velocity v along a direction forming an angle, α, with the fore-aft axis depends on eye-to-target distance, d, and eye position, θ, according to the equation (e.g., see Angelaki and Hess, 2001):

$$\dot{\theta} = \frac{v}{d} \sin(\alpha - \theta) \tag{17.1}$$

If the gaze dependence described by equation (17.1) is neurally encoded by premotor and motor cells, traditional Sherringtonian laws of reciprocal innervation of agonist–antagonist relationships could control the direction and amplitude of the ocular response to fore-aft motions. Alternatively, however, it has been proposed that the gaze dependence of the TVOR could also arise at the neuromuscular level through co-contraction of the agonist and antagonist muscles (Seidman et al., 1999). Accordingly, an imbalance in the co-contraction of the eye muscles (rather than a reversal in the firing rate modulation of premotor and motor neurons) could determine the direction and amplitude of the elicited eye movement during fore-aft motion. The latter control scheme would predict no phase change in neural firing rates as a function of gaze direction during fore-aft motion.

Data regarding the neural processing of otolith-ocular signals in alert primates has been limited and usually gathered during eccentric rotations when semicircular canals and otolith organs are activated simultaneously (Chen-Huang and

McCrea, 1999b; McConville et al., 1996; Snyder and King, 1996). Only recently have neural responses to pure translational motion been reported for the first time (Angelaki et al., 2001). In this chapter, we summarize the main conclusions of the study and specifically address the following issues in more detail. First, whether the dependence of firing rates of eye movement–sensitive vestibular nuclei neurons on vergence angle during pure translational motion is appropriate to generate the respective vergence dependence of the TVOR. Second, whether the gaze dependence of the TVOR, including the response reversal for rightward and leftward eye positions during fore-aft motion is neurally encoded. Such a result would support the existence of a Sherringtonian reciprocal innervation for the TVOR, rather than a dynamic co-contraction of agonist-antagonist relationships (Seidman et al., 1999).

17.2   Methods

Data reported here were collected from two juvenile rhesus monkeys that were prepared for chronic recording of bilateral eye movements and single-unit activity. Eye movements were measured with the magnetic search coil technique and calibrated as explained in detail elsewhere (Angelaki, 1998; Angelaki et al., 2000). During experiments, the monkey was seated in a primate chair that was secured inside the inner frame of a vestibular turntable consisting of a three- dimensional rotator on top of a linear sled that moved in an earth-horizontal plane (Acutronics, Inc.). Both stimulus presentation and data acquisition were controlled with custom scripts written within the Spike2 software environment using the Cambridge Electronics Device (CED, model 1401) data acquisition system. The behavioural performance of the animal was monitored through interactions with a second computer that provided a continuous on-line TTL pulse as long as both ocular positions were maintained within 1 deg of ideal target fixation. This "eye-in-window" signal was monitored by the CED for on-line juice reward delivery and was saved for off-line analyses. Behavioral windows for each eye were calculated on-line based on the geometrical relationships that should govern appropriate target fixation for a given motion of the target and/or head movement (e.g., Angelaki et al., 2000; McHenry and Angelaki, 2000). Juice rewards were typically given at a frequency of ~1-2/s as long as the gaze direction of both eyes were within the specified behavioral windows.

Animals were required to fixate one of two small target lights that were back-projected using a double laser/x-y mirror galvanometer system (General Scanning) onto a 2-plane screen that was specifically constructed to view 3-D targets. One of the laser-galvanometer systems was used to present targets on the earth-vertical plane of the screen (mounted parallel to the animal's frontal eye plane; vergence angle of ~6.4 deg). This target was used to classify the neurons according to their responses during horizontal and vertical fixations and pursuit, as well as RVOR and TVOR suppression (for details, see Angelaki et al., 2001). The second laser/galvanometer set was used for back-projection on the bottom (earth-horizontal) plane of the screen. This surface was mounted at ∼6 cm below the eyes and was used to present targets that could be varied in depth (see below).

Extracellular recordings from vestibular nuclei neurons were obtained with epoxy coated, etched tungsten microelectrodes inserted into the brain through a 26-gauge stainless steel guide tube. Neural activity was amplified, filtered (300 Hz–6 kHz) and passed both to an audio amplifier and a BAK Instruments dual time-amplitude window discriminator whose output was displayed on an oscilloscope. For each recorded cell, acceptance pulses from the BAK window discriminator were used to trigger the event channel of the CED data acquisition system. In addition, the eight voltage signals from the two eye coil assemblies (four for each dual eye coil; see Angelaki, 1998), the "eye-in-window" signal from the "slave" computer to confirm appropriate behavioural performance, the three output signals of a 3-D linear accelerometer (mounted on fibreglass members that firmly attached the animal's head ring to the inner gimbal of the rotator) and the velocity and position feedback signals from the rotator were antialias filtered (200 Hz, 6-pole Bessel), and digitized by the CED at a rate of 833.33 Hz (16-bit resolution).

During the first experiments in each animal, the abducens nuclei were identified based on the characteristic burst-tonic activity of neurons (Fuchs and Luschei, 1970). Penetrations concentrated in a relatively small area in the rostral part of the vestibular nucleus extending 3 mm posterior and 4 mm lateral from the center of the abducens nuclei (Angelaki et al., 2001). These areas, consisting mainly of the medial and ventral lateral nuclei, have been shown to contain eye movement–sensitive cells with many projecting directly to the abducens and oculomotor nuclei (Cullen and McCrea, 1993; McCrea et al., 1987; Scudder and Fuchs, 1992). Once a vestibular nuclei neuron was isolated, smooth pursuit (0.5 Hz, ±10 deg), visually guided saccades, as well as RVOR and TVOR suppression were used to classify cells into one of four main groups (Scudder and Fuchs 1992): (a) Position-Vestibular-Pause (PVP) neurons were characterized by sensitivities to angular head velocity and eye velocity in opposite directions such that these signals superimposed during gaze stabilization on an earth-fixed target. All PVP cells included here had activities that modulated in phase with ipsilateral head velocity during RVOR suppression and in phase with contralateral eye velocity during horizontal smooth pursuit (type I PVP). (b) Eye-head (E-H) neurons exhibited a sensitivity to head velocity during RVOR suppression and to eye velocity during smooth pursuit in the same direction, such that the two signals opposed each other during rotation while stabilizing an earth-fixed target. This group included cells with ipsilaterally directed eye and head velocity sensitivities (E_i-H_i) as well as cells with contralaterally directed eye and head velocity sensitivities (E_c-H_c). (c) Burst-tonic (BT) neurons exhibited eye movement sensitivity, but did not modulate during RVOR suppression. (d) Vestibular-only (VO) neurons included cells that did not exhibit any slow eye movement sensitivity, but modulated during either rotational or translational movements. The present analysis focuses specifically on eye movement-sensitive neurons (PVP, BT, and E-H cells). None of the

VO cell responses exhibited any dependence on vergence angle or eye position.

For these experiments, the specific experimental protocol consisted of the following: (i) The dependence of neural firing rate on vergence angle was evaluated during lateral translation while fixating earth-stationary targets at different viewing distances (range of 2–10 deg of vergence). Most of the data were collected at 0.5 and 1 Hz (±0.03 g). (ii) The dependence of neural firing rate on gaze was tested during fore-aft translation (0.5 Hz, ±0.1 g) while fixating earth-stationary and head-fixed targets at different target eccentricities (±25 deg; mean vergence angle of ~4 deg). A few cells were also tested at 5 Hz (±0.19 g).

The targets for these stimuli were back-projected onto the earth-horizontal screen that was mounted 6 cm below the eyes (vertical eye position ~10 deg). As a result, vertical eye position covaried with vergence angle (9–22 deg for 2–10 deg of vergence) and horizontal eccentricity. This did not pose a problem for the following reasons. First, the TVOR during lateral motion is relatively insensitive to vertical eye position (Angelaki et al., 2000; Paige and Tomko, 1991a; Telford et al., 1997), such that vertical eye movements did not exceed a peak of 0.5 deg/s during lateral motion. Second, vertical eye velocity modulation ($\sim5-7$ deg/s) was present during fore-aft translation, but did not vary for different horizontal eccentricities. Thus, any dependence of neural firing rate on horizontal eye position only reflected the underlying changes in horizontal eye movements. Since the laser-galvanometer-screen assembly was fixed to the inner gimbal of the rotator and sled superstructure, earth-fixed targets during translation were presented using an appropriately scaled position feedback signal from the sled to drive the galvanometers on-line during motion. Because of the underlying geometry (e.g., equation 17.1), the eye movements required to foveate a near target during large sinusoidal head translational movements must follow a nonsinusoidal profile (Figure 17.1a and 17.2a; see also Angelaki et al., 2001). This is particularly true for fore-aft motion (Figure 17.2a), where the argument in the sine function of equation (17.1) exhibits large changes for a small change in angle θ.

All data analyses were performed off-line using Matlab (Mathworks, Inc.). Eye positions were calibrated and expressed as 3-D rotation vectors, as described in detail elsewhere (Angelaki, 1998; Angelaki et al., 2000). For the neural activity, unit clock values were converted to instantaneous firing rate that was computed as the inverse of interspike interval. Ocular and cell responses were evaluated by selecting only portions of the data without saccades and fast phases (thus avoiding the need for subtracting eye position). Firing rates from at least 3 cycles were then folded into a single cycle (no averaging was performed). Only portions in which the positions of both eyes were within ±1 deg of the target were included in the folding and further analyses. The neural response sensitivity and phase during translation, rotation, and pursuit were determined by fitting a sine function (first and second harmonics and a DC offset) to the overlaid data using a nonlinear least squares algorithm based on the Levenberg–Marquardt method. For neurons whose firing rate fell silent during a portion of the cycle, silent portions were excluded from the least-squares optimization. Phase was expressed as the difference (in

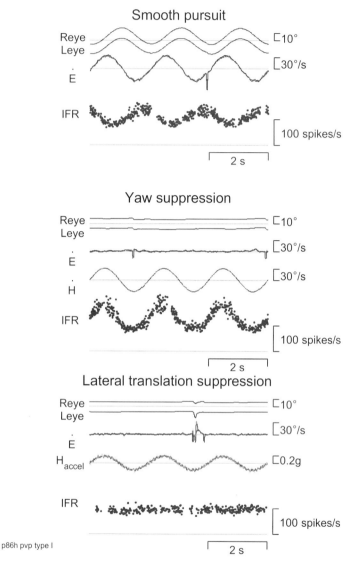

FIGURE 17.1. Responses of a position-vestibular cell during horizontal smooth pursuit, yaw RVOR suppression, as well as lateral translation (TVOR) suppression (0.5 Hz). From top to bottom, binocular (right and left eye) position, right eye velocity ($\dot{E}$), stimulus (head velocity, $\dot{H}$, for rotation and head acceleration, H_{accel}, for translation) and instantaneous firing rate (IFR) of the neuron (p86h). The cell increased its firing rate approximately in phase with contralateral eye velocity during smooth pursuit and ipsilateral head velocity during RVOR suppression, but it did not modulate during TVOR suppression.

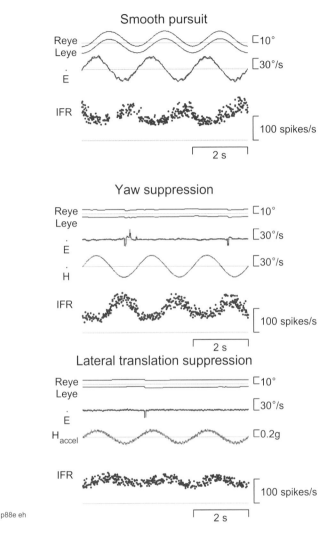

FIGURE 17.2. Responses of an eye-head cell during horizontal smooth pursuit, yaw RVOR suppression, as well as lateral translation (TVOR) suppression (0.5 Hz). From top to bottom, binocular (right and left eye) position, right eye velocity ($\dot{E}$), stimulus (head velocity, $\dot{H}$, for rotation and head acceleration, H_{accel}, for translation) and instantaneous firing rate (IFR) of the neuron (p88e). The cell increased its firing rate approximately in phase with contralateral eye velocity during smooth pursuit and contralateral head velocity during RVOR suppression. In contrast to PVP neurons, the majority of eye-head cells exhibited a small but consistent modulation during TVOR suppression.

degrees) between peak neural activity and peak head velocity (rightward velocity for lateral motion and backward velocity for fore-aft motion). That is, based on the coordinate system used to measure eye movements, lateral translation should elicit a TVOR with a phase of 0 deg. Fore-aft motion should elicit horizontal eye movements with a phase of 0 deg when looking to the right (negative eye positions) and 180 deg when looking to the left (positive eye positions; see also McHenry and Angelaki, 2000). The statistical significance of the dependence of firing rates on vergence angle was evaluated using linear regression analysis.

17.3   Results

Recent work in our laboratory has examined the activities of eye movement–sensitive neurons in the rostral vestibular nuclei during pure translational movements (Angelaki et al., 2001). Based upon their firing activities during fixations, smooth pursuit, and rotation, eye movement–sensitive neurons in the rostral vestibular nuclei were classified into eye-head (E-H), position-vestibular-pause (PVP) and burst-tonic (BT) cells (Angelaki et al., 2001; Cullen and McCrea, 1993; Scudder and Fuchs, 1992; Tomlinson and Robinson, 1984). We found that the firing rates of cells that carry both sensory head movement and motor like signals during rotation were more strongly related to the motor output than to the vestibular sensory signal during translation. In fact, the sensory/motor distinction appeared to vary for different classes of eye movement–sensitive vestibular nuclei cells. Specifically, the main secondary neuron in the disynaptic RVOR pathways (type I position-vestibular-pause cell) that exhibits a robust modulation during RVOR suppression did not modulate during TVOR suppression (Figure 17.1). In contrast, the majority of eye-head cells exhibited a clear response modulation during TVOR suppression in the absence of eye movements (Figure 17.2).

   In this chapter we will specifically focus on data from 15 eye movement-sensitive neurons in the rostral vestibular nucleus whose responses were characterized during translation while viewing earth-stationary targets at different distances and eccentricities. A more detailed description of neural responses for a larger cell population during translation has been reported elsewhere (Angelaki and Dickman, 2000; Angelaki et al., 2001).

17.3.1   Dependence on Vergence Angle

The firing rate modulation of all neurons tested during lateral motion at 0.5–5 Hz exhibited a significant dependence on vergence angle. An example of the responses of a PVP neuron for two different vergence angles is illustrated in Figure 17.3a. As the vergence angle increased from 4.3 deg to 9 deg, the amplitude of the elicited eye movement increased, and so did also the firing rate modulation of the cell (Figure 17.3a). This property is further illustrated in Figure 17.3b, where the dependence of firing rates on vergence angle during 0.5 Hz lateral translation has been summarized for a PVP (circles; same neuron as the one shown in Figure 17.3a), an $E_c–H_c$ and an $E_i–H_i$ neuron (up and down triangles, respectively). For all three cells illustrated, neural response amplitude when plotted as a function of vergence angle increased monotonically, similarly as horizontal eye velocity (Figure 17.3b, compare solid symbols with open circles used for plotting firing rates and eye velocity, respectively). Response phase did not exhibit any systematic dependence on vergence angle.

To evaluate whether the observed vergence-dependent change in firing rate modulation was sufficient to account for the respective dependence of the TVOR, the following analysis was performed. First, the dependencies of peak firing rate and eye velocity on vergence angle were quantified by fitting linear regressions, as illustrated in Figure 17.3b. Second, we estimated a *computed eye velocity-vergence slope* that would be elicited if each cell's modulation were directly linked to the actual eye movement according to its pursuit sensitivity (measured at 6.4° of vergence, i.e., approximately at the middle of the vergence range tested). If the vergence-dependent modulation of neural activity was sufficient to account for the corresponding eye movement relationship, actual and computed eye velocity-vergence slopes should be identical.

The results of this analysis are summarized in Figure 17.3c. Each filled black symbol corresponds to the slope of the regression analysis for a cell, with different symbols being used for BT, PVP, and eye-head cells. The neural firing rate-vergence slopes during 0.5 Hz translation increased linearly with the pursuit sensitivity of the cell ($r^2 = 0.82$). As a result of this dependence, the computed eye velocity-vergence slopes were similar to the actual slopes estimated directly from the eye velocity regression analysis (e.g., Figure 17.3c, compare open symbols with squares). Thus, as a population, the cells carry the appropriate signals to generate the viewing distance dependence of the TVORs.

In four cells that were also tested at 5 Hz, peak response amplitude also significantly increased as a function of vergence angle. In 3 of 4 cells, the 5 Hz slope was less than the 0.5 Hz. The fourth cell exhibited similar slopes for 0.5 and 5 Hz. Due to the ten-fold difference in frequency, 5 Hz translation data have not been directly compared with the 0.5 Hz pursuit.

17.3.2   Dependence on Gaze Direction

For neural signals to be appropriate to drive the oculomotor system and generate the gaze dependence of the TVOR (in the absence of co-contraction; see Section 17.4), peak neural firing rate and phase should also be a function of gaze, similarly as eye movements, during translation. Because gaze dependence is the strongest for fore-aft motion (e.g., Angelaki and Hess, 2001; McHenry and Angelaki, 2000; Paige and Tomko, 1991b; Seidman et al., 1999), this comparison focused on data collected during fore-aft motion. As previously reported, horizontal eye velocity is zero during straight-ahead gaze, whereas its dependence on gaze is manifested as a V-shaped curve for the response amplitude and a 180 deg-shift in phase for rightward and leftward eye positions (Figure 17.4b and c, open

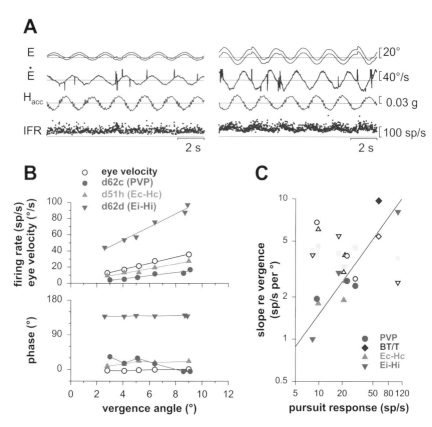

FIGURE 17.3. Dependence on vergence angle. (a) Responses of a horizontal type I PVP neuron recorded in the left vestibular nucleus during 0.5 Hz lateral translation at two vergence angles, 4.3 deg (left traces) and 9 deg (right traces). From top to bottom, binocular eye position (E), eye velocity ($\dot{E}$), stimulus (head acceleration, H_{acc}) and instantaneous firing rate (IFR) of the neuron (d62c). Positive acceleration is rightward, corresponding to leftward displacement of the sled and eliciting rightward (negative deflection) eye movements. Neural firing rate peaked approximately in phase with contralateral (rightward, negative) eye velocity. (b) Peak right eye velocity (open circles) and neural response and phase from a PVP (d62c, same as in a), an E_c-H_c (d51h) and an E_i-H_i (d62d) cell (solid symbols) are plotted as a function of vergence angle during 0.5 Hz (± 0.03 g) lateral motion. Solid lines for neural response and eye velocity are linear regressions through the corresponding data ($r^2 = 0.92 - 0.99$). Phase is expressed relative to linear head velocity. (c) The slopes of the linear relationship between neural firing rate and vergence angle (e.g., solid lines in b) are plotted as a function of the respective cell modulation during horizontal pursuit (0.5 Hz, ± 10 deg, 6.4 deg of vergence) as solid black symbols. The solid line is a linear regression, with coefficients: $y = -0.59 + 0.77x$, $r^2 = 0.82$. Solid grey squares plot the respective regression slopes of eye velocity. Open symbols illustrate the computed eye velocity slope that would be generated by each cell alone according to its slope and eye movement (pursuit) response sensitivity.

circles). All vestibular nuclei cells tested during fore-aft motion for different gaze directions changed their firing rate modulation phase for targets to the left and to the right. This is illustrated for an E_i-H_i cell in Figure 17.4a. When looking 15 deg to the left, neural firing rate peaked approximately in phase with forward head velocity (i.e., lagged negative linear acceleration by ∼90 deg; Figure 17.4a, left traces). When looking 6 deg to the right, neural firing rate peaked approximately in phase with backward head velocity (i.e., lagged positive linear acceleration by ∼90 deg; Figure 17.4a, right traces).

The amplitude and phase of the first harmonic description of these response modulations have been summarized in Figure 17.4b and c for three different cells. The BT and PVP cell responses exhibited an eye-position dependence that paralleled that of horizontal eye velocity, it was nearly zero for straight-ahead gaze and reversed ∼180 deg for leftward (positive) and rightward (negative) eye positions (Figure 17.4c). In contrast, the eye-head cell whose data are illustrated in Figure 17.4b, departed in several important ways from the eye position dependence of horizontal eye velocity. First, the phase shift for fixations to the left and to the right was not 180 deg, but closer to 90 deg (Figure 17.4b, bottom, solid triangles vs. open circles). Second, the cell exhibited a clear response modulation even in the absence of horizontal eye velocity during straight ahead gaze (0 deg eye position). Finally, a clear modulation was also observed during fore-aft TVOR suppression (Figure 17.4b, open triangles).

Even though the example illustrated in Figure 17.4b represents the largest departure from the corresponding oculomotor behaviour, eye-head responses typically deviated in one or the other way from the motor behaviour. The most pronounced difference was in the response phase. Of the eye-head cells that were tested sufficiently at different gaze directions, the change in response phase for rightward and leftward fixations averaged 98.4±50.5 deg for five E_c-H_c cells and 127.9±2.7 deg for three E_i-H_i cells (as compared to a shift of 169.0±8.5 deg for eye velocity). The nonlinear firing rate–eye position dependencies of eye-head cells might be partly responsible for this behaviour. As shown in Figure 17.4c, the two non-eye-head vestibular nuclei neurons, as well as three abducens motoneurons (not shown) that were also tested with an identical protocol, exhibited firing rates whose dependence on eye position more closely mirrored that of eye movements, including a phase shift of ∼180 deg for left and right targets. These results during fore-aft motion suggest that the gaze dependencies expected for the geometrical tuning of the TVOR, including a significant phase change for gaze directions to the left and to the right, are qualitatively present in the firing rates of eye movement sensitive neurons in the vestibular nuclei. However, at least some E-H neurons exhibited properties that were not quantitatively identical to the underlying geometrical dependence of eye movements. This observation supports previous results suggesting that at least a subpopulation of eye-head cells could not be characterized as carrying completely transformed "motor" signals during translation (Angelaki et al., 2001; see Section 17.4).

The dependence of neural firing rates on gaze direction was not only limited to fore-aft motion. Figure 17.5 plots the responses of an ipsilateral eye-head cell

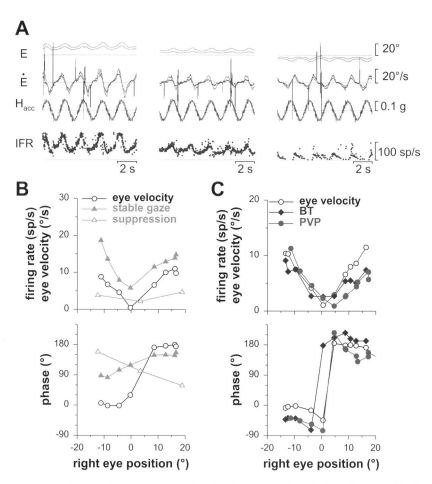

FIGURE 17.4. Dependence on eye position. (a) Responses of an E_i-H_i cell recorded in the left vestibular nucleus during fore-aft translation while fixating targets at three horizontal eccentricities, 15 deg to the left (left traces), 6.5 deg to the left (middle traces) and 6 deg to the right (right traces). From top to bottom, binocular eye position (E), eye velocity ($\dot{E}$), stimulus (head acceleration, H_{acc}) and instantaneous firing rate (IFR) of the neuron (d62d). Positive acceleration is backward. Neural firing rate peaked approximately in phase with ipsilateral (leftward, positive) eye velocity. Note the increase in mean firing rate with leftward (ipsilateral) eye positions. (b) Peak right eye velocity (open circles) and neural response amplitude and phase from an E_c-H_c cell (d51h; solid triangles) are plotted as a function of right eye position during 0.5 Hz fore-aft motion while fixating earth-stationary targets. Open triangles are the neural response during TVOR suppression. (c) Similar plots of the responses from a Bust-Tonic (d55f; solid diamonds) and PVP (d44n; solid circles) cell. Cells d55f and d44n did not modulate during TVOR suppression. Phase is plotted relative to backward linear head velocity.

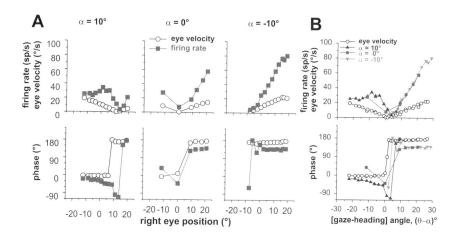

FIGURE 17.5. Dependence on eye position and heading direction. (a) Right eye velocity (open circles) and neural firing rate (solid squares) from an E_i-H_i cell (d62d; same as in Figure 17.4a) are plotted as a function of right eye position for three different heading directions; $\alpha = 0$ deg (along the animal's naso-occipital axis; middle), $\alpha = 10$ deg (along an axis rotated 10 deg clockwise; left) and $\alpha = -10$ deg (along an axis rotated 10 deg counter-clockwise; right). (b) The same data have been replotted, now relative to the difference angle between gaze and heading directions $(\theta - \alpha)$. The data for the different heading directions superimpose, suggesting that neural firing rates encode the respective dependence of eye velocity (see equation 17.1).

whose firing rate was recorded at different eye positions for three different directions of movement: fore-aft, as well as heading directions 10 deg to the right and to the left. As expected from equation (17.1), eye velocity followed the expected dependence on gaze and heading directions (see also Angelaki and Hess, 2001; Tomko and Paige, 1992). As explained above, during fore-aft motion eye velocity amplitude followed a V-shaped dependence on eye position, as well as a reversal in phase around an eye position of $\sim$0 deg (Figure 17.5a, middle; open circles). During heading directions to the left and to the right of the naso-occipital axis, the V-shaped curve and the phase reversal shifted accordingly along the eye position axis (as expected from equation 17.1). Neural firing rates paralleled this behaviour, as illustrated in Figure 17.5a (solid squares). In fact, the cell did not simply encode gaze but rather a function proportional to the difference angle $(\alpha - \theta)$, as expected from equation (17.1). This is further illustrated in Figure 17.5b, where the three heading direction data sets have been superimposed and plotted as a function of the difference angle $\alpha - \theta$ (rather than θ). Not only eye velocity, but also neural firing rates totally superimpose (Figure 17.5b, open circles and solid symbols, respectively).

17.4   Discussion

Three important issues for understanding the neural organization of the translational VOR have been addressed in these studies. First and in agreement with the conclusions of previous eccentric rotation studies (Chen-Huang and McCrea, 1999a, b; McConville et al., 1996), all neurons exhibited an increase in firing rate as vergence angle increased during pure translational motion. This dependence on vergence was sufficient and appropriate to account for the respective dependence of eye velocity in the TVOR. Second, all neurons tested also exhibited a dependence on heading and gaze directions, as expected from the corresponding dependence of eye velocity in the TVOR (Angelaki and Hess, 2001; Tomko and Paige, 1992). As will be addressed in more detail below, these results argue against dynamic co-contraction as a major mechanism to explain the gaze dependence of the TVOR (Seidman et al., 1999). Finally, the fact that eye-head cells exhibited a much more complex dependence on gaze, which was not identical to the respective dependence of eye velocity, further collaborates results during TVOR suppression (Angelaki et al., 2001) and suggests that eye-head cell responses are not mere replicas of the oculomotor behavior but might represent substrates in the sensorimotor transformations in the TVOR.

17.4.1   Sensorimotor Signal Transformations

Using a conceptual definition of "sensory" vs. "motor" for the vestibulo-ocular system, responses could be considered to be "motor" if they paralleled the functional dependencies of the eye movement and if their properties could directly account for the observed oculomotor behavior. For example, neurons carrying "motor" signals would be expected to exhibit a dependence on vergence angle and eye position as in the TVOR. In addition, neurons with "motor" signals should not modulate during TVOR suppression and would be expected to respond similarly during translation and pursuit, as long as the elicited eye movement was the same. On the other extreme, neurons (e.g., vestibular-only cells) would be considered to carry "sensory" signals if they were "afferent like," i.e., if their firing rate modulation exhibited neither a dependence on vergence angle nor a dependence on eye position and did not in any consistent way relates to the observed eye movements. Finally, neurons that modulate their firing rates with properties in-between "motor" and "sensory" could be considered as candidates of cells that might functionally lie within the sensori-motor processing that converts otolith signals into the TVORs.

The present results (see also Angelaki et al., 2001) suggest the following. First, there was no difference in neural firing rates when identical binocular eye movements (i.e., of the same amplitude, frequency and vergence angle) were elicited during head-stationary pursuit and lateral translation while viewing an earth-stationary target (Angelaki et al., 2001). Second, largely "motor" signals were observed when neural activity was tested during lateral motion while fixating targets at different viewing distances. Third, whereas neural firing rates exhibited a gaze dependence that was qualitatively similar to the respective gaze dependence of eye velocity, eye-head cell responses differed quantitatively from what would be expected if neural responses were mere replicas of the respective eye movements. Finally, eye-head cells differed from type I PVP neurons in the fact that only the former but never the latter modulated during TVOR suppression in the absence of eye movements (Angelaki et al., 2001).

Based on the results outlined above, we have concluded that type I PVP neurons could be considered to carry appropriately transformed "motor" signals during translation. In contrast, the majority of the eye-head cell responses seemed to represent intermediate processing stages in the TVOR in the sense that at least under some circumstances their responses appear to reflect a combination of "motor" and "sensory" signals. Specifically, a subset of these neurons exhibited a clear response modulation during TVOR suppression (Angelaki et al., 2001). Because of the small sensitivity during TVOR suppression, however, stable gaze responses during translation were similar to those during pursuit even for the neurons that exhibited a clear modulation during suppression (Angelaki et al., 2001). In addition, a clear departure from motor behaviour under stable gaze conditions was also observed here in the dependence on eye position during fore-aft motion (Figure 17.4).

The fact that translational but not rotational motion information appears to have been dynamically transformed into oculomotor-like signals at the level of type I PVP neurons has allowed us to form a hypothesis that type I PVP neurons receive direct canal but not otolith afferent signals (Angelaki et al., 2001). In contrast, Eye-Head cells may receive more direct otolith projections and might be functionally located within, rather than at the motor end, of the underlying sensorimotor transformations for the TVOR. Furthermore, neurons whose firing rates increased for ipsilaterally or contralaterally directed eye movements (eye-ipsi and eye-contra cells, respectively) exhibited distinct dynamic properties during TVOR suppression (Angelaki et al., 2001). Specifically, eye-ipsi neurons demonstrated relatively flat dynamics that were similar to those of the majority of vestibular-only neurons. In contrast, eye-contra cells were characterized by low-pass filter dynamics relative to linear acceleration and lower sensitivities than eye-ipsi cells (Angelaki et al., 2001).

These data, as well as distinct differences in the short-latency connections in the utriculo-abducens vs. semicircular canal-abducens pathways (Baker et al., 1969; Imagawa et al., 1995; Precht et al., 1969; Richter and Precht, 1968; Schwindt et al., 1973; Uchino et al., 1994, 1996, 1997), have been used to support the RVOR/TVOR model outlined in Figure 17.6. The model is based on a hypothesis originally proposed by Green and Galiana (1998) suggesting that the low-pass filter characteristics of the eye plant rather than central filtering provide the additional high-frequency temporal integration that is required in the TVOR as compared to RVOR pathways. Specifically, Green and Galiana (1998) proposed differential projections of semicircular canal and otolith signals within a shared premotor circuitry that generates the VORs. Figure 17.6b provides one of such realizations of the eye plant hypothesis, as recently supported by experimental observations (Angelaki et al., 2001).

17.4.2   The Floccular Lobe and the Translational VOR

If eye-head cells represent interneurons in the TVOR and since at least a subset of eye-head cells have been shown to receive inhibition from the cerebellar flocculus/ventral paraflocculus (FL/VPF; Lisberger and Pavelko, 1988; Lisberger et al., 1994; Partsalis et al., 1995; Zhang et al., 1995), it is possible that a cerebellar pathway though the FL/VPF is important for the TVOR. In support of this hypothesis, the early ensemble firing of FL/VPF Purkinje cells has been shown to be sensitive to translation, although not to viewing distance (Snyder and King, 1996). Deficits in the translational and eccentric rotation VOR have also been reported in patients with cerebellar dysfunction (Baloh et al., 1995; Crane et al., 2000). A direct, histologically verified FL lesion in monkeys to specifically investigate the importance of the FL/VPF in the properties of the reflex has not yet been reported.

A combination of the two hypotheses regarding Eye-Head cells, i.e., that they might receive primary otolith afferent input (see Angelaki et al., 2001) and direct inhibition from the FL/VPF raises the possibility that at least some of the secondary utriculo-ocular neurons might be floccular target neurons (FTNs). This idea, although totally speculative at this point, is consistent with the neuroanatomical and electrophysiological data that have identified FTNs in the ventrolateral vestibular nuclei (Langer et al., 1985b; Lisberger et al., 1994; Nagao et al., 1997; Sato et al., 1988), locations that at least partly overlap with those of secondary utriculo-ocular neurons (Uchino et al., 1994). Even though not directly investi-

A   Feedforward model of the RVOR/TVOR

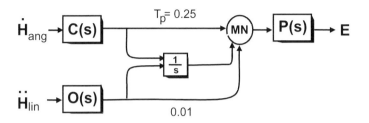

B   Feedback model of the RVOR/TVOR

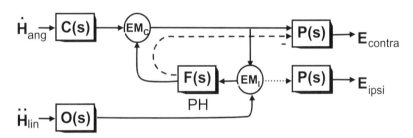

FIGURE 17.6. (a) Feedfoward and (b) feedback models of the RVOR/TVOR (modified after Green and Galiana, 1998; Angelaki et al., 2001). Top: The classical feedforward RVOR model consists of two parallel pathways: One pathway that conveys head velocity signals from the semicircular canals, $C(s)$, via vestibular neurons directly to extraocular motoneurons (MN) and an indirect pathway via a neural integrator (1/s, representing a low-pass filter with a long time constant; Robinson, 1981; Skavenski and Robinson, 1973). Linear acceleration signals from the otolith organs, $O(s)$, project to the eye plant though a set of parallel pathways where the projection via the neural integrator that is shared with the RVOR is stronger compared to a very weak direct otolith-ocular projection (set to 1 and 0.01, respectively). Bottom: In a feedback realization of the RVOR (Galiana and Outerbridge 1984), eye movement–sensitive vestibular neurons are proposed to be interconnected in positive feedback loops with a neural filter, $F(s)$, that represents an internal model of the eye plant, i.e., $F(s) = P(s)$ (presumed to exist in the nucleus prepositus hypoglossi, PH). Canal and otolith signals project onto a shared premotor network at unique sites such that a lumped premotor eye-contraversive sensitive cell type (EMC) receives direct canal projections but only indirect otolith projections via an ipsiversive neuron (EMI) and the neural filter $F(s)$. The EMI cell population is assumed to make weak excitatory projections to the ipsilateral abducens (dotted lines), as well as stronger projections to the ipsilateral PH (output of $F(s)$). Solid lines represent excitatory connections. Dashed lines from the PH to the contralateral abducens are used for inhibitory connections (-).

gated in primates (but see Scudder and Fuchs, 1992), it is generally thought that at least some FTNs are premotor cells that project directly to the ipsilateral abducens (Highstein, 1973; Ito et al., 1977; Sato et al., 1988). The otolith signals to the flocculus could either be direct (Vrabec et al., 1997) or indirect through the caudal medial and inferior subdivisions of the vestibular nuclei (Langer et al., 1985a). These FL-projecting areas considerably overlap with those that receive direct utricular afferent input (Imagawa et al., 1995; Siegborn and Grant, 1983).

17.4.3   Dependence on Gaze Direction

Whereas the TVOR dependence on the inverse of viewing distance could be implemented neurally by a simple scaling transformation, the reflex dependence on gaze is more complex and dependent on the direction of motion (see equation 17.1). Previous studies have postulated two possible mechanisms to account for the V-shaped curve of eye velocity as a function of gaze during fore-aft motion (Seidman et al., 1999). One possibility is that the nonlinear scaling described by equation (17.1) is neurally implemented, whereas muscle activation is based on the traditional Sherringtonian laws of reciprocal innervation of agonist–antagonist interactions. For this to be the case, neural activities during fore-aft motion must also exhibit a V-shape dependence on eye position and reverse modulation phase for gaze directions to the right and to the left. Alternatively, neural activities can maintain the same phase relationship to the head translational stimulus for leftward and rightward eye positions. According to this latter scheme, a dynamic co-contraction of the agonist and antagonist muscles for each eye could provide the flexibility to change the direction of the elicited eye movement as a function of gaze (Seidman et al., 1999).

The present results, showing that all neurons exhibited a change in response modulation phase (relative to head velocity) for leftward and rightward gaze directions, are not consistent with the co-contraction hypothesis. Thus, the present results would support a neural implementation of equation (17.1), although the details regarding such multivariate nonlinear scaling remain subject to future studies.

Acknowledgement

The work was supported by grants from NIH (EY12814 and DC04260).

References

Angelaki, D. E. (1998). Three-dimensional organization of otolith-ocular reflexes in rhesus monkeys. III. Responses to translation. *J. Neurophysiol.*, 80: 680–695.

Angelaki, D. E. and Dickman, J. D. (2000). Spatiotemporal processing of linear acceleration: Primary afferent and central vestibular neuron responses. *J. Neurophysiol.*, 84:
2113–2132.

Angelaki, D. E., Green, A. M. and Dickman, J. D. (2001). Differential sensorimotor processing of vestibulo-ocular signals during rotation and translation. *J. Neurosci.*, 21:
3968–3985.

Angelaki, D. E. and Hess, B. J. M. (2001). Direction of heading and vestibular control of
binocular eye movements. *Vis. Res.*, 4: 3215–3218.

Angelaki, D. E. and McHenry, M. Q. (1999). Short-latency primate vestibulo-ocular responses during translation. *J. Neurophysiol.*, 82: 1651–1654.

Angelaki, D. E., McHenry, M. Q. and Hess, B. J. M. (2000). Primate translational vestibulo-
ocular reflexes. I. High frequency dynamics and three-dimensional properties during
lateral motion. *J. Neurophysiol.*, 83: 1637–1647.

Baker, R. G., Mano, N. and Shimazu, H. (1969). Postsynaptic potentials in abducens
motoneurons induced by vestibular stimulation. *Brain Res.*, 15: 577–580.

Baloh, R. W., Yue, Q. and Demer, J. L. (1995). The linear vestibulo-ocular reflex in normal subjects and patients with vestibular and cerebellar lesions. *J. Vestib. Res.*, 5:
349-361.

Busettini, C., Masson, G. S. and Miles, F. A. (1996a). A role for stereoscopic depth cues
in the rapid visual stabilization of the eyes. *Nature*, 380: 342–345.

Busettini, C., Masson, G. S. and Miles, F. A. (1997). Radial optic flow induces vergence
eye movements with ultra-short latencies. *Nature*, 390: 512-515.

Busettini, C., Miles, F. A. and Krauzlis, R. J. (1996b). Short-latency disparity vergence
responses and their dependence on a prior saccadic eye movement. *J. Neurophysiol.*,
75: 1392–1410.

Busettini, C., Miles, F. A., Schwarz, U. and Carl, J. R. (1994). Human ocular responses to
translation of the observer and of the scene: dependence on viewing distance. *Exp.
Brain Res.*, 100: 484–494.

Chen-Huang, C. and McCrea, R. A. (1999a). Effects of viewing distance on the responses
of horizontal canal-related secondary vestibular neurons during angular head rotation. *J. Neurophysiol.*, 81: 2517–2537.

Chen-Huang, C. and McCrea, R. A. (1999b). Effects of viewing distance on the responses
of vestibular neurons to combined angular and linear vestibular stimulation. *J. Neurophysiol.*, 81: 2538–2557.

Crane, B. T., Tian, J.-R. and Demer, J. L. (2000). Initial vestibulo-ocular reflex during transient angular and linear acceleration in human cerebellar dysfunction. *Exp.
Brain Res.*, 130: 486–496.

Cullen, K. E. and McCrea, R. A. (1993). Firing behaviour of brain stem neurons during
voluntary cancellation of the horizontal vestibuloocular reflex. I. Secondary vestibular neurons. *J. Neurophysiol.*, 70: 828–843.

Fuchs, A. F. and Luschei, E. S. (1970). Firing patterns of abducens neurons of alert monkeys in relationship to horizontal eye movement. *J. Neurophysiol.*, 33: 382–392.

Galiana, H. L. and Outerbridge, J. S. (1984). A bilateral model for central neural pathways
in the vestibulo-ocular reflex. *J. Neurophysiol.* 51: 210–241.

Gianna, C. C., Gresty, M. A. and Bronstein, A. M. (1997). Eye movements induced by lateral acceleration steps: effect of visual context and acceleration levels. *Exp. Brain Res.*, 114: 124-129.

Green, A. M. and Galiana, H. L. (1998). A hypothesis for shared central processing of canal and otolith signals. *J. Neurophysiol.*, 80: 2222–2228.

Highstein, S. M. (1973). Synaptic linkage in the vestibulo-ocular and cerebello-vestibular pathways to the VIth nucleus in the rabbit. *Exp. Brain Res.*, 17: 301–314.

Imagawa, M., Isu, N., Sasaki, M., Endo, K., Ikegami, H. and Uchino, Y. (1995). Axonal projections of utricular afferents to the vestibular nuclei and the abducens nucleus in cats. *Neurosci. Let.*, 186: 87–90.

Ito, M., Nisimaru, N. and Yamamoto, M. (1977). Specific patterns of neuronal connexions involved in the control of the rabbit's vestibulo-ocular reflexes by the cerebellar flocculus. *J. Physiol.*, 265: 833–854.

Langer, T. P., Fuchs, A. F., Chubb, M. C., Scudder, C. A. and Lisberger, S. G. (1985b). Floccular efferents in the rhesus macaque as revealed by autoradiography and horseradish peroxidase. *J. Comp. Neurol.*, 235: 26–37.

Langer, T. P., Fuchs, A. F., Scudder, C. A. and Chubb, M. C. (1985a). Afferents to the flocculus of the cerebellum in the rhesus macaque as revealed by retrograde transport of horseradish peroxidase. *J. Comp. Neurol.*, 235: 1–25.

Lisberger, S. G. and Pavelko, T. A. (1988). Brain stem neurons in modified pathways for motor learning in the primate vestibulo-ocular reflex. *Science*, 242: 771–773.

Lisberger, S. G., Pavelko, T. A. and Broussard, D. M. (1994). Responses during eye movements of brain stem neurons that receive monosynaptic inhibition from the flocculus and ventral paraflocculus in monkeys. *J. Neurophsiol.*, 72: 909–927.

McConville, K. M. V., Tomlinson, R. D. and Na, E.-Q. (1996). Behavior of eye movement-related cells in the vestibular nuclei during combined rotational and translational stimuli. *J. Neurophysiol.*, 76: 3136–3148.

McCrea, R. A., Strassman, A., May, E. and Highstein, S. M. (1987). Anatomical and physiological characteristics of vestibular neurons mediating the horizontal vestibulo-ocular reflex of the squirrel monkey. *J. Comp. Neurol.*, 264: 547–570.

McHenry, M. Q. and Angelaki, D. E. (2000). Primate translational vestibulo-ocular reflexes. II Vergence and version responses during fore-aft motion. *J. Neurophysiol.* 83: 1648–1661.

Miles, F. A. (1993). The sensing of rotational and translational optic flow by the primate optokinetic system. In F. A. Miles and J. Wallman (eds), *Visual Motion and its Role in the Stabilization of Gaze*, pp. 393-403, Amsterdam: Elsevier.

Miles, F. A. (1998). The neural processing of 3-D visual information: Evidence from eye movements. *Eur. J. Neurosci.*, 10: 811–822.

Nagao, S., Kitamura, T., Nakamura, N., Hiramatsu, T. and Yamada, J. (1997). Differences of the primate flocculus and ventral paraflocculus in the mossy and climbing fiber input organization. *J. Comp. Neur.*, 382: 480–498.

Paige, G. D. and Tomko, D. L. (1991a). Eye movement responses to linear head motion in the squirrel monkey. I. Basic characteristics. *J. Neurophysiol.*, 65: 1170–1182.

Paige, G. D. and Tomko, D. L. (1991b). Eye movement responses to linear head motion in the squirrel monkey. II. Visual-vestibular interactions and kinematic considerations. *J. Neurophysiol.*, 65: 1183–1196.

Partsalis, A. M., Zhang, Y. and Highstein, S. M. (1995). Dorsal Y group in the squirrel monkey. I Neuronal responses during rapid and long-term modifications of the vertical VOR. *J. Neurophysiol.*, 73: 615–631.

Precht, W., Richter, A. and Grippo, J. (1969). Responses of neurones in cat's abducens nuclei to horizontal angular acceleration. *Pflügers Arch.*, 309: 285–309.

Richter, A. and Precht, W. (1968). Inhibition of abducens motoneurones by vestibular nerve stimulation. *Brain Res.*, 11: 701-705.

Robinson, D. A. (1981). The use of control systems analysis in the neurophysiology of eye movements. *Ann. Rev. Neurosci.*, 4: 463–503.

Sato, Y., Kanda, K.-I. and Kawasaki, T. (1988). Target neurons of floccular middle zone inhibition in medial vestibular nucleus. *Brain Res.*, 446: 225–235.

Schwarz, U. and Miles, F. A. (1991). Ocular responses to translation and their dependence on viewing distance: I Motion of the observer. *J. Neurophysiol.*, 66: 851–864.

Schwarz, U., Busettini, C. and Miles, F. A. (1989). Ocular responses to translation are inversely proportional to viewing distance. *Science*, 245: 1394–1396.

Schwindt, P. C., Richter, A. and Precht, W. (1973). Short latency utricular and canal input to ipsilateral abducens motoneurons. *Brain Res.*, 60: 259–262.

Scudder, C. A. and Fuchs, A. F. (1992). Physiological and behavioral identification of vestibular nucleus neurons mediating the horizontal vestibuloocular reflex in trained rhesus monkeys. *J. Neurophysiol.*, 68: 244–264.

Seidman, S. H., Paige, G. D. and Tomko, D. L. (1999). Adaptive plasticity in the nasoocipital linear vestibulo-ocular reflex. *Exp. Brain Res.*, 125: 485–494.

Siegborn, J. and Grant, G. (1983). Brainstem projections of different branches of the vestibular nerve. An experimental study by transganglionic transport of horseradish peroxidase in the cat. I. The horizontal ampullar and utricular nerves. *Arch. Italiennes de Biologie*, 121: 237–248.

Skavenski, A. A. and Robinson, D. A. (1973). Role of abducens neurons in the vestibuloocular reflex. *J. Neurophysiol.* 36: 724–738.

Snyder, L. H. and King, W. M. (1996). Behavior and physiology of the macaque vestibuloocular reflex response to sudden off-axis rotation: Computing eye translation. *Brain Res. Bull.*, 40: 293–302.

Telford, L., Seidman, S. H. and Paige, G. D. (1997). Dynamics of squirrel monkey linear vestibuloocular reflex and interactions with fixation distance. *J. Neurophysiol.*, 78: 1775–1790.

Tomko, D. L. and Paige, G. D. (1992). Linear vestibuloocular reflex during motion along axes between nasooccipital and interaural. *Ann. New York Acad. Sci.*, 656: 233–241.

Tomlinson, R. D. and Robinson, D. A. (1984). Signals in vestibular nucleus mediating vertical eye movements in the monkey. *J. Neurophysiol.*, 51: 1121–1135.

Uchino, Y., Ikegami, H., Sasaki, M., Endo, K., Imagawa, M. and Isu, N. (1994). Monosynaptic and disynaptic connections in the utriculo-ocular reflex arc of the cat. *J. Neurophysiol.*, 71: 950–958.

Uchino, Y., Sasaki, M., Sato, H., Imagawa, M., Suwa, H. and Isu, N. (1996). Utriculoocular reflex arc of the cat. *J. Neurophysiol.*, 76: 1896–1903.

Uchino, Y., Sasaki, M., Sato, H., Imagawa, M., Suwa, H. and Isu, N. (1997). Utricular input to cat extraocular motoneurons. *Acta Otolaryngol. Suppl.*, 528: 44–48.

Vrabec, J. T., Perachio, A. A. and Purcell, I. M. (1997). Transganglionic labeling of utricular primary afferent neuron projections to the brainstem and cerebellum in the macaque. *ARO Midwinter*, Tampa, FL.

Zhang, Y., Partsalis, A. M. and Highstein, S. M. (1995). Properties of superior vestibular nucleus flocculus target neurons in the squirrel monkey. I General properties in comparison with flocculus projecting neurons. *J. Neurophysiol.*, 73: 2261–2278.

18

Influence of Rotational Cues on the Neural Processing of Gravito-Inertial Force

Daniel M. Merfeld and Lionel H. Zupan

Sensory systems often provide ambiguous information. For example, otolith organs measure gravito-inertial force (GIF), the sum of gravitational force and inertial force due to linear acceleration. According to Einstein's equivalence principle, no set of linear accelerometers alone can distinguish gravitational force (which changes with head orientation during head tilt) from inertial force (which changes with linear acceleration of the head). Therefore, the central nervous system (CNS) must use other sensory cues to distinguish tilt from translation. For example, the CNS can use rotational cues provided by the semicircular canals and vision. Much of this chapter provides a brief review of studies showing the influence of rotational cues on the neural processing of tilt and translation. However, we also include preliminary unpublished data. We begin by discussing the underlying physics (and associated neural processes) and neural representations. We then present studies that measure the influence of rotational cues on tilt responses before presenting studies of translation responses. We finish by reviewing modeling approaches to sensory integration for both tilt and translation responses.

18.1   Introduction

Like all linear accelerometers, the otolith organs measure gravito-inertial force (GIF), which is the vector sum of gravitational force and inertial force due to linear acceleration (Figure 18.1). Since responses to tilt with respect to gravity (e.g., posture control) must differ from responses to linear acceleration (e.g., translational vestibulo-ocular responses), this gravito-inertial ambiguity presents a problem for the nervous system. No set of linear accelerometers alone can resolve this ambiguity. This is true regardless of how many linear accelerometers are in the set. How can the nervous system, which has no other way to determine the state of the external world but through its imperfect sensory systems, determine what portion of the otolith cue is due to gravity and what portion is due to linear acceleration? Data presented herein show that other sensory cues are used to help perform the neural processes necessary to estimate tilt and translation.

Which cues are available to assist with the resolution of the gravito-inertial ambiguity? The semicircular canals provide rotational cues measuring angular head movements; we will show that these cues assist with the neural processing of GIF cues. Visual rotational motion cues also provide rotational self-motion cues; we will show that these cues assist with the neural processing of GIF cues. Rotation about an axis aligned with gravity (e.g., shaking one's head to signal "no") does not influence the orientation of gravity relative to the head. But, whenever head rotation includes a component that is not aligned with gravity (e.g., nodding one's head to signal "yes"), the relative orientation of gravity with respect to the head changes. These "physics" show that rotation influences the relative orientation of gravity. Mimicking these real physical effects, the nervous system also appears to utilize rotational cues, provided via the semicircular canals and visual system, to help estimate the relative orientation of gravity. In Section 18.3 we will present studies showing the physiological influence of rotational cues on tilt responses.

As mentioned above, all linear accelerometers measure GIF, which is gravity minus linear acceleration (Figure 18.1). Mimicking this real physical effect, the nervous system appears to separate the ambiguous measurement of GIF into neural representations of gravity and linear acceleration with the difference between the two, estimated gravity minus estimated linear acceleration, equaling the estimated (or measured) specific GIF (Figure 18.2). Such processing is essential because GIF must be separated into gravitational and linear acceleration components in order to utilize the ambiguous measurement. For example, linear acceleration can be calculated (estimated) by subtracting the measurement of GIF from the calculated (estimated) gravitational vector. If such neural calculations are performed, any direct influence of rotational cues on tilt responses (as discussed in the previous paragraph) will indirectly influence translation responses. In Section 18.4 we will present studies showing the physiological influence of rotational cues on translation responses.

It is worth noting that much research has addressed the issue of how gravitational cues influence rotational responses (e.g., Angelaki and Hess, 1994; Benson, 1974; Guedry, 1965; Merfeld et al., 1993b; Raphan et al., 1981). The work presented in this chapter complements this work, since it focuses on the influence of rotational cues on tilt (gravity) and translation (linear acceleration) responses as opposed to the influence of gravity on rotation responses.

In addition, we will not draw a distinction between those studies that utilize experimental conditions to elicit illusory responses (e.g., Dichgans et al., 1972; Merfeld et al., 1999; Merfeld et al., 2001; Zupan et al., 2000) from those that utilize experimental conditions to elicit veridical responses (e.g., Angelaki et al., 1999; Merfeld and Young, 1995; Stockwell and Guedry, 1970; Zacharias and Young, 1981). These approaches offer complementary methodologies.

Much of this chapter provides a brief review of studies showing the influence of rotational cues on the neural processing of tilt and translation, but we also include preliminary unpublished data as well. We begin by presenting a discussion of physiological rotation cues available to the nervous system followed by a discussion of the underlying physics of rotation and GIF as well as the associated neural processes and representations. We then present studies that measure the influence of rotational cues on tilt responses before presenting studies of translation responses.

18.2   Background

18.2.1   Rotation Cues

In the two-decade frequency range from several hundredths to several Hz, which covers much of the physiologic range, the response of the canals well represents angular velocity (Wilson and Melvill Jones, 1979). Therefore, these sensors are often considered angular velocity sensors. At lower frequencies, the response of the canals begins to more closely represent angular acceleration. Furthermore, the canals show no response during extended constant velocity rotation.

The canal dynamics provide an experimental tool, utilized by many of the studies presented herein, allows the presentation of rotation cues from the canals in the absence of actual rotation. In brief, subjects are rotated at a constant velocity for an extended period of time, then rapidly brought to a stop. Transiently, in the absence of actual rotation, the canals provide a cue indicating rotation in the direction opposite the preceding rotation for more than 10 s. The actual strength of this postrotary canal cue depends primarily upon the duration and speed of the constant-velocity portion of the rotation as well as the rapidity of the deceleration.

Visual rotational motion cues yield sensations of rotation in the direction opposite the visual field rotation (Brandt et al., 1973, 1974; Howard and Heckmann, 1989; Schor et al., 1984). These illusory motion sensations are called "vection." Complementary to the canal dynamics described above, these responses are strongest at lower frequencies, where the canal responses are small, and fall off at higher frequencies (e.g., Mergner and Becker, 1990). Since rotational cues provided by the semicircular canals and those provided by the visual system combine in the vestibular nuclei (Waespe and Henn, 1977), very early in the neural processing of these cues, we will not treat the visual rotational cues differently than the canal cues.

The influence of visual orientation cues on perceived tilt has long been known (e.g., Asch and Witkin, 1948; Howard and Childerson, 1994; Howard and Heckmann, 1989). The work presented herein differs from this previous work, since we focus on the issue of how visual rotational motion cues, without orientation cues, influence the neural processing of gravity (tilt) and linear acceleration (translation). We acknowledge that visual orientation cues also play a key role in these neural processes; the influence of these cues is addressed elsewhere (Oman, 2002).

18.2.2 Physics

Before proceeding with an analysis of how the nervous system processes otolith cues, it is crucial to have an understanding of the underlying physics. First, we review our approach to understanding gravity and inertial forces. (Alternate approaches, which equivalently represent the physics, are briefly discussed in the appendix to this chapter.) Then we review how rotation influences the relative orientation of gravity with respect to the head.

Gravito-Inertial Force (GIF)

Our approach derives from the French mathematician d'Alembert who recognized that Newton's 2nd Law ($m\mathbf{a} = m\mathbf{g} + \sum \mathbf{F}$, where m is mass, $\mathbf{a}$ is linear acceleration *relative to an inertial reference frame*, $\mathbf{g}$ is gravity and $\sum \mathbf{F}$ is the sum of all nongravitational forces), could be applied in noninertial (rotating or accelerating) environments by defining fictional forces, which have come to be known as inertial (or d'Alembert) forces.

We define linear acceleration of the head (**a**) *relative to an inertial frame of reference*; inertial force per unit mass is defined as the negative of this linear acceleration ($\mathbf{f}_i = -\mathbf{a}$). GIF per unit mass (**f**) is then defined (Figure 18.1a) as the sum of gravitational force per unit mass (**g**) and the inertial (or d'Alembert) force per unit mass ($\mathbf{f}_i$), $\mathbf{f} = \mathbf{g} + \mathbf{f}_i$. Simple substitution yields $\mathbf{f} = \mathbf{g} - \mathbf{a}$, which we use to represent the relationship between gravity, linear acceleration, and GIF.[1]

We proceed using centrifugation as an example. For simplicity, we first present the analysis for fixed-radius constant angular velocity centrifugation, neglecting the "Coriolis" and "tangential" accelerations generally present. During fixed-radius constant-velocity centrifugation, the subject is always linearly accelerating toward the center of rotation; this linear acceleration is generally referred to as centripetal acceleration. The representation of this linear acceleration varies with the choice of reference frame. Figure 18.1 shows the representation of centripetal acceleration in a non-inertial head-fixed reference frame (Figs. 18.1a and b) as well as in an inertial reference frame (Figure 18.1c).

The inertial force that corresponds to centripetal acceleration is called centrifugal force and is experienced in the head-fixed reference frame. (This is perhaps the most widely known inertial force.) Its magnitude per unit mass is the distance from the center of rotation (r) times the magnitude of angular velocity (ω) squared ($|\mathbf{a}_C| = r\omega^2$). D'Alembert showed that Newton's laws can be applied in this head-fixed noninertial environment if we replace gravity with GIF (**f**): $m\mathbf{a} = m\mathbf{f} + \sum \mathbf{F}$, where **a** is the linear acceleration *relative to the head*[2] (fixed relative to the centrifuge arm) of an object having mass (m).

---

[1]Our approach matches that used previously in the *Handbook of Physiology* (Young, 1984). Young's approach, defining specific force equal to gravity minus linear acceleration, was derived from inertial guidance (Wrigley et al., 1969).

[2]**a** is linear acceleration relative to the head. This is distinct from linear acceleration relative to an inertial frame (**a**).

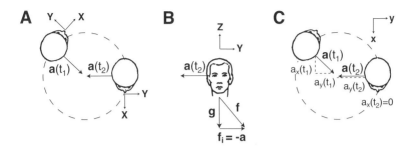

FIGURE 18.1. Inertial stimulation and gravito-inertial force (GIF).

In the subject's head-fixed reference frame (X, Y, Z), the head acceleration is represented by a constant interaural centripetal acceleration vector ($\mathbf{a}$) pointing toward the center of rotation. (a) View from above; (b) view from facing in the (Y, Z) plane. At two different head locations in space corresponding to two different instants, t_1 and t_2, we have $\mathbf{a}(t_1) = \mathbf{a}(t_2)$ in the head-fixed reference frame (X, Y, Z). This provides a simple representation of the linear acceleration of the head, but this rotating reference frame is a noninertial environment. Since Newton's 2nd Law ($m\mathbf{a} = m\mathbf{g} + \sum \mathbf{F}$, where m is mass, $\mathbf{a}$ is linear acceleration, $\mathbf{g}$ is gravity and, $\sum \mathbf{F}$ is the sum of all nongravitational forces) is only applicable in inertial reference frames, D'Alembert defined "fictional" inertial forces that allow the use of Newton's 2nd Law in noninertial frames of reference. The inertial force in this eccentric rotation example is called centrifugal force. Inertial force per unit mass ($\mathbf{f}_i$) is defined to be equal and opposite the linear acceleration of the head ($\mathbf{f}_i = -\mathbf{a}$). GIF per unit mass ($\mathbf{f}$) is defined as the sum of gravity ($\mathbf{g}$) plus the inertial (centrifugal) force per unit mass, $\mathbf{f} = \mathbf{g} + \mathbf{f}_i$. Substituting the definition of inertial force, we obtain $\mathbf{f} = \mathbf{g} - \mathbf{a}$. With GIF per unit mass ($\mathbf{f}$) replacing gravity ($\mathbf{g}$), D'Alembert showed that we can now apply Newton's 2nd Law in the non-inertial head-fixed frame ($m\mathbf{a} = m\mathbf{f} + \sum \mathbf{F}$) to determine the acceleration of an otoconia (or any other object) *relative to the head* ($\mathbf{a}$). (c) In the earth-fixed coordinate system (x,y,z), centripetal acceleration is a rotating vector that can be represented by $a_x = |\mathbf{a}| \sin \omega t$, $a_y = |\mathbf{a}| \cos \omega t$ and $a_z = 0$, where $|\mathbf{a}|$ represents the magnitude of the head linear acceleration. These equations describe the subject's motion in a circular path. Like planetary motion, the circular path results from the subject's constant acceleration toward the center of rotation. At different head locations, we have different linear acceleration in the earth-fixed reference frame. For example, for the head locations shown at t_1 and t_2, $a_x(t_1) \neq a_x(t_2)$ and $a_y(t_1) \neq a_y(t_2)$. In principle, we can apply Newton's 2nd law to calculate the acceleration of an individual otoconia *relative to the inertial coordinate system* ($\mathbf{a}_{oto}$ as $m\mathbf{a}_{oto} = m\mathbf{g} + \sum \mathbf{F}$). We can then calculate the acceleration of the otoconia relative to the head ($\mathbf{a}_{oto}$) by determining and subtracting the acceleration of the head with respect to the inertial frame of reference (a_x, a_y and a_z). Often, as in this example, the introduction of inertial forces (panels a and b) simplifies the analysis by eliminating time-varying elements such as the oscillating linear acceleration (a_x, a_y) found in the inertial reference frame.

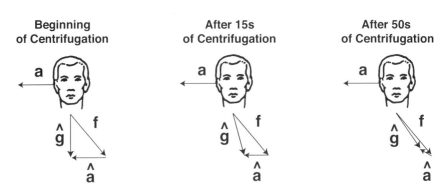

FIGURE 18.2. GIF resolution during fixed-radius centrifugation (e.g., Figure 18.1). According to the GIF resolution hypothesis, the nervous system estimates gravity ($\hat{\mathbf{g}}$) and linear acceleration ($\hat{\mathbf{a}}$) such that their difference approximately matches the GIF ($\mathbf{f} \approx \hat{\mathbf{g}} - \hat{\mathbf{a}}$). At the very beginning of the centrifugation when canal cues are still present, the nervous system correctly estimates both gravity ($\hat{\mathbf{g}} \approx \mathbf{g}$) and linear acceleration ($\hat{\mathbf{a}} \approx \mathbf{a}$). After 10s of centrifugation, the canal cues have begun to decay, and the estimate of gravity ($\hat{\mathbf{g}}$)) has begun to shift toward alignment with the measured GIF ($\mathbf{f}$); the amplitude of the estimate of linear acceleration ($\hat{\mathbf{a}}$) decreases accordingly. After 50 s of centrifugation, canal cues have nearly completely decayed, and the estimate of gravity ($\hat{\mathbf{g}}$) has nearly aligned with the measured GIF ($\mathbf{f}$). Concomitantly, the amplitude of the estimate of linear acceleration ($\hat{\mathbf{a}}$) becomes small.

Consider the acceleration of a single otoconia. In the steady-state, the acceleration of an otoconia relative to the head ($\mathbf{a}$) equals zero, so the sum of the external forces (e.g., tissue stress including hair cell bending forces) exactly balances the inertial force $0 = m\mathbf{f} + \sum \mathbf{F}$. Studies suggest that this balancing occurs with a time constant of around 2 ms (Grant and Best, 1986), so the relative acceleration of the otoconia (i.e., otolith organ dynamics) can be ignored for frequencies less than around 70 Hz. At higher frequencies, GIF can be considered *the* driving force causing motion of the otoconia, but otolith dynamics (e.g., viscous and other mechanical forces) also influence the otoconial motion. If Grant's analysis (1986) is correct, otoconial displacement is proportional to GIF for frequency components less than 70 Hz.

During transient rotational motions, centrifugation might also include "Coriolis" ($\mathbf{a}_{co}$) and "tangential" ($\mathbf{a}_t$) accelerations in addition to centripetal acceleration ($\mathbf{a}_c$). If so, GIF can be represented as $\mathbf{f} = \mathbf{g} - (\mathbf{a}_c + \mathbf{a}_{co} + \mathbf{a}_t)$, where $\mathbf{a}_c + \mathbf{a}_{co} + \mathbf{a}_t$ is simply the total linear acceleration of the head. Similarly, during purely translational motion, the linear acceleration of the head ($\mathbf{a}_{tr}$) is often known and controlled. GIF can simply be calculated as $\mathbf{f} = \mathbf{g} - \mathbf{a}_{tr}$. Again, both of these relationships are specific applications of the general definition that $\mathbf{f} = \mathbf{g} - \mathbf{a}$, where $\mathbf{a}$ is the linear acceleration of the head relative to an inertial reference frame.

Rotation and Gravity

When the head rotates, the motion of gravity relative to the head can be represented by the differential equation, $d\mathbf{g}/dt = -\boldsymbol{\omega} \times \mathbf{g}$, where $\boldsymbol{\omega}$ is the angular velocity of the head relative to the inertial frame of reference. For example, this equation dynamically describes how gravity pitches forward and backward relative to the head-fixed frame of reference when the head nods signaling "yes." (It is actually the head that is moving with respect to gravity, but in the head-fixed reference frame the opposite appears true.) This equation was directly obtained from the mathematical relationship between time derivative operators in rotating (head) and fixed (external world) frames of reference (Spiegel, 1972). If angular velocity is known over a period of time and the initial orientation of gravity is known, this equation can be integrated to calculate the relative orientation of gravity at any time, $\mathbf{g} = \int (d\mathbf{g}/dt)dt = \int (-\boldsymbol{\omega} \times \mathbf{g})dt$.

18.2.3   Internal Models and Neural Representations of Physical Quantities

To avoid confusion, it is important to distinguish neural quantities or representations from the physical quantities they represent. The equations $d\mathbf{g}/dt = -\boldsymbol{\omega} \times \mathbf{g}$ and $\mathbf{f} = \mathbf{g} - \mathbf{a}$ presented above mathematically represent real-world physics. If the nervous system includes internal models[3] of these physical effects (e.g., Angelaki et al., 1999, 2001; Merfeld et al., 1993a, 1999, 2001; Merfeld and Young, 1995; Zupan et al., 2000), these internal models can be mathematically represented as $\hat{\mathbf{f}} = \hat{\mathbf{g}} - \hat{\mathbf{a}}$ and $d\hat{\mathbf{g}}/dt = -\hat{\boldsymbol{\omega}} \times \hat{\mathbf{g}}$, where $\hat{\mathbf{f}}$, $\hat{\mathbf{g}}$, $\hat{\mathbf{a}}$, and $\hat{\boldsymbol{\omega}}$ are neural representations of the physical variables ($\mathbf{f}$, $\mathbf{g}$, $\mathbf{a}$, and $\boldsymbol{\omega}$) described above. The carets simply distinguish neural representations from real physical variables, since the nervous system only has access to neural representations or to sensory measurements of the physical variable, not the physical variables per se.

We can integrate the differential equation $d\hat{\mathbf{g}}/dt = -\hat{\boldsymbol{\omega}} \times \hat{\mathbf{g}}$ to obtain the neural representation of gravity as a function of time, $\hat{\mathbf{g}} = \int (d\hat{\mathbf{g}}/dt)dt = \int (-\hat{\boldsymbol{\omega}} \times \hat{\mathbf{g}})dt$. Furthermore, $\hat{\mathbf{f}} = \hat{\mathbf{g}} - \hat{\mathbf{a}}$ and $d\hat{\mathbf{g}}/dt = -\hat{\boldsymbol{\omega}} \times \hat{\mathbf{g}}$ can be combined, eliminating $\hat{\mathbf{g}}$, to yield $d\hat{\mathbf{a}}/dt = -\hat{\boldsymbol{\omega}} \times \hat{\mathbf{a}} - d\hat{\mathbf{f}}/dt - \hat{\boldsymbol{\omega}} \times \hat{\mathbf{f}}$, which, notational differences aside, is similar to the equation derived by Angelaki and colleagues (1999). Finally, we assume that the neural representation of GIF approximately equals the GIF itself ($\hat{\mathbf{f}} \approx \mathbf{f}$). We will not distinguish the neural representation of GIF, from actual GIF, though this may provide a fruitful endeavor in the future.

The GIF resolution equation $\hat{\mathbf{f}} = \hat{\mathbf{g}} - \hat{\mathbf{a}}$ can be rewritten as $\hat{\mathbf{a}} = \hat{\mathbf{g}} - \hat{\mathbf{f}}$, which implies that the nervous system can calculate ("estimate") linear acceleration from

---

[3] An internal model is a neural system that mimics a physical process (e.g., a physical relationship, sensory dynamics, or motor dynamics). When a physical process can be described by a mathematical operation (e.g., $\mathbf{f} = \mathbf{g} - \mathbf{a}$), an internal model signifies that a neural process equivalent to this mathematical operation occurs ($\hat{\mathbf{f}} = \hat{\mathbf{g}} - \hat{\mathbf{a}}$) within the neurons that calculate and encode the neural representations (e.g., $\hat{\mathbf{f}}$, $\hat{\mathbf{g}}$, $\hat{\mathbf{a}}$, or $\hat{\boldsymbol{\omega}}$) of the physical variables (e.g., $\mathbf{f}$, $\mathbf{g}$, $\mathbf{a}$, or $\boldsymbol{\omega}$).

the neural representations of GIF ($\hat{\mathbf{f}}$) and gravity ($\hat{\mathbf{g}}$). Furthermore, since the neural representation of GIF is assumed approximately equal to GIF, we can calculate linear acceleration as $\hat{\mathbf{a}} \approx \hat{\mathbf{g}} - \mathbf{f}$. This implies that we, as investigators, can calculate a putative representation of linear acceleration from the difference between a measure of the neural representation of gravity ($\hat{\mathbf{g}}$) and GIF ($\mathbf{f}$). If the neural representation of gravity is misaligned with GIF, a non-zero estimate of linear acceleration should result (Figure 18.2).

In the following section we focus on the effect that rotational cues ($\hat{\boldsymbol{\omega}}$) have on tilt responses ($\hat{\mathbf{g}}$). In our models, this influence appears relatively direct, with the neural representation of gravity influenced by rotational cues. Mathematically, this direct influence is written as $\hat{\mathbf{g}} = \int (-\hat{\boldsymbol{\omega}} \times \hat{\mathbf{g}}) dt$.

Next, in Section 18.4 we focus upon the effect that rotational cues have on translation responses. In our models, this influence of rotational cues is less direct for translation responses that for tilt. In this section, we focus upon the separation of linear acceleration from gravity, which can be represented mathematically as $\hat{\mathbf{a}} = \hat{\mathbf{g}} - \hat{\mathbf{f}}$. Rotational cues indirectly influence translation responses via their direct influence ($\hat{\mathbf{g}} = \int (-\hat{\boldsymbol{\omega}} \times \hat{\mathbf{g}}) dt$) on the neural representation of gravity.

18.3   Influence of Rotational Cues on Tilt Responses

A number of different experimental paradigms and measurement techniques have been used to demonstrate the influence of rotational cues on tilt responses. A partial coverage of these investigations follows. We begin with studies that utilize perceptual measures of tilt, then proceed to studies that use manual control measures, and end by presenting eye movements.

18.3.1   Perceptual Measures of Tilt

It has long been known that rotational cues influence the perception of tilt. Von Holst and Grisebach (1951) rotated subjects about an earth-horizontal axis of rotation, with the head positioned such that the rotation primarily provided roll stimulation to the head (Figure 18.3). This rotation was performed for several revolutions; then the subjects were decelerated to a stop, initiating a transient postrotational canal response. Following deceleration, the data show a transient illusory roll tilt measured using a visual task. Similarly, we recently found that some subjects verbally reported illusory yaw tilt when postrotational yaw canal cues were present (Merfeld et al., 1999). Our yaw tilt findings were consistent with the roll tilt findings of von Holst; the illusory yaw tilt was almost always in the direction that gravity would have tilted if the subject had truly been rotating in the direction indicated by the semicircular canals.

Another study showed that veridical canal cues helped subjects rapidly and accurately estimate their orientation relative to gravity (Figure 18.4) when they were rapidly roll tilted 30 deg (Stockwell and Guedry, 1970). These data were

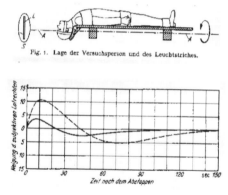

FIGURE 18.3. Perceived vertical following rotation about an earth-horizontal axis. (a) After two to three revolutions (duration of the last revolution taking 4-5s) about the subject's head-fixed naso-occipital axis, subject's indicated their perception of earth-vertical with a luminous bar L attached to a disk S they could rotate by use of hand gears. As indicated on the cartoon, the subject's head is bent backward at the neck. (b) Perceived deviation from physical vertical as a function of time after the subject has been stopped with subject's forehead either up (solid line) or down (dashed line). Reprinted from von Holst, E. and Grisebach, E. (1951). Einfluss des Bogengangssystems auf die "subjektiv" Lotrechte beim Menschen. *Naturwissenschaften*, 38: 67–68, with permission.

contrasted to earlier data (Figure 18.4) showing that illusory roll tilt took 30 – 60 sec to develop fully (Clark and Graybiel, 1966; Graybiel and Brown, 1951), when subjects were tested using fixed-radius constant-velocity centrifugation (similar to that demonstrated in Figures 18.1 and 18.2). The subjects were always upright with respect to gravity during the centrifugation paradigm, with the centrifugal force aligned with the interaural axis (y-axis) yielding roll tilt of the GIF (Figure 18.1). These centrifugation data were originally interpreted and modeled (Mayne, 1974) to indicate that human perception of tilt (gravity) could be explained by some form of low-pass filtering of the otolith cues, which measure the tilted GIF. But the influence of the semicircular canal cue, present during Stockwell and Guedry's rapid roll tilt, allowed these subjects to rapidly and accurately estimate the relative orientation of gravity. This rapid change in perceived roll tilt was inconsistent with the notion that the otolith cues were simply low-passed filtered to yield tilt and showed that roll canal cues could help accurately and rapidly calculate the relative orientation of gravity.

The above studies demonstrate the direct influence of canal cues on tilt in the same plane. In other words, yaw rotation cues influenced yaw tilt and roll rotation cues influenced roll tilt. We (Merfeld et al., 2001) and other colleagues (Seidman et al., 1998) have recently performed studies showing that yaw canal cues can influence the perception of roll tilt and pitch tilt, respectively. Both studies utilized fixed-radius centrifugation, similar to that discussed above, to confirm previously reported results (Clark and Graybiel, 1963; Graybiel and Brown, 1951) that the illusory tilt took 20 sec or longer to develop. In our study, the subjects were oriented

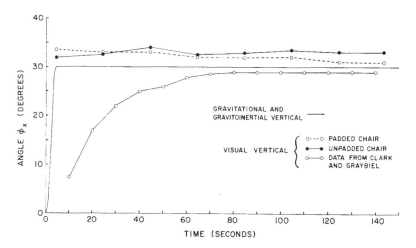

FIGURE 18.4. Mean estimates of visual vertical for subjects tilted at a rate of 6 deg/s^2. Data were combined for both right and left directions of tilt. On the ordinate, angular displacements are expressed in degrees from the subject's z-axis. The solid line represents approximate angular displacement from the subject's z-axis of the gravitational vertical in the present experiment and of the gravito-inertial vertical in (Clark and Graybiel, 1966). Plotted circles (open, closed, or dotted) represent angular displacement from the subject's z-axis of the lighted column used to indicate the visual vertical. Reproduced from Stockwell and Guedry (1970) with permission from *Acta Otolaryngologica*, 70: 170-175.

such that the centrifugal force was aligned with the interaural axis (Merfeld et al., 2001), yielding roll tilt of the GIF. The time course of the perceptual response (Figure 18.5) closely matches that which would occur if the GIF were processed with a low-pass filter[4] having a time constant of 14 s (cut-off frequency of 0.01 Hz).

Both investigations also used another motion paradigm — variable-radius centrifugation. In this condition, the subjects were rotated at a constant velocity for several minutes while near the center of rotation. Then, after the canal cues had decayed, the subjects were moved outward radially, such that the centrifugal force that they experienced was nearly identical to that experienced during the fixed-radius trials. An illusory roll tilt, somewhat like that found during fixed-radius trials, was measured, but the illusory tilt developed much more rapidly during the variable-radius trials. The time course of perceived roll tilt in the absence of canal cues closely matches that which would occur if the GIF were processed with a low-pass filter having a time constant of 6 s (Figure 18.5). This time-constant corresponds to a cut-off frequency of about 0.03 Hz. (It is interesting to note that this time course roughly mimics the dynamics of the semicircular canals which have a time constant around 5 to 6 s. This may indicate that the time course of the roll tilt response during variable-radius centrifugation is limited by the absence of a roll velocity cue from the four vertical canals which measure roll rotation.) Similar results were reported in the pitch plane by Seidman et al. (1998). Since the only significant difference between the fixed-radius and variable-radius paradigms is the presence (fixed-radius) or absence (variable-radius) of yaw canal cues, and since the time course of the roll tilt responses was quite different for these two motion paradigms, the presence of the yaw canal cues has been shown to influence the perception of tilt in both the roll and pitch planes. Possible mechanisms are discussed elsewhere (Merfeld and Zupan, 2002; Merfeld et al., 2001; Zupan et al., 2002).

Visual rotational motion cues have also been shown to influence perceived tilt. In one study roll rotation cues, provided via a random dot display that precluded edge orientation cues, were shown to elicit illusory tilt as measured by having the subjects align a rod with the perceived vertical (Dichgans et al., 1972). The steadystate illusory tilt increased with the angular velocity of the visual cue up to an average tilt saturation of about 15 deg that was reached for angular velocities above 30 deg/s. Similar findings have been reported (e.g., Held et al., 1975; Zupan and Merfeld, 2002) by several other studies including one by Ian Howard (Howard and Childerson, 1994) who also demonstrated that the illusory tilt became greater when the visual stimuli included orientation cues in addition to the rotational motion cues. We extended these various findings by showing that the illusory tilt does not extinguish immediately when the visual cues are eliminated (Zupan

---

[4]We are not suggesting that the nervous system performs low-pass filtering; such a low-pass filter is inconsistent with available data. This does not negate the fact that the low-pass filter metaphor provides a good phenomenological fit to these data.

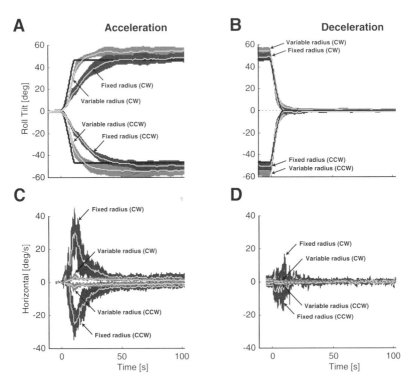

FIGURE 18.5. Perceived tilt and translational VOR during *fixed radius* and *variable radius* centrifugation. Solid gray lines show average ($N = 6$) perceptual tilt response (a) during and after angular acceleration and (b) during and after deceleration. Shading, dark and light, indicates standard deviation for fixed-radius and variable-radius, respectively. CW and CCW refer to the direction of actual centrifuge rotation as observed from above. Positive indicates perceived tilt toward the left-ear-down orientation; negative indicates right-ear-down. Measurement made using visual bar. Putative translational horizontal VOR (c) during and after angular acceleration and (d) during and after deceleration. Shading convention as above. Positive indicates slow phase velocity toward the subject's left; negative indicates response to right. Data from Merfeld et al. (2001), with permission.

and Merfeld, 2002). When lights were extinguished, the illusory tilt decayed back toward zero over a period of roughly 30 s (Figure 18.6).

18.3.2   Manual Control Measures of Tilt

Manual control has been used to investigate sensory processing and sensory integration (Huang and Young, 1988; Merfeld, 1996; Stephenson, 1993; Zacharias and Young, 1981). Some of these studies investigated canal-otolith interactions using a paradigm in which the subjects were tilted in roll in complete darkness (Merfeld, 1996; Stephenson, 1993) and asked to keep themselves upright via a control signal (e.g., joystick) by which they controlled their motion. Sum-of-sine motion disturbances were used with disturbance frequencies ranging between 0.014 and 0.668 Hz. Typically, the subjects accurately kept themselves upright at frequencies below about 0.4 Hz, with the ability to "null" the motion falling off at higher frequencies.[5]

As discussed previously, the data during variable-radius centrifugation (Figure 18.5) were phenomenologically consistent with a low-pass filter having a cutoff frequency of about 0.03 Hz. This means that changes in the otolith cues do not contribute by themselves to perception of tilt at frequencies above about 0.03 Hz. Therefore, the significant nulling responses above 0.03 Hz demonstrate the influence of the semicircular canals on tilt estimation and control. A different analysis of similar manual control data, focusing on subject's manual control describing function,[6] came to a similar conclusion that canal cues contribute substantially to the processing of tilt at frequencies above roughly 0.1 Hz (Huang and Young, 1988).

Similar manual control methods have also been used to investigate the influence of visual rotational motion cues. Motion trials were accompanied by visual rotation cues that included constant-velocity visual rotational motion (Huang and Young, 1988). A significant nonzero tilt was measured during the trials that included the constant velocity visual stimulation, while the mean tilt without constant velocity visual stimulation was not significantly different from zero.

These data appear to show that the visual rotational cues induced rotational vection, which, in turn, induced illusory tilt. The illusory tilt, in turn, elicited a manual control response by the subject to counter ("null") their perceived tilt. This response acted to tilt the subject opposite their direction of perceived tilt. Therefore, these measurements confirm the finding that visual rotation cues influence tilt responses, since significant mean tilt responses were not present with the subjects in complete darkness but were present when constant-velocity rotational

---

[5]The response fall-off at higher frequencies is due to dynamic limitations of the motion device combined with limitations in the ability to move the limbs at high frequencies because of biomechanical dynamics and neural transmission delays. The highfrequency fall-off does not indicate limitations in the ability to process the sensory information at these frequencies.

[6]A describing function provides the frequency response of a human operator performing a manual control task (e.g., McRuer and Weir, 1969).

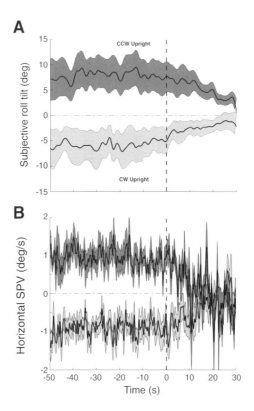

FIGURE 18.6. Subjective roll tilt and induced VOR responses during roll optokinetic stimulation in an upright orientation. CW and CCW refer to the direction of drum rotation from the subject's perspective. (a) Average roll tilt measurement during and after CW (light shading) and CCW (dark shading) drum rotation. Positive indicates roll tilt to the subject's left; negative indicates roll tilt to the subject's right. Shading indicates the standard error of the average response ($N = 17$). The vertical dashed line (time "zero") indicates "lights off." (b) Horizontal slow phase eye velocity (SPV) during and after CW (light shading) and CCW (dark shading) drum rotation. The SPV is the derivative of eye position after saccade removal. Positive indicates horizontal eye movements to the subject's left; negative indicates horizontal eye movements to the subject's right. Data from Zupan and Merfeld (2002), with permission.

cues were provided.

The above findings have been extended by manual control studies performed before and after spaceflight by human astronaut subjects (Merfeld, 1996). These preflight/postflight adaptation results show that astronauts were not able to perform the roll task in the dark as well 6–10 hrs after landing as before flight. The responses returned to near normal within 24 to 48 hrs. Furthermore, the response changes appeared predominantly in the mid-frequency range where the canals contribute substantially to tilt estimation. These data suggest that extended exposure to microgravity leads to a degradation in the neural processes by which the canals help estimate (or perceive) tilt. This is probably because the canals no longer help calculate the relative orientation of gravity during spaceflight in the absence of the gravitational cue normally provided by the otoliths.

18.3.3   Eye Movement Measures of Tilt

In one recent study (Hess and Angelaki, 1999), primary eye position was measured to change sinusoidally during off-vertical axis rotation (OVAR) but not during pure translation providing similar otolith cues. The observed modulation was enhanced at the beginning of OVAR when the rotational cues from the semicircular canals were veridical. Since a primary difference between the incoming sensory information during these motion conditions was the contribution of the semicircular canals, the authors concluded that the cues from the semicircular canals helped discriminate between gravity and linear acceleration.

Shifts in the axis of eye rotation in monkeys are also used as an indicator of tilt. Monkeys demonstrate a large, robust axis shift[7] when canal responses are not aligned with gravity as during post-rotational tilt ("dumping") stimulation or following constant-velocity off-vertical axis rotation (OVAR). One interpretation of these data is that the measured axis shift represents a neural miscalculation that shifts the estimated rotational axis from that indicated by the semicircular canals toward alignment with gravity, reducing the sensory conflict between the incoming cues (Merfeld, 1995b; Merfeld et al., 1993a, b; Merfeld and Young, 1995; Zupan et al., 2002). Another interpretation is that the nervous system performs sensory integration calculations in an inertial (as opposed to head-fixed) reference frame (Angelaki and Hess, 1994). While these hypotheses appear to be distinct and potentially conflicting, both groups seem to agree that the axis of eye rotation in monkeys provides an indicator of gravity as represented (and/or

---

[7]An axis shift refers to a shift in the eye rotation axis from alignment with the rotation indicated by the semicircular canals toward alignment with gravity or gravito-inertial force. Such shifts occur during motion paradigms that misalign semicircular canal cues and gravity (e.g., postrotational tilt or "dumping"). Indications of a planar shift in the eye responses were first demonstrated using two-dimensional eye movement recordings (Harris, 1987; Harris and Barnes, 1987; Raphan et al., 1981). However, since measurement of the axis of eye rotation requires full three-dimensional (3-D) eye recordings, measurement of an axis shift awaited 3-D recordings and analyses (e.g., Angelaki and Hess, 1994; Merfeld and Young, 1992; Merfeld et al., 1991, 1993b).

estimated) by the nervous system.

Studies with monkeys have shown that, during fixed-radius centrifugation, the axis of eye rotation aligns rapidly with the tilted orientation of GIF (Merfeld and Young, 1995; Wearne et al., 1999). This contrasts with the slow gradual perceptual tilt measured in humans during similar centrifugation paradigms (Clark and Graybiel, 1966; Curthoys, 1996; Graybiel and Brown, 1951; Merfeld et al., 2001) as demonstrated in Figs. 18.4 and 18.5. The rapid shift in the axis of eye rotation might indicate that the influence of rotational cues on tilt estimation is less pervasive for monkeys than has been demonstrated in humans. This substantial difference in the time course of tilt responses might also indicate that monkeys more readily utilize measured GIF to provide the direction of gravity. However, it remains to be proven that the rapid tilt in the axis of eye rotation is directly correlated with a rapid tilt in the monkey's neural representation of tilt (or perceived tilt).

Finding other eye movements clearly indicative of "tilt" is challenging. Torsion has been considered an indicator of roll "tilt," but a recent study (Merfeld et al., 1996) suggested that human torsion responses include two components, one indicating roll tilt and a second that is a simple reflexive response to interaural shear of the otoliths. This finding is consistent with findings in monkeys (Angelaki, 1998; Paige and Tomko, 1991) and humans (Merfeld et al., 1996) showing significant torsional responses when subjects were accelerated along their interaural axis while in the supine orientation. Since there is no roll stimulation in this paradigm (the tilt of GIF is in yaw) and since humans do not experience illusory roll tilt during this motion (Merfeld et al., 1996), the torsion evident during this paradigm seems unlikely to be a response elicited in conjunction with a neural estimate of roll tilt. In addition, since torsion may not be a clear and simple tilt indicator, its analogs in the other two dimensions (i.e., vertical and horizontal "doll reflexes" Citek and Ebenholtz, 1996), may also be questionable tilt indicators.

18.4   Influence of Rotational Cues on Translation Responses

A few experimental paradigms and measurement techniques have been used to demonstrate the influence of rotational cues on translation responses. A partial coverage of these investigations follows. We begin by describing a study that utilizes perceptual measures of translation, then proceed to studies that use eye movements. All studies will include some form (visual or canal) of rotational cue that influences the neural representation of gravity and linear acceleration. As discussed previously, the influence of rotational cues on translation processing ($\hat{\mathbf{a}} = \hat{\mathbf{g}} - \hat{\mathbf{f}}$) is somewhat indirect, acting through the direct influence of rotational cues on the neural representation of gravity ($\hat{\mathbf{g}} = \int (-\hat{\boldsymbol{\omega}} \times \hat{\mathbf{g}})dt$).

18.4.1   Perceptual Measures of Translation

There is little data showing that perception of translation is influenced by rotational cues. The lack of evidence could indicate that such an influence doesn't exist. However, we were not aware of a single study investigating this influence until we recently completed a preliminary investigation. For our study, we utilized a methodology that had previously been developed by Benson (1986) to investigate human translation perception in the dark. Our subjects were asked to indicate what direction (left or right) they moved when they were translated to the left or right via a single cycle of sinusoidal acceleration. The single cycle of acceleration had a duration of 4 s ($f = 0.25$ Hz), and the magnitude of acceleration level was between 0 and 20 mG (0, ±2, ±4, ..., ±20 mG). The actual linear translation displacement was directly proportional to the acceleration level.[8] If subjects were not certain which direction they translated, they were instructed to do their best or simply guess. At the end of each motion trial, a tone indicated that the subject must push one of two buttons ("forced choice"). Each subject pushed a button held in their left hand when they experienced motion to the left, and a button in their right hand for motion experienced to the right.

When subjects perform this task in the dark, they are able to detect large amplitude motions with nearly 100% accuracy (Benson et al., 1986). At very low levels of acceleration, the subject's ability to correctly determine the actual direction of translation becomes 50%, no better than simply guessing. Our data (Figure 18.7) confirm these findings.

We then added a roll visual rotation cue at a speed of 60 deg/s, similar to the vection cues discussed previously that have been shown to induce a small illusory tilt. With the visual motion cue continuously rotating about the subject's line of sight, we presented single cycle acceleration profiles, like those previously experienced in the dark. Each subject was tested in the dark prior to optokinetic stimulation.

Our preliminary data indicate that the perception of translation is affected by the presence of the visual rotation cue (Figure 18.7). First, the ability to detect the direction of linear motion became much more variable in the presence of the visual roll rotation cue. This increase in variability may be because the vection experienced by the subject is highly variable (e.g., Brandt et al., 1974). Second, the mean number of trials in which the subject indicated left or right was biased by the direction of rotation of the visual cue. In the dark, the subject's reported that they moved to the right on 50.0% of the trials. Clockwise (from the subject's perspective) rotation of the visual display led to reported motion to the right for 48.6% of the trials. With counterclockwise rotation of the visual display, the subjects reported motion to the right for 57.0% of the trials. This small effect is in the direction consistent with our predicted influence of visual rotation cues on estimated linear acceleration as illustrated in the figure inserts.

---

[8] Integrating a single cycle of sinusoidal acceleration yields a bell-shaped linear velocity. Integrating again yields a translation displacement.

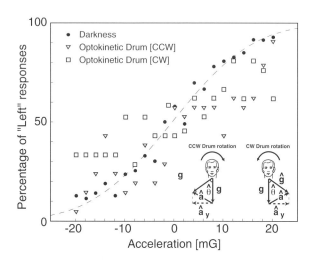

FIGURE 18.7. Perception of subjective direction during passive interaural translation. Subjects were passively translated along their interaural axis in either darkness (black circle) or in light while viewing an optokinetic dome rotating about their line of sight in either CW (open square) or CCW (open triangle) directions. The translation displacement P followed the time course $P(t) = A/\omega \cdot [t - 1/\omega \cdot \sin(\omega t)]$ where A is the maximum acceleration and $\omega = 2\pi f$ with $f = 0.25$ Hz. Interaural acceleration to the subject's left is positive. The percentage of responses where subjects perceived being passively translated "to the left" is plotted as a function of the maximum acceleration. A three-parameter psychometric function assumed to have the form of a normal probability integral (Carpenter-Smith et al., 1995; Foster and Bischof, 1991) was fitted through least-square optimization to the data in darkness (dotted line). Two cartoons illustrate the GIF resolution during optokinetic stimulation. Optokinetic stimulation induces illusory roll tilt. This is represented diagramatically by showing the estimate of gravity ($\hat{g}$) tilted in roll away from gravity (g). According to the GIF resolution hypothesis, the discrepancy between measured and estimated gravity leads to a nonzero interaural estimate of linear acceleration ($\hat{a}_y$) to the subject's left for CW dome rotation (to the right for CCW dome rotation).

18.4.2   Eye Movement Measures of Translation

In order to keep an object foveated while translating to the left, the eye must rotate toward the right. The translational VOR is a widely known and accepted eye movement response to translational motion in the dark (e.g., Baarsma and Collewijn, 1975; Baloh et al., 1988; McCabe, 1964; Niven et al., 1966; Paige, 1989; Paige and Tomko, 1991; Schwarz et al., 1989; Schwarz and Miles, 1991; Skipper and Barnes, 1989). This compensatory eye response will be the translational eye movement measure discussed throughout this section.

A simple early demonstration of the influence of rotational cues on translation responses was obtained using simple roll tilts, like those presented previously (Figure 18.4) for measures of tilt perception. Squirrel monkeys were rapidly tilted in roll while 3-D eye movements were measured (Merfeld and Young, 1995). The tilts were rapid so that gravity provided significant high-frequency interaural shear force to the otolith organs. Despite the presence of high-frequency interaural otolith cues, no consistent horizontal translational VOR was measured after the roll tilt. In contrast, in the absence of roll canal cues, similar high frequency interaural otolith cues had been shown to elicit significant horizontal translational VOR responses in the same monkeys (Paige and Tomko, 1991) during sinusoidal interaural linear acceleration. Since the horizontal VOR was present when roll canal cues were absent and absent when canal cues were present, we concluded that canal cues influence the neural processing of the otolith cues.

These findings have been confirmed and extended using high-frequency sinusoidal roll tilt stimulation (Angelaki et al., 1999). In this study, rhesus monkeys were sinusoidally roll tilted, sinusoidally translated along their inter-ural axis, or simultaneously roll tilted and translated. The interaural shear force during the roll tilt alone and translation alone stimuli were nearly identical. Yet, little or no horizontal VOR response was observed during the roll tilt stimulation, while a substantial horizontal VOR was observed during the inter-aural translation. This clearly demonstrates the influence of the canal cues on the translation VOR response, since the response pattern was clearly altered by the presence (roll tilt only) or absence (translation only) of the canal cue.

During combined roll tilt and linear acceleration (Angelaki et al., 1999), the horizontal eye responses remained compensatory for the applied stimuli, even when the relative phase between the linear acceleration and roll tilt was such that little or no inter-aural shear force was measured by the otolith organs. Essentially, the nervous system appeared able to estimate the motion accurately and to elicit appropriate reflexive responses when the canal cues were present. However, when the canal cues were altered via canal plugging, the horizontal responses no longer compensated for the applied stimuli, clearly and convincingly demonstrating the influence of the semicircular canals on the neural processing of translation.

We (and Angelaki et al.) presume that the nervous system accomplished this by using the rotational cue from the canals to help correctly estimate the relative orientation of gravity. Linear acceleration can then be estimated as the difference between this estimate of gravity and the otolith measurement of GIF (e.g.,

Figure 18.2). Unfortunately, no measure of tilt[9] was available for either of these studies (Angelaki et al., 1999; Merfeld and Young, 1995).

Other recent investigations and explanations (Merfeld et al., 1999; Zupan et al., 2000) take these findings one step further, showing that translational responses can be elicited even in the absence of actual linear acceleration. Our studies utilized the well-studied postrotational tilt ("dumping") protocol. As mentioned in the introduction, the semicircular canals provide a postrotatory response that gradually decays to zero following deceleration from extended constant velocity yaw rotation. Since the orientation of the canals is fixed in the head, the horizontal canals continue to signal rotation in the head-fixed reference frame even after head tilts. Therefore, when a subject is tilted 90 deg during this postrotatory period, the canals continue to indicate rotation, but this rotation is now about an axis perpendicular to gravity. If yaw rotation were truly to occur about an earth-horizontal axis then the relative orientation of gravity would rotate with respect to the head as in barbecue spit rotation. However, in the postrotatory condition no actual movement is occurring, so the otolith organs accurately provide a cue indicating the unchanging orientation of gravity.

It has long been known that human responses following postrotational tilt depend upon subject orientation following the tilt. For example, the human horizontal VOR is greater when the subject is tilted "nose-up" than when tilted "nose-down" (Benson and Bodin, 1966; Merfeld et al., 1999; Zupan et al., 2000). We have also measured a similar asymmetry following constant velocity "barbecue-spit" rotation when the subjects are brought to a stop in the nose-up or nose-down orientation (Merfeld et al., 1999; Zupan et al., 2000). Other paradigms appear to induce a similar position-dependent asymmetry. For example, horizontal OKN and OKAN responses of humans induced by yaw rotation of an optokinetic surround, with subjects and optokinetic surround aligned with the earth-horizontal, have been shown to be greater when the subject is oriented nose-up than when oriented nose-down (Wall et al., 1999), and a similar asymmetry has been observed during caloric stimulation of canal-plugged monkeys (Minor and Goldberg, 1990; Paige, 1985).

Our data and analysis (Merfeld et al., 1999; Zupan et al., 2000; Wall et al., 1999) suggest that the response in each of the above paradigms includes a horizontal angular response (elicited by the canals or vision) and a horizontal linear VOR (compensatory to an interaural estimate of linear acceleration) induced by the rotational cue even in the absence of linear acceleration. Based on our data (Figure 18.8), the underlying angular responses appear independent of subject orientation in humans, while the induced linear VOR depends on subject orientation (e.g., nose-up or nose-down).

Specifically, the rotational cues from the semicircular canals are independent of head orientation following postrotatory tilt, since the canal cue is head-fixed.

---

[9]The torsional VOR is primarily an angular response to the roll angular velocity, though it may include a tilt component. See Section 18.3.3, "Eye Movement Measures of Tilt," for related discussion.

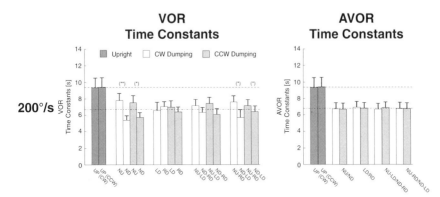

FIGURE 18.8. Mean time constants (+/- 1SE) of postrotatory VOR and angular VOR for 200 deg/s dumping protocols and for eight orientations (nose-up: NU; nose-down: ND; left-ear-down: LD; right-ear-down: RD; and the four intermediate orientations: NU-LD, NU-RD, ND-LD, ND-RD). (a) Histograms show mean time constants of post-rotatory VOR following clockwise (white) and counterclockwise (light shading) rotations about an earth-vertical axis followed by a tilt ("dumping") for eight subjects. Mean time constants of postrotatory VOR following rotations about an earth-vertical axis without postrotatory tilt are indicated by dark shading. The dash-dotted line indicates the VOR mean time constant for all dumping protocols. The dashed line indicated the VOR mean time constant for all upright rotations without tilt. The asterisks indicate the level of statistical significance (∗ for $p < 0.05$; ∗∗ for $p < 0.005$) in the difference between two mean time constants for orientations separated by 180 deg. (b) Mean time constants of post- rotatory angular VOR (AVOR) eye movements for 200 deg/s dumping protocols with standard error bars for eight subjects. The AVOR is computed as the average of VOR responses separated by 180 deg (e.g., nose-up and nose-down). Histograms show mean time constants of postrotatory AVOR following clockwise (white) and counterclockwise (light shading) dumping protocols. Mean time constants of postrotatory VOR following rotations about an earth-vertical axis without postrotatory tilt are indicated by dark shading. The dash-dotted line indicates the mean time constant for all dumping protocols. The dashed line indicated the mean time constant for all upright rotation protocols followed by no tilt. Data are from the same subjects and trials as presented in Zupan et al., 2000.

In addition, data show that psychophysical reports about the duration of rotation sensation following 90 deg tilts do not depend on subject orientation (Benson and Bodin, 1966). Based on this evidence, we calculated the linear VOR response component by subtracting VOR responses for final orientations separated by 180 deg (divided by 2). We used subtraction because the angular VOR was hypothesized to be independent of head orientation, and we predicted (Merfeld et al., 1999; Zupan et al., 2000) that the direction of the estimated linear acceleration should reverse for subject orientation separated by 180 deg (e.g., nose-up vs. nose-down). Performing this calculation, we found that the linear VOR component varied sinusoidally with head orientations following both CW and CCW rotations (but shifted for each rotation direction) as we predicted. Complementary to these calculations, we found the angular VOR response component after a postrotatory tilt by averaging VOR responses for final orientations separated by 180 deg (e.g., nose-up and nose-down). Time constants for the eight different angular VOR responses obtained using this averaging method (Figure 18.8) were not significantly different ($p > 0.05$) after 50 deg/s, 100 deg/s and 200 deg/s rotations, supporting the original assumption (Zupan et al., 2000).

These predictions were made using logic similar to that discussed in previous sections. An illusory tilt is induced when the canal cue is misaligned with gravity. The illusory tilt demonstrates that the neural representation of gravity ($\hat{\mathbf{g}}$) differs from the otolith measure of gravity ($\mathbf{g}$), which is the total GIF present during this paradigm since there is no linear acceleration ($\mathbf{f} = \mathbf{g}$). As discussed above, such a difference should be interpreted as linear acceleration ($\hat{\mathbf{a}} = \hat{\mathbf{g}} - \mathbf{f}$) if the nervous system implements an internal model of GIF resolution. The nervous system should respond to this estimate of linear acceleration in much the same manner as a true linear acceleration, since it cannot distinguish one from the other because of the inherent measurement ambiguity. In the nose-up and nose-down orientations, this neural representation of linear acceleration is substantially aligned with the interaural axis. The interaural component of estimated linear acceleration should elicit a horizontal translational VOR, that reverses between nose-up and nose-down orientations even in the absence of actual linear acceleration.

This GIF resolution hypothesis ($\hat{\mathbf{a}} = \hat{\mathbf{g}} - \hat{\mathbf{f}}$) was first developed to explain centrifugation responses (Merfeld, 1990). As discussed previously (Figure 18.4), illusory tilt develops much more gradually during acceleration on a fixed radius centrifuge than the tilt sensation dissipates during deceleration (Clark and Graybiel, 1963; Clark and Graybiel, 1966; Graybiel and Brown, 1951). The gradual build up of the illusion of tilt, much slower than the actual GIF tilt, meant that a substantial difference between the measured GIF and the neural representation of gravity existed during acceleration. The GIF resolution hypothesis suggested that this difference should be interpreted as linear acceleration ($\hat{\mathbf{a}} = \hat{\mathbf{g}} - \hat{\mathbf{f}}$). Since a horizontal translational VOR is elicited by linear acceleration, the hypothesis suggested that a horizontal translational VOR should be present transiently during centrifuge acceleration with little or no translational VOR during deceleration.[10] Furthermore, the time course of the horizontal translational VOR should match the time course with which estimated tilt aligns with GIF.

These predictions were consistent with earlier studies using centrifugation showing that the horizontal VOR putatively, including both translational and angular components, was dramatically different with humans facing-motion than with back-to-motion (Lansberg et al., 1965; Young, 1967). The facing-motion responses, with the linear and angular responses summing, were substantially larger than the back-to-motion responses, with the linear and angular responses opposing each other. We have recently confirmed these findings in a comprehensive centrifugation study with human subjects (Merfeld et al., 2001). Also consistent with model predictions, we found that the time course of the translational VOR component (Figure 18.5) was approximately equal to the time course with which the illusory tilt aligns with GIF.

While monkey responses during centrifugation are quantitatively different from the human responses, we had previously reported compatible results in squirrel monkeys (Merfeld and Young, 1995). We found a facing-motion and back-to-motion asymmetry in the horizontal VOR response, similar though smaller than in humans. Furthermore, the axis of eye rotation lagged (slightly) behind the actual GIF tilt. This provides evidence for a small lag in the alignment of the neural representation of gravity with GIF in squirrel monkeys, since as discussed previously the axis of eye rotation is an indicator for the neural representation of gravity. The lag in alignment of the axis of eye rotation with the stimulus was approximately the same as the time course over which the translational VOR component decayed to zero, again consistent with the GIF resolution hypothesis.

Based on the internal model hypotheses discussed previously, responses during off-vertical axis rotation (OVAR) also support the influence of rotational cues on translation responses. Specifically, when a subject is rotated about an off-vertical rotation axis, the orientation of gravity relative to the head rotates. If perfect rotational cues were available, the nervous system could, in principle, keep precise track of the relative orientation of gravity using the rotational mechanism discussed previously ($\hat{g} = \int(-\hat{\omega} \times \hat{g})$. Under such perfect conditions, the estimate of gravity would equal true gravity as measured by the otoliths. Furthermore, because there would be no difference between true gravity and the estimate of gravity, the estimate of linear acceleration would equal zero. Conversely, as the canal cue decays to zero the available rotational cue is less effective in helping estimate the relative orientation of gravity, and the neural estimate of gravity should lag farther and farther behind the otolith measurement of gravity. Therefore, a larger estimate of linear acceleration should result as the canal cue decays.

Data show that the horizontal sinusoidal VOR modulation builds up gradually

---

[10]The estimate of gravity ($\hat{g}$) and the measurement of GIF ($\hat{f}$) were nearly equal during deceleration, but not acceleration, of the centrifuge. Hence the estimate of linear acceleration, the difference between measured GIF and estimated gravity, would be near-zero during angular deceleration and larger during angular acceleration.

in humans as the canal cues decay toward zero (Wall, 1987) during yaw rotation about an earth-horizontal axis. This is particularly evident in human subjects because in humans the horizontal angular VOR response decays toward zero, maintaining a very small bias component. Assuming that the majority of the horizontal modulation component is a translational VOR, these data precisely match the predictions outlined above, with the horizontal translational VOR component becoming larger as the angular response decays toward zero. The smaller modulation component and larger bias component found in monkeys (e.g., Goldberg and Fernandez, 1982; Raphan et al., 1981) are also consistent with this putative mechanism. If the bias component is presumed to represent a reflexive response to rotational motion, then a large bias component suggests that the nervous system has a more accurate rotational estimate to help keep track of the relative orientation of gravity relatively well. A smaller difference between the estimated orientation of gravity and the measured orientation of gravity would lead to a smaller estimate of linear translation and a smaller sinusoidal modulation component (translational VOR).

18.5   Sensory Integration Modeling

Interactions between the sensory systems that contribute to spatial orientation have long been modeled. One of the earliest vestibular sensory integration models was developed by Mayne (1974). To a large extent, this model relied on low-pass and high-pass filtering of the cues from the otolith organs to elicit tilt and translation responses, respectively, but this model also included two-dimensional influences of semicircular canal cues on the processing of cues from the otolith organs. The two-dimensional implementation of this model as well as the general reliance of the model on simple filtering of gravito-inertial cues are the primary limitations of this model.

Other early models of the angular vestibulo-ocular reflex (Raphan et al., 1977; Robinson, 1977) focused on the prolongation of the VOR compared to the activity of the semicircular canal first-order afferent. To do so, one model used a single positive feedback loop (Robinson, 1977). Another used two parallel paths (Raphan et al., 1977), including a hypothesized "leaky integrator" that is part of a "velocity storage mechanism." Robinson's model (Robinson, 1977) was modified to implement the influence of the otolithic information on visual-vestibular interactions (Hain, 1986). Hain's model implements the influence of otolith information on rotational cues, but this model does not implement the influence of canal cues on self-orientation. The model developed by Raphan and colleagues (Raphan et al., 1977) has been modified to implement the influence of the otolithic information on monkey reflexive eye movements (Wearne et al., 1999). In this new model, the otolith inputs modulate the "velocity storage" dynamic characteristics and a direct translational VOR pathway is added. Wearne's model implements the tendency of the eye movement axis of rotation to align with gravity. But this model does not implement the influence of canal cue on the neural processing of otolith cues. Several other models (Galiana and Outerbridge, 1984; Green and Galiana, 1998) successfully match characteristics of angular VOR responses but do not include sensory interactions between two or more sensory systems.

Other models utilized techniques borrowed from optimal estimation to perform multisensory integration (Borah et al., 1988; Ormsby and Young, 1977). Ormsby's model included the influence of rotational cues on the orientation of gravity using a mechanism that resembles an internal model of a physical relationship. But the primary estimation processes were carried out for each sensory system individually. This differs from the internal model approach in which the primary estimation processes are carried out via the internal models. Borah's model, based on Kalman filtering, did not explicitly include internal models. Both of these models were important incremental steps toward the development of the more recent models that include explicit internal models.

A family of models that include explicit internal model representations have been developed to help explain sensory interactions between vestibular cues from the semicircular canals and otolith organs (Glasauer and Merfeld, 1997; Glasauer and Mittelstaedt, 1992; Merfeld, 1990, 1995a, b; Merfeld et al., 1993a; Merfeld and Zupan, 2001; Zupan et al., 2002). Crucial elements include the internal models of the influence of rotational cues on the neural representation of gravity $(d\hat{\mathbf{g}}/dt = -\hat{\boldsymbol{\omega}} \times \hat{\mathbf{g}})$ and the gravito-inertial resolution processing $(\hat{\mathbf{f}} = \hat{\mathbf{g}} - \hat{\mathbf{a}})$ discussed previously. These models of vestibular interactions simulate the dominant components of squirrel monkey (Merfeld, 1990, 1995b; Merfeld et al., 1993a) and human responses (Glasauer, 1992; Merfeld, 1995a, b; Merfeld and Zupan, 2001; Zupan et al., 2002) during complex, three-dimensional motion stimuli (e.g., off-vertical axis rotation and eccentric rotation). These models, along with those discussed below, are an example of modeling predictions preceding, and in fact leading to, experimental findings.

Another model describing three-dimensional, sensory interactions between visual and vestibular cues was developed in parallel with these efforts. This model is based on the concept of "coherence constraints" (Droulez and Darlot, 1989). This model also includes internal models of sensory dynamics, body dynamics and physical relationships but differs in its implementation of these internal models. The coherence constraint model of visual-vestibular interactions simulates reflexive eye movements induced by complex, three-dimensional motion stimuli in darkness and in light (Zupan, 1995; Zupan et al., 1994) and has also been used to model motor control of an eye and a forearm (Darlot et al., 1996).

Recently, the observer (Merfeld, 1990, 1995b; Merfeld et al., 1993a; Merfeld and Zupan, 2001) and coherence constraints (Droulez and Darlot, 1989; Zupan, 1995; Zupan et al., 1994) models have been combined in a general model of human sensory integration, referred to as a sensory weighting model (Zupan et al., 2002). This new model uses sensory weighting as the main mechanism to centrally estimate physical variables. This model also includes internal models of sensory dynamics, body dynamics, and physical relationships but differs in its implementation of these internal models when compared to observer and coherence constraint models. This model accurately simulates reflexive eye movements and perceptual responses during complex three-dimensional visual and vestibular stimulations.

Another model of human perception during vestibular-only stimulation (Holly, 2000) is also based on the assumption that the central nervous system has somehow integrated the laws of physics governing combined rotation and translation of the head. This model does not differ philosophically from the observer, coherence constraint and sensory weighting models which explicitly implement internal models of the laws of physics. A primary difference is that Holly's model does not include the sensory system dynamics that define motion transduction by the vestibular system. Recently, another VOR model based on optimal observer theory (originally used to model posture control Kuo, 1997) accurately simulates reflexive eye movements during one-dimensional canal or visual stimulation. By defining the required noise levels in both canal and visual sensory channels for the design of an optimal Kalman filter (Gelb, 1974), Kuo is able to model the learning process for both vestibulo-ocular and optokinetic response dynamics by use of adaptive filters (Kuo and Henry, 1997). Extending this model to include gravity and linear acceleration cues and the related neural estimation processes will advance our understanding the role that noise plays when rotational cues influence the neural processing of tilt and translation.

18.6   Discussion

As presented above, studies have shown that rotational cues from the canals and vision influence tilt responses. These influences have been shown with perceptual, eye movement, and manual control measures of tilt. Therefore, the influence of rotational cues on these behavioral responses is well established. On the other hand, the neural basis underlying these responses has just begun to be investigated (Zhou et al., 1998, 2000). Understanding these neural processes is essential to fully understand the sensory interactions involved.

Furthermore, eye movements are the only behavioral measure of tilt that exists for monkeys, and these measures are somewhat indirect. Therefore, in order to build a bridge between human responses (perceptual, eye movement, and manual control) and monkey responses, including neural recordings, other monkey behavioral measures (e.g., psychophysics, manual control) are essential.

On the other hand, it is worth noting that the perceptual pathways are not identical to the eye movement pathways. Despite this distinction, we compare eye movement and perceptual responses in several places above. We do this because these are the only available measures that allow us to compare tilt and translation responses; we do not believe that the neural processing of eye movements and perceptual responses necessarily utilize identical neural elements, though some common neural pathways likely contribute to both types of responses (Zupan et al., 2002).

Other studies have begun to show that translation responses are also influenced by rotational cues. However, this influence of rotational cues on translation responses is less established than for tilt. There are several reasons for this. One is the indirect influence of rotational cues on translation responses discussed previously. The second reason is that eye movements are almost the only translational measure available. This is problematic because the horizontal translational VOR is not trivial to distinguish from a horizontal angular VOR. This is especially problematic for the paradigms that directly elicit horizontal angular VOR responses, but is a potential problem even during the combined tilt translation paradigm, since the observed horizontal response could include an axis shift component. Demonstrating that the translational VOR components demonstrate characteristics like variation with target distance and gaze dependence would add to the credibility of this explanation. However, several of these responses have fairly low frequency dynamics (e.g., Merfeld et al., 1999, 2001; Zupan and Merfeld, 2002; Zupan et al., 2000), and it has been shown that the dependence of the translational VOR diminishes for low-frequency responses (Telford et al., 1997). Neural and psychophysical recordings of translational responses will provide a resolution to this issue

Appendix A: GIF Notation Sets

Alternative notation sets are used in other published reports (e.g., Angelaki et al., 1999; Wearne et al., 1999) that investigate the neural processing of ambiguous gravito-inertial cues. Since there is agreement on the underlying physics, these approaches simply represent different ways to represent the physics mathematically . To minimize confusion due to notational differences, we briefly present the alternative notations below.

Some papers (e.g., Wearne et al., 1999) define "gravito-inertial acceleration" as the quantity measured by the otolith organs. The true linear acceleration (A) used by these authors is defined using the same definition as us $(A = \mathbf{a})$. In vector notation, the vector points in the direction of the linear acceleration and has a magnitude equal to the size of the acceleration. The authors also define a term called "gravitational acceleration" (A_g), which is opposite our "gravity" $(A_g = -\mathbf{g})$. It is simply the equivalent linear acceleration, having a magnitude of 9.81 m/s^2, which would yield a force equal to gravitational force. This approach complements d'Alembert's approach, which defines a fictional force due to linear acceleration, by defining a fictional linear acceleration with an effect equivalent to gravity. Gravito-inertial acceleration (GIA) is simply the vector sum of linear acceleration and gravitational acceleration, which yields a vector that has the same magnitude as our specific GIF but points in the opposite direction (GIA $= -\mathbf{f}$ or GIA $= -$GIF).

Another approach has been borrowed from Vieville and Faugeras (1990). In this approach, Angelaki and colleagues (1999) define "resultant linear acceleration"

(α), as the vectorial sum of the "gravity" (g) and "translational acceleration" (f), $\alpha = g + f$. The similarity of this approach to our notation was briefly discussed in a joint article (Angelaki et al., 2001). Differences result from the use of different reference frames.

Acknowledgements

We gratefully acknowledge the assistance provided by numerous coworkers, who helped develop, design, and perform several of the published studies presented herein. Support provided by NIH/NIDCD grants DC03066, DC04158 and DC04644 and NASA grant 99-HEDS-03-370.

References

Angelaki, D. (1998). Three dimensional organization of otolith-ocular reflexes in rhesus monkeys. III. Responses to translation. *J. Neurophysiol.*, 80: 680–695.

Angelaki, D. and Hess, B. (1994). Inertial representation of angular motion in the vestibular system of rhesus monkeys. I. Vestibuloocular reflex. *J. Neurophysiol.*, 71: 1222–1249.

Angelaki, D., McHenry, M., Dickman, J. D., Newlands, S. and Hess, B. (1999). Computation of inertial motion: neural strategies to resolve ambiguous otolith information. *J. Neurosci.*, 19 :316–327.

Angelaki, D., Wei, M. and Merfeld, D. (2001). Vestibular discrimination of gravity and translational acceleration. *Ann. New York Acad. Sci.*, 942: 114–127.

Asch, S. and Witkin, H. (1948). Studies in space orientation: I. Perception of the upright with displaced visual fields. *J. Exp. Psych.*, 38: 325–337.

Baarsma, E. and Collewijn, H (1975). Eye movements due to linear accelerations in the rabbit. *J. Physiol.*, 245: 227–247.

Baloh, R., Beykirch, K., Honrubia, V. and Yee, R. (1988). Eye movements induced by linear acceleration on a parallel swing. *J. Neurophysiol.*, 60 :2000–2013.

Benson, A. (1974). Modification of the response to angular accelerations by linear accelerations. In H. Kornhuber (ed.), *Handbook of Sensory Physiology. Volume VI, Vestibular System Part 2: Psychophysics, Applied Aspects and General Interpretations*, pp. 281–320. New York: Springer-Verlag.

Benson, A. and Bodin, M. (1966). Comparison of the effect of the direction of the gravitational acceleration on post-rotational responses in yaw, pitch, and roll. *Aerospace Med.*, 37: 889–897.

Benson, A. J., Spencer, M. B. and Stott, J. R. (1986). Thresholds for the detection of the direction of whole-body, linear movement in the horizontal plane. *Aviat. Space Environ. Med.*, 57 :1088–96.

Borah, J., Young, L. R. and Curry, R. E. (1988). Optimal estimator model for human spatial orientation. *Annals New York Acad. Sci.*, 545: 51–73.

Brandt, T., Dichgans, J. and Buchle, W. (1974). Motion habituation: inverted self-motion perception and optokinetic after-nystagmus. *Exp. Brain Res.*, 21: 337–352.

Brandt, T., Dichgans, J. and Koenig, E. (1973). Differential effects of central versus peripheral vision on egocentric and exocentric motion perception. *Exp. Brain Res.*, 16: 476–491.

Carpenter-Smith, T. R., Futamura, R. G. and Parker, D. E. (1995). Inertial acceleration as a measure of linear vection: an alternative to magnitude estimation. *Percept. Psychophys.*, 57: 35–42.

Citek, K. and Ebenholtz, S. (1996). Vertical and horizontal eye displacement during static pitch and roll postures. *J. Vestib. Res.*, 6: 213–228.

Clark, B. and Graybiel, A. (1963). Contributing factors in the perception of the oculogravic illusion. *Am. J. Psych.*, 76: 18–27.

Clark, B. and Graybiel, A. (1966). Factors contributing to the delay in the perception of the oculogravic illusion. *Am. J. Pysch.*, 79: 377–388.

Curthoys, I. (1996). The delay of the oculographic illusion. *Brain Res. Bull.*, 40: 407-412.

Darlot, C., Zupan, L., Etard, O., Denise, P. and Maruani, A. (1996). Computation of inverse dynamics for the control of movements. *Biol. Cybern.*, 75: 173–186.

Dichgans, J., Held, R., Young, L. R. and Brandt, T. (1972). Moving visual scenes influence the apparent direction of gravity. *Science*, 178: 1217–1219.

Droulez, J. and Darlot, C. (1989). The geometric and dynamic implications of the coherence constraints in three-dimensional sensorimotor interactions. In M. Jeannerod (ed.), *Attention and Performance XIII*, pp. 495–526. New York: Erlbaum.

Foster, D. and Bischof, W. (1991). Thresholds from psychometric functions: superiority of bootstrap to incremental and probit variance estimators. *Psych. Bull.*, 109: 152–159.

Galiana, H. and Outerbridge, J. (1984). A bilateral model for central neural pathways in the vestibulo-ocular reflex. *J. Neurophysiol.*, 51: 210–241.

Gelb, A. (1974). *Applied Optimal Estimation*. Cambridge, MA: MIT Press.

Glasauer, S. (1992). Interaction of semicircular canals and otoliths in the processing structure of the subjective zenith. *Ann. New York Acad. Sci.*, 656: 847-849.

Glasauer, S. and Merfeld, D. (1997). Modelling three dimensional vestibular responses during complex motion stimulation. In M. Fetter, H. Misslisch, D. Tweed and T. Halswanter (eds.), *Three-Dimensional Kinematics of Eye, Head, and Limb Movements*, pp. 389-400, Amsterdam: Harwood Academic Publishers GMBH.

Glasauer, S. and Mittelstaedt, H. (1992). Determinants of orientation in microgravity. *Acta Astronautica*, 27: 1–9.

Goldberg, J. and Fernandez, C. (1982). Eye movements and vestibular-nerve responses produced in the squirrel monkey by rotations about an Earth-horizontal axis. *Exp. Brain Res.*, 46: 393–402.

Grant, J. and Best, W. (1986). Mechanics of the otolith organ - dynamic response. *Ann. Biomed. Engineer.*, 14: 241–256.

Graybiel, A. and Brown, R. (1951). The delay in visual reorientation following exposure to a change in direction of resultant force on a human centrifuge. *J. Gen. Psych.*, 45: 143–150.

Green, A. and Galiana, H. (1998). Hypothesis for shared central processing of canal and otolith signals. *J. Neurophys.*, 80: 2222–2228.

Guedry, F. (1965). Orientation of the rotation-axis relative to gravity; its influence on nystagmus and the sensation of rotation. *Acta Otolaryngol.*, 60: 30–49.

Hain, T. C. (1986). A model of the nystagmus induced by off vertical axis rotation. *Biol. Cyber.*, 54: 337–350.

Held, R., Dichgans, J. and Bauer, J. (1975). Characteristics of moving visual scenes influencing spatial orientation. *Vis. Res.*, 15: 357–365.

Hess, B. J. and Angelaki, D. E. (1999). Oculomotor control of primary eye position discriminates between translation and tilt. *J. Neurophysiol.*, 81: 394–398.

Holly, J. E. (2000). Baselines for three-dimensional perception of combined linear and angular self-motion with changing rotational axis. *J. Vestib. Res.*, 10: 163–178.

Howard, I. P. and Childerson, L. (1994). The contribution of motion, the visual frame, and visual polarity to sensations of body tilt. *Perception*, 23: 753–762.

Howard, I. P. and Heckmann, T. (1989). Circular vection as a function of the relative sizes, distances, and positions of two competing visual displays. *Perception*, 18: 657–665.

Huang, J. and Young, L. R. (1988). Visual field influence on manual roll and pitch stabilization. *Aviation, Space, and Environ. Med.*, 59: 611–619.

Kuo, A. D. (1997). An optimal control model of human balance: can it provide theoretical insight to neural control of movement? *Proc. American Control Conference, FA08*, pp. 4–8. Albuquerque, NM: IEEE.

Kuo, A. D. and Henry, S. M (1997). Adaptive filter model of vestibulo-ocular reflex accounts for velocity storage. *Soc. for Neurosci.*, 18: 509.

Lansberg, M., Guedry, F. and Graybiel, A. (1965). Effect of changing resultant linear acceleration relative to the subject on nystagus generated by angular acceleration. *Aero. Med.*, 36: 456–460.

Mayne, R. (1974). A systems concept of the vestibular organs. In H. Kornhuber (Ed.), *Handbook of Sensory Physiology, Vol. VI. Vestibular System, Part 2. Psychophysics, Applied Aspects and General Interpretations*, pp. 493-580. New York: Springer-Verlag

McCabe, B. (1964). Nystagmus response of the otolith organs. *Laryngoscope*, 74: 372–381.

McRuer, D. and Weir, D. (1969). Theory of manual vehicular control. *Ergonomics*, 12: 599–633.

Merfeld, D. M. (1990). Spatial orientation in the squirrel monkey: an experimental and theoretical investigation. *Ph.D. Thesis*, MIT.

Merfeld, D. M. (1995a). Modeling human vestibular responses during eccentric rotation and off vertical axis rotation. *Acta Oto-Laryngologica (Supp.)*, 520: 354–359.

Merfeld, D. M. (1995b). Modeling the vestibulo-ocular reflex of the squirrel monkey during eccentric rotation and roll tilt. *Exp. Brain Res.*, 106: 123–134.

Merfeld, D. (1996). Effect of space flight on ability to sense and control roll tilt: human neurovestibular studies on SLS-2. *J. Appl. Physiol.*, 81: 50–57.

Merfeld, D., Teiwes, W., Clarke, A., Scherer, H. and Young, L. R. (1996). The dynamic contributions of the otolith organs to human ocular torsion. *Exp. Brain Res.*, 110: 315–321.

Merfeld, D. M. and Young, L. R. (1995). The vestibulo-ocular reflex of the squirrel monkey during eccentric rotation and roll tilt. *Exp. Brain Res.*, 106: 111–122.

Merfeld, D. M., Young, L. R., Oman, C. and Shelhamer, M. (1993a). A multi-dimensional model of the effect of gravity on the spatial orientation of the monkey. *J. Vest. Res.*, 3: 141–161.

Merfeld, D. M., Young, L. R., Paige, G. Tomko, D. (1993b). Three dimensional eye movements of squirrel monkeys following post-rotatory tilt. *J. Vestib. Res.*, 3: 123–139.

Merfeld, D. M. and Zupan, L. H. (2002). Neural Processing of Gravito-Inertial Cues in Humans. III. Modeling tilt and translation responses. *J. Neurophysiol.*, 87: 819–833.

Merfeld, D. M., Zupan, L. H. and Gifford, C. A. (2001). Neural Processing of gravito-inertial cues in humans. II. Influence of the semicircular canals during eccentric rotation. *J. Neurophysiol.*, 85: 1648–1660.

Merfeld, D. M., Zupan, L. and Peterka, R. (1999). Humans use internal models to estimate gravity and linear acceleration. *Nature*, 398: 615–618.

Mergner, T. and Becker, W. (1990). Perception of horizontal self-rotation: multisensory and cognitive aspects. In R. Warren and A. Wertheim (eds.), *Perception & Control of Self-Motion*, pp. 219–263. Hillsdale: Lawrence Erlbaum Associates.

Minor, L. and Goldberg, J. (1990). Influence of static head position on the horizontal nystagmus evoked by caloric, rotational and optokinetic stimulation in the squirrel monkey. *Exp. Brain Res.*, 82: 1–13.

Niven, J., Hixson, W. and Correia, M. (1966). Elicitation of horizontal nystagmus by periodic linear acceleration. *Acta Otolaryngologica*, 62: 429–441.

Oman, C. (2002). Human Visual Orientation in Weightlessness. In L. Harris and M. Jenkin (eds.), *Levels of Perception*. New York: Springer Verlag. This volume.

Ormsby, C. M. and Young, L. R. (1977). Integration of semicircular canal and otolith information for multisensory orientation stimuli. *Math. Biosci.*, 34: 1–21.

Paige, G. (1985). Caloric responses after horizontal canal inactivation. *Acta Otolaryngol*, 100: 321–327.

Paige, G. (1989). The influence of target distance on eye movement responses during vertical linear motion. *Exp. Brain Res.*, 77: 585–593.

Paige, G. and Tomko, D. (1991). Eye movement responses to linear head motion in the squirrel monkey. I. Basic characteristics. *J. Neurophysiol.*, 65: 1170–1182.

Raphan, T., Cohen, B. and Henn, V. (1981). Effects of gravity on rotatory nystagmus in monkeys. *Ann. NY Acad. Sci.*, 374: 44–55.

Raphan, T., Matsuo, V. and Cohen, B. (1977). A velocity storage mechanism responsible for optokinetic nystagmus (OKN), optokinetic after-nystagmus (OKAN) and vestibular nystagmus. In R. Baker and A. Berthoz (eds.), *Control of Gaze by Brain Stem Neurons, Developments in Neuroscience*, pp. 37–47. Elsevier/North Holland Biomedical Press.

Robinson, D. (1977). Vestibular and optokinetic symbiosis: an example of explaining by modelling. In R. Baker and A. Berthoz (eds.), *Control of Gaze by Brain Stem Neurons, Developments in Neuroscience*, pp. 49–58. Elsevier/North-Holland Biomedical Press.

Schor, C. M., Lakshminarayanan, V. and Narayan, V. (1984). Optokinetic and vection responses to apparent motion in man. *Vis. Res.*, 24: 1181–1187.

Schwarz, C., Busettini, C. and Miles, F. (1989). Ocular responses to linear motion are inversely proportional to viewing distance. *Science*, 245: 1394–1396.

Schwarz, C. and Miles, F. (1991). Ocular responses to translation and their dependence on viewing distance. I. Motion of the observer. *J. Neurophysiol.*, 66: 851–864.

Seidman, S., Telford, L. and Paige, G. (1998). Tilt perception during dynamic linear acceleration. *Exp. Brain Res.*, 119: 307–314.

Skipper, J. and Barnes, G. (1989). Eye movements induced by linear acceleration are modified by visualisation of imaginary targets. *Acta Otolaryngologica (Supp.)*, 468: 289–293.

Spiegel, M. (1972). Moving coordinate systems. In *Theory and Problems of Theoretical Mechanics with an Introduction to Lagrange's Equations and Hamiltonian Theory*, pp. 144–159. New York: Schaum Publishing Company.

Stephenson, S. B. (1993). Influence of the visual field on manual roll and lateral stabilization. *M.Sc. Thesis*, MIT, Boston, MA.

Stockwell, C. and Guedry, F. (1970). The effect of semicircular canal stimulation during tilting on the subsequent perception of the visual vertical. *Acta Oto-Laryngologica*, 70: 170–175.

Telford, L., Seidman, S. and Paige, G. (1997). Dynamics of squirrel monkey linear vestibuloocular reflex and interactions with fixation distance. *J. Neurophysiol.*, 78: 1775–1790.

Vieville, T. and Faugeras, O. (1990). Cooperation of the inertial and visual systems. In T. Henderson (ed.), *Traditional and Non-traditional Robotic Sensors*, Berlin: Springer-Verlag.

von Holst, E. and Grisebach, E. (1951). Einfluss des Bogengangssystems auf die "subjektiv" Lotrechte beim Menschen. *Naturwissenschaften*, 38: 67–68.

Waespe, W. and Henn, V. (1977). Neuronal activity in the vestibular nuclei of the alert monkey during vestibular and optokinetic stimulation. *Exp. Brain Res.*, 27: 523–538.

Wall, C. (1987). Eye movements induced by gravitational force and by angular acceleration: their relationship. *Acta Otolaryngologica*, 104: 1–6.

Wall, C., Merfeld, D. and Zupan, L. (1999). Effects of static orientation upon human optokinetic afternystagmus. *Acta Otolaryngologica*, 119: 16–23.

Wearne, S., Raphan, T. and Cohen, B. (1999). Effects of tilt of the gravito-inertial acceleration vector on the angular vestibuloocular reflex during centrifugation. *J. Neurophysiol.*, 81: 2175-2190.

Wilson, V. and Melvill Jones, G. (1979). *Mammalian Vestibular Physiology*. New York: Plenum.

Wrigley, W., Hollister, W. and Denhard, W. (1969). *Gyroscopic Theory, Design, and Instrumentation.* Cambridge, MA: MIT Press.

Young, L. R. (1967). Effects of linear acceleration on vestibular nystagmus. Proc. 3rd Symposium on the Role of the Vestibular Organs in Space Exploration, NASA SP-152, pp. 383-391. US Government Printing Office: Washington, DC.

Young, L. R. (1984). *Handbook of Physiology*, Oxford: Oxford University Press.

Zacharias, G. L. and Young, L. R. (1981). Influence of combined visual and vestibular cues on human perception and control of horizontal rotation. *Exp. Brain Res.*, 41: 159–171.

Zhou, W., King, W. M., Tang, B. and Newlands, S. (1998). Characteristics of angular and linear motion signals in the macaque vestibular nuclei. *Society for Neurosci.*, 24: 1744.

Zhou, W., Tang, B. and King, W. M. (2000). Vestibular neurons encode linear translation and head tilt with respect to gravity. *Society for Neurosci.*, 26: 1491.

Zupan, L. (1995). Modélisation du Réflexe Vestibulo-Oculaire et Prédiction des Cinétoses. Ph.D. thesis, Ecole Nationale Supérieure des Télécommunications.

Zupan, L., Droulez, J., Darlot, C., Denise, P. and Maruani, A. (1994). Modelization of vestibulo-Ooular reflex (VOR) and motion sickness prediction. In M. Marinaro and P. G. Morasso (Eds.), *International Congress on Application of Neural Networks*, pp. 106-109. Sorrento, Italy: Springer-Verlag.

Zupan, L. and Merfeld, D. M. (2002). Neural processing of gravito-inertial cues in humans: IV. Influence of visual rotational cues during roll optokinetic stimulation. *J. Neurophysiol.*, submitted.

Zupan, L., Merfeld, D. M. and Darlot, C. (2002). Using sensory weighting to model the influence of canal, otolith and visual cues on spatial orientation and eye movements. *Biol. Cyber.*, 86: 209–230.

Zupan, L., Peterka, R. and Merfeld, D. M. (2000). Neural processing of gravito-inertial cues in humans: I. Influence of the semicircular canals following post-rotatory tilt. *J. Neurophysiol.*, 84: 2001–2015.

19

Human Visual Orientation in Weightlessness

Charles M. Oman

An astronaut's sense of self-orientation is relatively labile, since the gravitational "down" cues provided by gravity are absent, visual cues for orientation are often ambiguous, and familiar objects can be difficult to recognize when viewed from an unfamiliar aspect. This chapter surveys the spatial orientation problems encountered in weightlessness including 0-G inversion illusions, visual reorientation illusions, EVA height vertigo, and spatial memory problems described by astronauts. We consider examples from the Space Shuttle, Mir, and the International Space Station. A vector model for sensory cue interaction is synthesized which includes gravity, gravireceptor bias, frame (architectural symmetry), and polarity cues, and an intrinsic "idiotropic" tendency to perceive the visual vertical in a footward direction. Experimental evidence from previous studies and recent research by our York and MIT teams in orbital flight is summarized.

19.1   Introduction

Understanding how humans maintain spatial orientation in the absence of gravity is of practical importance for astronauts and flight surgeons. It is also of fundamental interest to neurobiologists and cognitive scientists, since the force of gravity is a universal constant in normal evolution and development. Gravireceptor information plays a major role in the coordination of all types of body movement, and anchors the coordinate frame of our place and direction sense, as neurally coded in the limbic system.

This chapter reviews four related types of spatial orientation problems, as described by crew members on the US Shuttle and Russian and international space stations. We synthesize a set of working hypotheses which account for static orientation illusions in 0-G and 1-G, their relationship to height vertigo and spatial memory, and the role of visual cues. We then summarize supporting evidence from ground, parabolic, and orbital flight experiments. There is evidence that astronauts are more susceptible to dynamic (circular- and linear-vection) self-motion illusions during the first weeks of spaceflight, but for reasons of brevity, these dynamic illusions are not considered here.

This year's symposium honors Professor Ian Howard, who has made so many contributions to the understanding of human perception. Human spatial orientation has been a longstanding interest of Ian's. His 1982 book *Human Visual Orientation*, though out of print, remains the student's best introduction to this subject. Over the subsequent two decades, he and his students built a set of unique stimulus devices in the basements of three buildings: the now legendary rotating sphere, the vection sled, the mirrored bed and two tumbling rooms[1]. They did a series of experiments on static and dynamic visual orientation that are landmarks in this field. Ian has always been fascinated by the orientation illusions reported by astronauts, and has done experiments in parabolic flight. In the early 1990s, he accepted my challenge to help me write the first NASA proposal for what has since become a series of continuing space flight investigations on human visual orientation on the Shuttle and the International Space Station, employing virtual reality technology in space for the first time. Both in the laboratory and in the field, Ian's discipline, intellect, curiosity, creativity, infectious scientific passion, and adaptability to Tex-Mex food inspired everyone, including our astronauts. Some of the results from Neurolab — our first flight — are included here. Our laboratories also continue to collaborate in ground-based research sponsored by the NASA National Space Biomedical Research Institute.

19.2   Human Orientation Problems in Space Flight

Vision plays a critical role in maintaining spatial orientation in weightlessness. One of the most striking things about entering 0-G is that if the observers are in a windowless cabin, usually no one has any sensation of falling. Obviously "falling" sensations are visually and cognitively mediated. If the observers make normal head movements, the visual surround seems quite stable. Oscillopsia (apparent motion of the visual environment), so common among patients who have inner ear disease, is only rarely reported in weightlessness. What can change — often in dramatic fashion — is one's perception of static orientation with respect to the cabin and the environment beyond:

19.2.1   0-G Inversion Illusions

Ever since the second human orbital space flight by the late Gherman Titov in 1961, crew members in both the U. S. and Russian space programs have described a bizarre sensation of feeling continuously inverted in 0-G, even though in a familiar "visually upright" orientation in the cabin (Gazenko, 1964; Oman et al., 1986). "The only way I can describe it," some say, "is that though I'm floating upright in the cabin in weightlessness. Both the spacecraft and I seem to somehow be flying upside down." Visual cues clearly play a role in the strength of the illusion, but in contrast with visual reorientation illusions (Section 19.2.2), inversion illusions are

---

[1] A Quicktime video of the Tumbling Room can be found on the CD-ROM.

relatively persistent, and continue after eyes are closed. Some report the illusion is stronger in the visually symmetrical mid-deck area of the Shuttle than when on the flight deck, or in the asymmetrical Spacelab module. Inversion illusion is sometimes reversible by belting or pulling yourself firmly into a seat, or looking at yourself in a mirror. The illusion is quite common among Shuttle crew members in the first minutes of weightlessness, continuing or recurring for minutes to hours thereafter, but reports are rare after the second day in orbit. It is almost universal in parabolic flight among blindfolded volunteers entering weightlessness for the first time (Lackner, 1992). As detailed later, inversion illusion in 0-G has been attributed to the combined effects of gravitational unloading of the inner ear otolith organs, elevation of viscera, and also to the sensations of facial fullness and nasal stuffiness caused by sitting with feet elevated prior to launch, launch accelerations, and 0-G fluid shift.

Many astronauts are familiar with "aerobatic" inversion illusion, a sensation of inversion resulting from the "eyeballs up" acceleration component involved in an aerobatic pushover or inverted flight. Since the Shuttle thrusts into orbit into an inverted attitude, and crew members experience "eyeballs-in and up" acceleration, it is not surprising crew members experience aerobatic inversion illusion during launch. Perhaps the aerobatic inversion illusion due to the launch profile primes the onset of 0-G inversion illusion after entering weightlessness.

19.2.2   Visual Reorientation Illusions

Unlike their predecessors in the Mercury, Gemini, and Apollo programs, Skylab and Shuttle astronauts no longer routinely worked in their seats. Instead, their tasks frequently required them to move around and to work in orientations relative to the spacecraft interior, that were physically impossible to practice in simulators beforehand. Fundamental symmetries in the visual scene can create an ambiguity in the perceived identity of surrounding surfaces. When floating horizontally or upside down, they discovered that the spacecraft floor, ceiling, and walls would frequently exchange identities: "You know intellectually what is going on, but somehow whichever surface is seen beneath your feet seems like a floor"; "surfaces parallel to your body axis are walls"; "surfaces overhead are ceilings" (Figure 19.1).

Interior architectural asymmetries and familiar objects in fixed locations provided important landmarks which tended to prevent or reverse the illusion. However, the human body is also a familiar form, viewed on Earth primarily in a gravitationally upright position. Astronauts found that catching sight of another crew member floating inverted nearby would sometimes make themselves suddenly feel upside down (Figure 19.2). The Earth can provide a powerful "down"-orienting stimulus when viewed out a porthole or when on a spacewalk. In crew debriefings, other examples abounded: Astronauts working inverted on the flight deck, photographing the Earth through the overhead windows, felt they were looking "down" through windows in the floor of a gondola. Crew members working close to the canted upper racks in the Spacelab module were surprised to look

FIGURE 19.1. Crew member with feet toward Spacelab ceiling seems right side up. Note canted "upper" racks in the lower part of the picture. NASA photo.

down and see the lower racks tilting outward beneath them. Astronauts in the nodes and laboratory modules of the U. S. portions of the International Space Station sometimes find it difficult to distinguish walls from ceiling from floor, since the modules have a square cross section, and interchangable rack systems. Crew members passing headfirst through the horizontal tunnel connecting Space-lab with the Shuttle mid-deck sometimes feel as if they are ascending inside a vertical tube. Encountering another crew member coming the other way can make them suddenly feel as if they are upside down, descending headfirst. Looking backward at their own feet makes them feel upright again.

After these illusions were described by Skylab crew members (Cooper, 1976) and in more detail by the crew of Spacelab-1, we decided to name them "visual reorientation illusions" (Oman et al., 1984, 1986; Oman 1986), since they differed from 0-G inversion illusions in several important respects: First, the sensation was not necessarily of being "upside down" — rather, the subjective vertical was frequently beneath your feet. Second, whereas inversion illusions were difficult to reverse and continued when eyes were closed, VRIs were easily reversed, and depended typically on what you were looking at. Though VRIs usually occurred spontaneously, they could be manipulated cognitively in much the same way one can reverse a figure/ground illusion, or the perceived orientation of a Necker cube. "I can make whichever way I want to be down become down" was the frequent comment. When one slowly rolls inside a spacecraft, the moment of interchange of the subjective identity of the walls, ceilings, and floors is perceptually quite a distinct event, just as is the reversal of the corners of a Necker cube, or a figure-ground illusion. Lastly, most crew members experienced VRIs, and susceptibility continued throughout even long-duration Skylab and Mir missions, whereas 0-G inversion illusions are rare after the first day or two in weightlessness. VRIs

FIGURE 19.2. Seeing a crew member in an inverted position can make an observer himself feel "upside down." NASA photo.

have also been described in parabolic flight (Graybiel and Kellogg, 1967; Lackner and Graybiel, 1983) though the distinction between inversion and reorientation illusions was not made in the older literature. Astronauts now sometimes refer to VRIs as "the downs." Actually, it is possible to have a VRI right here on Earth, as when you leave an underground subway station labyrinth, and upon seeing a familiar visual landmark, realize, for example, that you are facing east, not west. On Earth, gravity constrains our body orientation, and provides an omnipresent "down" cue, so we normally only experience VRIs about a vertical axis. However, VRIs can be easily created about the gravitational horizontal in a 1-G laboratory using real or virtual tumbling rooms (Howard and Childerson, 1994; Oman and Skwersky, 1997).

19.2.3 Inversion Illusions, VRIs, and Space Sickness

There is relatively strong circumstantial and scientific evidence (reviewed by Oman and Shubentsov, 1992) that head movements made about any axis, particularly in pitch, are the dominant stimulus causing space sickness. However, it is clear from crew member reports that inversion illusions and VRIs — when they occur — often increase nausea. Crew members experiencing inversion illusions are reportedly continually aware of the sensory cue discrepancy. Apparently it is the onset of a VRI — and the sudden change in perceived self-orientation without a concurrent change in semicircular canal or otolith cue — that provides the nauseogenic stimulus. For example, one Shuttle pilot awoke, removed the sleep shades from the flight deck windows, saw the Earth above instead of below where he had previously seen it, and vomited immediately after. Other crew members described vomiting attacks after seeing other crew members — or even just space suits — floating inverted nearby, and suddenly feeling tilted or uncertain about their orientation. One astronaut who was feeling nauseous described "getting it over with" simply by deliberately cognitively inducing VRIs. This causal relationship makes sense in terms of what we know about the role of vestibular sensory conflict in motion sickness (Reason, 1978; Oman, 1982, 1990). Once we recognized the etiologic role of VRIs in space sickness (Oman et al., 1984, 1986; Oman, 1986), we suggested that whenever anyone on board was suffering from space sickness, everyone — not just the afflicted — should try to work "visually upright" in the cabin. This advice has since been broadly accepted by Shuttle crews.

19.2.4   EVA Height Vertigo

Over the past decade, there have been anecdotal reports from several crew members that while working inverted in the Shuttle payload bay, or while standing in foot restraints on the end of the Shuttle robot arm (Figure 19.3), or hanging at the end of a pole used as a mobility aid, they experienced a sudden attack of height anxiety and fear of falling toward Earth somewhat resembling the physiological height vertigo many people experience on Earth when standing at the edge of a cliff or the roof of a tall building. Some report experience enhanced orbital motion awareness, and a sensation of falling "down." The associated anxiety is disturbing, or in some cases even disabling, causing crew members to "hang on for dear life." A NASA astronaut flying on Mir published a vivid account (Linenger, 2000; see also Richards et al., 2001). We do not yet have prospective or retrospective statistical data on the incidence of the phenomenon. However, height vertigo is clearly a potential problem that will become more important during the ISS construction era, when many more EVAs are being made.

19.2.5   3D Spatial Memory and Navigation Difficulties

The U.S. and Russian space programs gradually evolved to using larger vehicles, with more complex three-dimensional architectures. For practical reasons, the local visual verticals in different modules are not universally coaligned. Ground trainer modules are not always physically connected in the same way as they are in the actual vehicle. Therefore, occupants say that they have difficulty visualizing the spatial relationships between landmarks on the interiors of the two modules. They cannot point in the direction of familiar interior landmarks in other modules the way they say they could when in their homes on Earth. They often do not know instinctively which way to turn when moving between modules through symmetrical multi-ported nodes. Shuttle crew members visiting the Mir station (Figure 19.4) often had difficulty finding their way back, without assistance from Mir crew members, or arrows fashioned and positioned to help them (Richards et al., 2001) Comparable problems have not been described within the Shuttle itself, probably because the flight deck, mid-deck, and payload bay research modules have coaligned and less ambiguous internal visual verticals. Maintaining spatial

FIGURE 19.3. Spacewalking Shuttle crew member looking "down" while standing in foot restraints on the end of the Canadian robotic arm. NASA photo.

orientation during EVA activity on the outside of the Mir and International Space Station was sometimes also difficult, particularly during the dark half of each orbit, due to the lack of easily recognizable visual landmarks.

Several operational crises that occurred in 1997 aboard the Russian Mir station convinced crew members and human factors specialists that the ability to make three-dimensional spatial judgements is important in emergency situations and critical if an emergency evacuation is necessary in darkness, or when smoke obscures the cabin. Twice when collisions with Progress spacecraft were imminent, crew members moved from module to module and window to window, unsuccessfully trying to locate the inbound spacecraft. Another emergency required the crew to reorient the entire station using thrusters on a docked Soyuz spacecraft. Members of the crew in the Mir base block module discovered they had great difficulty mentally visualizing the orientation of another crew member in the differently oriented Soyuz cockpit, and verbally relaying the appropriate commands (Burrough, 1998). Related difficulties are being encountered on the new International Space Station. Egress routes to the Shuttle or Soyuz spacecraft require turns in potentially disorienting nodes. Emergency egress is complicated by the limited capacity of rescue vehicles, so different crew members are assigned different vehicles and egress routes. One early station crew placed emergency "Exit" signs beside the node hatches, but subsequently discovered that one of the signs had been misplaced, probably as a result of a visual reorientation illusion. Improved egress signs are in development, and "you are here" maps, inflight practice, and preflight virtual reality-based spatial memory training are under consideration.

FIGURE 19.4. Russian Mir space station had four research modules connected to a central node. Visual verticals of some modules were not coaligned. NASA photo.

19.3   A Model for Human Visual Orientation

Based on prior research on human visual orientation in 1-G (reviewed by Howard, 1982), and synthesizing more recent theories and experiments of Mittelstaedt (1983, 1988), Young et al. (1986), Oman (1986), Oman et al. (1986), and Howard and Childerson (1994), the following heuristic model for static orientation perception emerges:

19.3.1   Beginning with a 1-G Model

On Earth in 1-G, the direction of the subjective vertical (SV) is the nonlinear sum of three vectors:

G the gravitational stimulus to the otoliths, cardiovascular, and kidney gravireceptors.

B a net gravireceptor bias acting in the direction of the body's major axis. The magnitude and headward vs. footward direction is presumed to be an individual characteristic.

V the perceptual visual vertical, is normally determined by:

F "frame" (architectural symmetry) visual cues, disambiguated by

P "polarity" cues, associated with the recognition of top/bottom of familiar objects in view, and

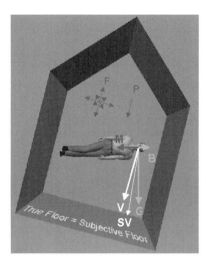

FIGURE 19.5. Model for 1-G "Tilted Room" illusion.

M an "idiotropic" tendency to perceive the visual vertical as oriented along the body axis in a footward direction.

Note that as is the convention in engineering and physics, the G vector defining the gravitational "vertical" is depicted pointing "down," as are the corresponding V, P, and M vectors. (Mittelstaedt has adopted the opposite convention). The idiotropic vector is denoted "M" in recognition of Mittelstaedt's many contributions (Young et al., 1986).

The SV in complete darkness (sometimes called the postural vertical) is determined only by the G and B vectors. The SV of gravitationally horizontal observers who have a headward gravireceptor bias is tilted slightly in a headward direction, that is they report feeling tilted slightly head down, and conversely. Measurement of the postural vertical provides a convenient way to assess a person's gravireceptor bias B — at least in 1-G.

The "idiotropic" tendency M affects all judgements of SV when any visual cues are present. The idiotropic effect a usually stronger than gravireceptor bias, even when the latter is in a headward direction. Hence the SV of a horizontally recumbent subject is deviated footward. When no F or P cues are present, the resultant of M and B deviates the SV footward. Hence an observer perceives a dimly lit gravitationally vertical line as rotated in the opposite direction to body tilt — the well-known Aubert illusion.

Figure 19.5 shows a horizontally recumbent observer viewing the interior of a tilted, barnlike room in 1-G. The major and minor axes of symmetry of the visual environment are depicted with the array of bidirectional vectors F. Since the room interior has a familiar shape, and readily distinguishable ceiling (top) and floor (bottom), it is also said to possess visual polarity, depicted by the vector P. The visual vertical V lies along one of the major the axes of symmetry in a direction closest to P and M. Here V points in the direction of the true floor, so

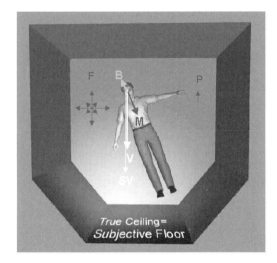

FIGURE 19.6. Model for 0-G Visual Reorioentation Illusion. Crew member inverted in a Spacelab module feels right side up.

it is subjectively perceived as a floor. The direction of the subjective vertical SV is determined by a nonlinear interaction of the visual V and gravireceptor (G+B) vectors. How the vectors combine depends on the orientation of the subject. For relatively small static tilts of the subject or the environment as shown in the figure — up to a limit of perhaps 45 degrees — the SV lies in a direction intermediate between V and (G+B). However, if the subject is not in the normal erect position, but instead recumbent, supine, or prone with respect to gravity, and V aligns with M, the SV can be "captured" by (i.e., align with) the V and M vectors. Thus, a supine subject feels gravitationally upright if the environment is tilted so P and V align with the body axis M.

19.3.2   Extending the Model to 0-G

How the model applies in weightlessness is shown in Figure 19.6. The physical stimulus to the body's gravireceptors G is absent, but a headward or footward bias B remains. As in 1-G, the direction of the visual vertical V is determined by the interaction of environmental frame F and polarity P cues, and the idiotropic vector M. Depending on the relative weighting the SV is captured by the visual vertical V or the resultant of the idiotropic vector M and the gravireceptor bias vector B. Unlike the near-upright 1-G case, the SV never lies in an intermediate direction between V and (M+B). It is always captured by one or the other. In Figure 19.6, the observer is depicted inside a Spacelab module, which has canted overhead racks. The structured environment provides a strong set of symmetry cues F. Here, the observer's feet are oriented toward the canted ceiling, and the footward idiotropic bias overcomes relatively weak polarity cues available from the visual scene. The perceptual visual vertical and the SV point toward the true

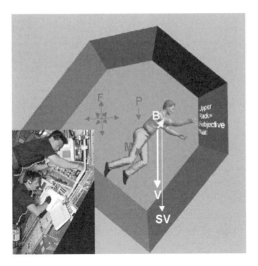

FIGURE 19.7. Model for VRI when working close to a canted upper rack in Spacelab. Nasa photo inset.

ceiling, which the observer perceives as a subjective floor. The observer experiences a visual reorientation illusion.

It is important to understand that frame and polarity cues are not physical properties of the entire visual environment. Both depend on the observer's viewpoint and gaze direction. For example, Figure 19.7 shows a crew member working on equipment mounted in the upper Spacelab racks. Working close to the upper racks, the dominant frame cue in the scene is aligned with the upper rather than lower racks. Written labels on rack-mounted equipment enhance the strength of downward polarity cues. As a result, V is parallel to the plane of the upper rack, which is perceived as a subjective wall. Unless the subject has a strong idiotropic bias M, the SV is also in the plane of the upper rack. If the observer momentarily looks "down" at the lower rack, he is surprised that it seems to tilt outward at the bottom.

Figure 19.8 illustrates the factors that likely contribute to a 0-G inversion illusion. This observer is shown floating with his feet in the general direction of the true floor. The frame, polarity and idiotropic cues F, P, and M align the visual vertical V toward the floor. Hence the true floor is perceived as a floor, and the subjects report being "visually upright" in the cabin. However, unlike the individuals depicted in previous figures, this person has an abnormally large headward gravireceptor bias, so though visually upright with respect to the cabin, he feels that he and the entire spacecraft are somehow upside down.

Figure 19.9 provides a plausible explanation for the onset of EVA height vertigo. In the left panel, the crew member is working "visually upright" in the payload bay of the Space Shuttle. The Earth is perceived as being "above." However, if the crew member rolls inverted, and sees the Earth beneath his feet, rather than feeling upside down, idiotropic M and Earth view polarity cues reverse the direc-

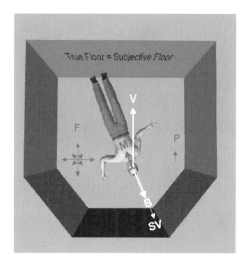

FIGURE 19.8. Model for 0-G inversion illusion.

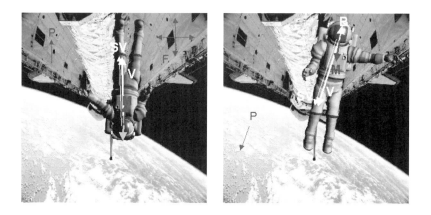

FIGURE 19.9. Model for EVA height vertigo.

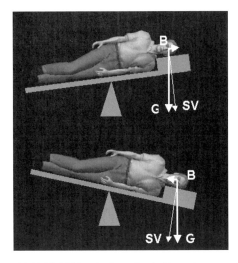

FIGURE 19.10. Tilting bed test for 1-G gravireceptor bias.

tion of the visual and subjective verticals, as shown in the right panel. Suddenly the crew member perceives he is hanging by one hand beneath an inverted space-craft.

19.4   Related Experiments

19.4.1   Gravireceptor Bias

Laboratory evidence for the existence of a gravireceptor bias comes from the experiments of Mittelstaedt (1986), who asked observers lying on a tilting bed to set themselves gravitationally horizontal in darkness. More than 40 normals and five previously flown astronauts were tested. The tilt angle of the entire group averaged almost perfectly horizontal, but there were consistent differences between individuals. As shown in Figure 19.10, some tended to set the bed a few degrees head down, while others set it a few degrees head up. It was a personal characteristic htat remained stable over periods of more than three years. Mittelstaedt hypothesized that those who set the bed slightly head up did so to effectively cancel out a headward gravireceptor bias, and noted that the two astronauts who had experienced inversion illusion in orbital flight had head up bias, whereas the other three did not. Pursuing the origin of the bias, he conducted experiments on a short-radius centrifuge where the observers could adjust their position relative to the axis of rotation until they felt horizontal subjectively . Normal observers felt horizontal when the rotation axis passed through their upper chest. Presumably the effect on tilt perception of the centrifugal stimulus to the vestibular otoliths was being balanced by centrifugal stimulation of previously unknown gravireceptors located on the other side of the axis of rotation. In further tests on paraplegics

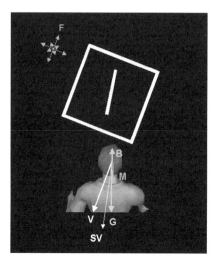

FIGURE 19.11. Rod and Frame test.

and nephrectomized patients (Mittelstaedt, 1996), he found evidence that the effect was mediated by mechanoreceptors in the kidneys and large blood vessels of the abdomen.

It remains to be verified how well 1-G tilting bed tests of individual graviceptor bias predict 0-G inversion illusion under operational conditions. Also, B is a multisensory bias which could conceivably be influenced by 0-G and launch acceleration induced fluid shift, facial edema, and nasal stuffiness not present in the 1-G tilting bed tests. If so, graviceptor bias measured in 1-G may be somewhat different than that found in flight.

19.4.2   Visual Frame Effects

In their classic "rod and frame" experiments, Witkin and Asch (1948) asked erect observers in a dark room to set a dimly lit pivoting rod to the SV. The rod was surrounded by a similarly lit square frame, which was tilted 28 deg clockwise or counterclockwise with respect to G. As depicted in Figure 19.11, the observer's SV indications deviated consistently in the direction of frame rotation. There were consistent differences between observers in the size of the effect, with group average being about 6 deg. The effect diminished with larger frame tilts, probably because the square was perceived as an upright diamond, so the diagonals became the perceptually dominant axes. Ebenholtz (1977) later showed that larger frames induced greater rod tilt than smaller ones, showing that field of view is important in producing a frame effect. Singer et al. (1970) and Howard and Childerson (1994) extended this result by having gravitationally upright observers view the interior of an unfurnished cubic chamber. The SV was consistently deviated toward the nearest axis of room symmetry, either the floor-ceiling-wall directions, or the room diagonals.

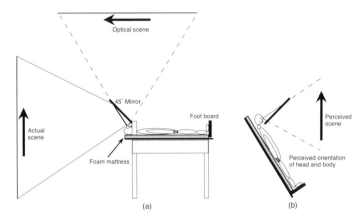

FIGURE 19.12. Mirror bed apparatus of Hu et al. Used with permission.

19.4.3   Visual Polarity Effects

Howard (1982) noted that in daily life, there is a class of common objects that we almost always encounter in an "upright" orientation with respect to gravity. Examples include tables, chairs, rugs, doors, houses, trees, cars, or human figures. These objects all have a readily identifiable "top" and "bottom," with mass distributed approximately equally on either side of an axis of symmetry, so they do not tip over. Howard refers to these as "intrinsically polarized" objects. Their relative orientation of conveys information about the direction of gravity, and can help disambiguate frame cues. Many other objects such as coins, pencils, books, etc., which are not usually seen in a consistent gravitational orientation, are described as "nonpolarized."

In the context of orientation in weightlessness, it is important to note that large surfaces, including those which extend beyond the immediate field of view, establish the major planes of visual space, but if their visual identity is ambiguous, they can provide only frame cues. We believe that in weightlessness the perceptual floor/ceiling/wall ambiguity of such surfaces is resolved by the relative orientation of the surface with respect to the body axis, or polarized visual details on the surface itself.

To experimentally measure object polarity, Hu, Howard and Palmisano (1999) had observers lying supine on an elevated bed (Figure 19.12) look upward into a wide mirror angled at 45 deg so they saw a left-right reversed view of the laboratory beyond the head of the bed. If the scene was a blank wall, observers perceived it as a ceiling. However, when intrinsically polarized objects were placed in view, the observers perceived their heads as upright, and their bodies tilted by an amount which varied depending on the characteristics of the objects in the scene. The extent of perceived body tilt was used as a measure of visual polarity. Polarized objects placed in the background appeared more potent than in the foreground. It was also confirmed that nonpolarized objects can inherit a form of "extrinsic" polarity if they appear to be lying on or hanging from other objects.

A second type of "extrinsic" polarity derives from the conventional location of certain types of objects. For example, doors and simple window and picture frames are often up-down symmetrical. However their placement relative to adjacent surfaces provides extrinsic polarity cues for surfaces in a vertical plane. We do not expect to see a door in the middle of a wall, or a picture frame positioned close to a floor. It makes sense to think that object polarity depends little on the relative orientation of the object and observer. We must only recognize "what" type of object it is. Relative orientation is probably more important for distinguishing details that allow an observer to distinguish "which" specific member of a class an object is.

19.4.4   Interaction Between Gravity, Polarity, Frame, and Idiotropic Cues

The rules describing how G, F, P, and M cues in various directions are combined under 1-G conditions have been defined in experiments where observers have viewed the interior of tilted furnished rooms. The interiors were fitted with anchored tables, desks, bookshelves, and other props so as to provide strong frame and polarity cues. For practical reasons, most of the testing has been done with the observers and rooms tilted less than 30 deg from the gravitational vertical (Kleint, 1936; Asch and Witkin, 1948; Singer et al., 1970).

As in the Rod and Frame experiments, the indicated subjective vertical represents a compromise between the gravitational and frame/polarity directions. Howard and Childerson (1994) tested at larger room tilt angles, and found that the SV was deviated toward the floor-ceiling-wall closest to being beneath their feet, but not to the diagonals (as in their frame experiments described in Section 19.4.2 above). The subjects were not asked whether the subjective identities of the floor-wall-ceiling surfaces exchanged as the room rotated into various positions, but in retrospect, and after trying it ourselves, Ian and I are almost certain they did, and thus experienced VRIs analogous to those of astronauts. Subsequently, Howard and Hu (2001) also tested at the 90 and 180 deg extremes of body tilt. We knew from earlier experiments (e.g. Young, Oman and Dichgans, 1975) that pitch and roll angular self-motion illusions (vection) was enhanced when the observer's head and body were supine or inverted. But we were still surprised to discover that when Howard and Hu's observers were gravitationally supine or inverted, and the room polarity vector was aligned with their body axis M vector, a substantial fraction felt gravitationally upright in the motionless room! It was as if gravireceptor information was being discounted when the head-body axis was not in the familiar gravitationally upright position. The subjective vertical seemed to be closely aligned with the coaligned idiotropic, visual polarity, and visual frame axes. This sort of "capture" was reminiscent of what we think happens to the astronauts. Not all subjects felt this, of course. Some still felt oriented with respect to gravity, and others said their perceptions seemed to switch back and forth in a confusing way between the two rival interpretations. It was also amusing that if gravitationally supine but subjectively upright observers extended their arms gravitationally upward, the arms felt oddly levitated, as if floating. It felt different than extending your arms while lying supine in bed at home in a gravitationally upright visual environment. Ian refers to this special sensation as a "levitation" illusion. Howard, Jenkin and Hu (2000) also showed that the incidence of "levitation" illusion increases as a function of age. We cannot be sure whether the latter is due to increased experience with polarity cues as one ages, senescent loss of vestibular receptor sensitivity, or both.

19.4.5   Animal and Human Visual Orientation Experiments in Weightlessness

Many astronauts have asked us, "Isn't it strange that we still have a vertical in weightlessness, even though dropped objects don't fall?" Certainly, but since an astronaut's job requires knowing whether they are facing forward or aft, port, or starboard in the spacecraft, everyone maintains an exocentric (allocentric) reference frame. This frame is the anchor for our hierarchically organized set of knowledge and visual memories for where things are, the latter is sometimes called a "spatial framework." The framework lets us remember where things are, look and reach for things in the correct direction, and mentally visualize unseen parts of the vehicle in correct relative orientation. Based on recordings from place and direction cells in the limbic system of animals on Earth, O'Keefe (1976) and Taube et al. (1990) believe that the human sense of place and direction is neurally coded in a gravitationally horizontal plane. Taube showed that prominent visual landmarks can reorient our sense of direction within this horizontal plane. Normally, the orientation of this plane is anchored by gravity. Taube et al. (1999) recently monitored rat head direction cells in parabolic flight, and Knierim et al. (2000) studied place cell behaviour in orbital flight. Both experiments confirmed that place and direction cells usually continue to maintain allocentric place and directional coding when the animals walk on the floor or walls of the test chamber. However, in both experiments, there was evidence that the allocentric reference frames sometimes — but not always — reoriented onto the surface the animal was walking on. Apparently humans are not the only animals who experience VRIs in weightlessness.

These animal experiments strongly support the notion that the human CNS also maintains an allocentric reference direction at the neural level, represented by the SV direction in the present model. It makes sense to think that the CNS uses this SV direction to determine the perceptual identity of ambiguous nonpolarized surfaces in the visual surround. However, since the SV direction is not "anchored" by gravity, idiotropic and gravireceptor bias and visual polarity cues can cause the orientation of the horizontal reference plane to shift suddenly. Depending on

FIGURE 19.13. Neurolab crew member wearing head-mounted display and spring harness. NASA photo.

the individual, "down" is either along the body axis, or perpendicular to the subjective floor (Figure 19.6). However, if gravireceptor bias is strongly headward, in conflict with the visual vertical V, the observer experiences a 0-G Inversion Illusion (Figure 19.8) by assuming that the SV is no longer associated with the local visual vertical V, but with an unseen outside coordinate frame, and describes himself as right side up in an upside down vehicle.

In 1998, we had the opportunity to quantify how frame and polarity cues affected the SV in four astronauts on the STS-90 Neurolab mission (Oman et al., 2000). For practical reasons, we could not use real tilted visual environments, so instead our observers wore a wide field of view (65 deg × 48 deg), color stereo head-mounted display (Figure 19.13) and viewed a sequence of virtual spacecraft interior scenes (Figure 19.14), presented at random angles with respect to their body axis. Subjects indicated the SV using a joystick-controlled pointer. Responses were categorized as to whether they were aligned within 5 deg of one of the scene visual axes, the body axis, or in between. We defined a metric that gave us a measure of average visual vs. idiotropic dominance across all angles of scene tilt. We tested the subjects preflight and postflight in both a gravitationally upright and supine position. Inflight, we tested them on the third or fourth day of the mission both free floating and while "standing" in a spring harness that pulled them down to the deck with a 70-lb force. As we expected, we usually saw "in between" responses only in the 1-G conditions. Inflight, responses aligned with either the body axis or one of the scene axes. Based on results of previous and concurrent 0-G linear- and circular-vection experiments (Young et al., 1996; Oman et al., 2000), which showed increased sensitivity to moving visual scenes, we expected that our observers might also rely more on the orientation of frame in polarity cues in motionless visual scenes. One observer who was

FIGURE 19.14. Stereogram of polarized visual scene used in Neurolab visual orientation experiments.

moderately visually independent on the ground became more visually dependent in flight, and then recovered postflight, after a short period of carryover. He responded to the scene polarity manipulation inflight. But the other three observers — two of whom were strongly "idiotropic" and one of whom was strongly "visual" — showed little overall change during or after the flight. The "down" cues from the spring harness did reduce the visual category responses of the two visual observers. None of these observers showed any signs of "inversion illusion" during these flight day 4 tests, although one reported a brief inversion illusion in darkness while a subject in another experiment. Obviously these results are only preliminary. Ultimately we cannot be sure that subjects respond exactly the same way to our virtual environments as they would if we could use real ones. But our results do confirm the notion — suggested by Young et al. (1986) and Reschke et al. (1994) based on astronaut debriefs — that crew members differ markedly in terms of whether they adopt a "visual" or "idiotropic" reference frame in making subjective vertical judgements. We expect to have the opportunity to test more observers over a longer flight duration on International Space Station missions starting in about two years.

19.4.6 EVA Height Vertigo

Height vertigo, experienced by many people when standing on top of a high structure, is generally seen as a normal physiological aversive response to a potentially dangerous situation. Symptoms include subjective instability of posture and locomotion, coupled with a fear or sensation of falling, and autonomic symptoms. Brandt et al. (1980) found the intensity of symptoms was greatest when the subject was standing, intermediate when sitting, and least when lying. It was strong when there were no stationary objects in front of the subject within 15-20 m. They noted that when a standing observer looks out over a distant vista, the subtle visual cues resulting from small translations of the body's centre of mass fall below visual threshold. The observer must depend on other vestibular and proprioceptive sources of information to be sure his center of gravity does not slip forward of his point of support. If the subject increases postural reflex gains in response to this uncertainty, his postural sway amplitude may actually increase, increasing his anxiety further. Of course, height vertigo is not limited to situations in which subjects stand erect. The training director of a major New England area telephone company has estimated that fully one-third of lineman trainees drop out due to height vertigo experienced while learning to climb telephone poles (personal communication). Height workers generally say that habituation usually occurs after repeated, graded exposures. It makes sense to think that EVA height vertigo is triggered by visual reorientation illusions resulting from seeing the Earth "below," as described in Section 19.3 and Figure 19.9. If subjects feel they are "standing" on the end of the Shuttle robot arm looking down at Earth, the lack of visual cues from nearby Shuttle-stationary objects in response to body movement may seem disturbing. Based on this interpretation obvious EVA height vertigo countermeasures include immediately rotating the body to face the spacecraft, and, if possible, working "right side up" relative to the spacecraft with the Earth nadir is in the upper visual vield. Use of body and hand restraints in addition to foot restraints may be helpful. Preflight practice with these techniques or even graded preflight habituation of the susceptible is possible, but the use of virtual reality techniques may be required since when using conventional underwater EVA training techniques, the pool walls nearby are readily visible.

19.4.7   3D Spatial Memory and Navigation Difficulties

Given that the interior architectures of space station modules and nodes are so symmetrical, and VRIs happen often, it is not that surprising crew members occasionally have difficulty maintaining a exocentric reference frame veridically aligned with the vehicle. However there is a second problem that relates to the way we establish local spatial frameworks, and the difficulty we apparently have in vehicles like the Mir station or ISS when we have to turn the spatial frameworks — originally learned in 1-G simulators — over in our minds, connect them together, and make spatial judgements. It is not so easy. Humans appear to choose salient spatial reference points to define a "spatial framework" and use this to remember the location of other objects and places in hierarchical fashion (Sadalla et al., 1980; McNamara, 1986; Franklin and Tversky, 1990), often employing their body axes to help establish referent directions. Observers can use mental imagery to change viewpoint location and direction. Creem et al. (2001) recently found that observers can more easily rotate memories of previously seen external object arrays about their body axis — perhaps because we have do it in everyday life — though the relative orientation of the gravity vector was unimportant. We recently studied how observers establish a spatial framework inside a cubic virtual room and recognize targets after the room had been rotated 90 or 180 deg about any of the three axes, not just the body axis. Observers had to memorize the relative directions of objects at the center of each wall, and correctly deduce the direction of an unseen target object after the objects located ahead and below were shown as a relative orientation cue. As in Creem et al.'s study, performance had little to do with the relative direction of gravity. Those who performed best also performed significantly better on traditional card and cube paper-and-pencil tests of mental rotation ability. Most observers could do the 3-D task robustly within 60 trials, but many said they memorized the cube in a particular reference orientation, and employed rules (e.g., remembering opposite pairs, and/or learning the three objects in a specific corner) to assist themselves in determining relative target directions. Taken together, these studies imply that astronauts should anticipate difficulty in situations where they have to rotate mentally the spatial framework of their current module or adjacent modules in order to make spatial judgements, and even greater difficulty making spatial judgements between modules, if the spatial frameworks must be mentally rotated from the orientation learned on the ground in 1-G trainers in order to connect the frameworks together. Further experiments on this question are currently underway in our laboratory. Potential inflight countermeasures for 3-D spatial memory and navigation difficulties now under consideration include route and emergency egress path signs; the use of easily remembered icons and colored surfaces to establish spatial reference landmarks and directions in a station rather than module-centric coordinate system; "you-are-here" maps with the major spatial reference landmarks on the interior of each module clearly shown; inflight egress practice sessions; and preflight training using virtual reality techniques so that crew members learn how to establish a consistent hierarchical spatial framework for the entire assembly of modules and nodes.

19.5   Conclusion

There is still a great deal that we do not understand about human visual orientation, both on Earth and in weightlessness. Our current models are useful in parsing and understanding the different types of 0-G illusions, but the models cannot yet be used to make quantitiative predictions for individual subjects, since they are largely heuristic and incomplete. For example, we need to better understand the effects of fluid shift and otolith unweighting on the gravireceptor bias terms in our models, and have reliable ways of predicting or measuring their magnitude and time course in 0-G. The orientation model presented in this paper is a simple one, and does not include the effects of surface contact forces, which can have a major effect when present. We also know that visual and vestibular angular velocity cues influence the SV, and in certain situations can cause static illusions such as "aviator's leans," but these effects are omitted from the current model. Why does susceptibility to "levitation" illusion gradually increase with age on Earth? The stability of the Aubert illusion in individuals suggests idiotropic bias is relatively constant in 1-G, but does it change after months of living in 0-G, in an environment where a "floor" is no longer consistently beneath us? Can we develop models for the way humans represent 3-D spatial frameworks, and validate them? After living in space for many months, will humans develop a more robust ability to establish 3-D spatial frameworks, and turn them over in our minds? My hope is that continued scientific research in weightlessness aboard the space station and its successors will ultimately help provide answers to these questions.

Acknowledgments

Supported by NASA Cooperative Agreement NCC9-58 with the National Space Biomedical Research Institute, and NASA Grant NAG9-1004 from Johnson Space Center.

References

Asch, S. E. and Witkin, H. A. (1948). Studies in space orientation: I. Perception of the upright with displaced visual fields. *J. Exp. Psych.*, 38: 325–337.

Brandt, T., Arnold, F., Bles, W. and Kapteyn, T. S. (1980). The mechanism of physiological height vertigo I. Theoretical approach and psychophysics. *Acta Otolaryngol.*, 89: 513–523.

Burrough, B. (1998). *Dragonfly. NASA and the Crisis Aboard Mir*. New York: Harper Collins.

Cooper, H. (1976). *A House in Space*. New York: Holt, Rinehart and Winston.

Creem, S. H., Wraga, M. and Prottitt, D. R. (2001). Imagining physicaly impossible self-rotations: geometry is more important than gravity. *Cognition*, 81: 41–64.

Ebenholtz, S. M. (1977). Determinants of the Rod and Frame Effect: The role of retinal size. *Percept. and Psychophys.*, 22: 531–538.

Franklin, N. and Tversky, B. (1990). Searching imagined environments. *J. Exp. Psych.: Gen.*, 119: 63–76.

Gazenko, O. (1964). Medical studies on the cosmic spacecrafts "Vostok" and "Voskhod".

Graybiel, A. and Kellogg, R. S. (1967). Inversion illusion in parabolic flight: its probable dependence on otolith function. *Aviation, Space, and Environ. Med.*, 38: 1099–1013.

Howard, I. P. (1982). *Human Visual Orientation*. Toronto: Wiley.

Howard, I. P. and Childerson, L. (1994). The contribution of motion, the visual frame, and visual polarity to sensations of body tilt. *Percept.*, 23: 753–762.

Howard, I. P. and Hu, G. (2001). Visually inducd reorientation illusions. *Perception*, 30: 583–600.

Howard, I. P., Jenkin, H. L. and Hu, G. (2000). Visually-induced reorientation illusions as a function of age. *Aviation, Space, and Environ. Med.*, 71S: A87–A91.

Hu, G., Howard, I. P. and Palmisano, S. (1999). The role of intrinsic and extrinsic polarity in generating reorientation illusions. *Invest. Ophthal. and Vis. Sci.*, 40: S801.

Kleint, H. (1936). Versuche über die Wahrnehmung. *Zeitschrift fur Psychologie*. 138: 1–34.

Knierim, J. J., McNaughton, B. L. and Poe, G. R. (2000). Three-dimensional spatial selectivity of hippocampal neurons during space flight. *Nature Neurosci.*, 3: 209–210.

Lackner, J. (1992). Spatial orientation in weightless environments. *Percept.*, 21: 803–812.

Lackner, J. and Graybiel, A. (1983). Perceived orientation in free fall depends on visual, postural, and architectural factors. *Aviation, Space and Environ. Med.*, 54: 47–51.

Linenger, J. M. (2000). *Off the Planet: Surviving Five Perilous Months Aboard the Space Station Mir*. New York: McGraw-Hill.

McNamara, T. P. (1986). Mental representations of spatial relations. *Cog. Psych.*, 18: 87–121.

Mittelstaedt, H. (1983). A new solution to the problem of subjective vertical. *Naturwissenschaften*, 70: 272–281.

Mittelstaedt, H. (1988). Determinants of space perception in space flight. *Adv. Oto-Rhino-Laryng.*, 42: 18–23.

Mittelstaedt, H. (1996a). Somatic gravireception. *Biolog. Psych.*, 42: 53–74.

Mittelstaedt, H. (1996b). Inflight and postflight results on the causation of inversion illusions and space sickness. *Scientific Results of the German Spacelab Mission D1*, Norderney, Germany, Wissenshaftliche Projecktfuhrung D1/DFVLR, Koln, Germany.

Mittelstaedt, H. and Glasauer, S. (1993). Crucial effects of weightlessness on human orientation. *J. Vest. Res.*, 3: 307–314.

O'Keefe, J. (1976). Place units in the hippocampus of the freely moving rat. *Exp. Neurol.*, 51: 78–109.

Oman, C. M. (1982). A heuristic mathematical model for the dynamics of sensory conflict and motion sickness. *Acta Otolaryngologica*, Stockholm (Suppl 392), 1–44.

Oman, C. M. (1986). Etiologic role of head movements and visual cues in space motion sickness on Spacelabs 1 and D-1. *Proc. 7th IAA Man in Space Symposium: Physiologic Adaptation of Man in Space*, Houston, TX.

Oman, C. M. (1987). The role of static visual orientation cues in the etiology of space motion sickness. *Proc. Symposium on Vestibular Organs and Altered Force Environment*. Houston, TX, NASA/Space Biomedical Research Institute, pp. 25–37.

Oman, C. M. (1990). Motion sickness: a synthesis and evaluation of the sensory conflict theory. *Can. J. Physiol. Pharmacol.*, 68: 294–303.

Oman, C. M., Howard, I. P., Carpenter-Smith, T., Beall, A. C., Natapoff, A., Zacher, J. E. and Jenkin, H. L. (2000). Neurolab experiments on the role of visual cues in microgravity spatial orientation. *Aviation Space and Environ. Med.*, 71: 293.

Oman, C. M., Lichtenberg, B. K., Money, K. E. and McCoy, R. K. (1984). Symptoms and signs of space motion sickness on Spacelab-1. *Proc. NATO-AGARD Aerospace Medical Panel Symposium on Motion Sickness: Mechanisms, Prediction, Prevention and Treatment*, Williamsburg, Va, NATO AGARD CP-372. Later republished as Oman, C. M., Lichtenberg, B. K. et al. In G. H. Crampton (Ed.) *Symptoms and Signs of Space Motion Sickness on Spacelab-1. Motion and Space Sickness*, pp. 217–246. Boca Raton, FL: CRC Press.

Oman, C. M., Lichtenberg, B. K., Money, K. E. and McCoy, R. K. (1986). MIT/Canadian vestibular experiments on the Spacelab-1 mission: 4. Space motion sickness: symptoms, stimuli, and predictability. *Exp. Brain Res,*, 64: 316–334.

Oman, C. M. and Shubentsov, I. (1992). Space sickness symptom severity correlates with average head acceleration. In A. L. Bianch, L. Grelot, A. D. Miller and G. L. King (Eds.), *Mechanisms and Control of Emesis*, Colloque INSERM/Libbey Eurotext, Ltd. 233: 185–194.

Oman, C. M. and Skwersky, A. (1997). Effect of scene polarity and head orientation on illusions in a tumbling virtual environment. *Aviation, Space, and Environ. Med.*, 68: 649.

Reason, J. T. (1978). Motion sickness adaptation: a sensory mismatch model. *J Roy. Soc. Med.*, 71: 819–829.

Reschke, M. F., Bloomberg, J. J. et al. (1994). Neurophysiological Aspects: Sensory and Sensory-Motor Function. *Space Physiology and Medicine*. A. E. Nicogossian, Lea and Febiger.

Richards, J. A., Clark, J. B. et al. (2001). Neurovestibular effects of long-duration space-flight: a summary of Mir phase 1 experiences. *NASA Johnson Space Center National Space Biomedical Research Institute*.

Sadalla, E. K., Burroughs, W. J. and Staplin, L. J. (1980). Reference points in spatial cognition. *J. Exp. Psych.: Human Learn. and Mem,*, 6: 516–525.

Singer, G., Purcell, A. T. et al. (1970). The effect of structure and degree of tilt on the tilted room illusion. *Percept. and Psychophys.*, 7: 250–252.

Taube, J. S., Muller, R. U. and Ranck, J. B. Jr. (1990). Head direction cells recorded from the postsubiculum in freely moving rats. *J. Neurosci.*, 10: 436–447.

Taube, J. S., Stackman, R. W. et al. (1999). Rat head direction cell responses in 0-G. *Soc. Neurosci. Abstr.*, 25: 1383.

Witkin, H. A. and Asch, S. E. (1948). Studies in space orientation: IV. Further experiments on perception of the upright with displaced visual fields. *J. Exp. Psych.*, 38: 762–782.

Young, L. R., Mendoza, J. C., Groleau, N. and Wojck, P. W. (1996). Tactile influences on astronaut visual spatial orientation: Human neurovestibular experiments on Space-lab Life Sciences — 2. *J. Applied Physiol.*, 81: 44-49.

Young, L. R., Oman, C. M. and Dichgans, J. M. (1975). Influence of head orientation on visually induced pitch and roll sensation. *Aviation, Space, and Environ. Med.*, 46: 264–268.

Young, L. R., Oman, C. M., Watt, D. G. C., Money, K. E., Lichtenberg, B. K., Kenyon, R. V. and Arrott, A. P. (1986). MIT/Canadian vestibular experiments on the Spacelab-1 mission: 1. Sensory adaptation to weightlessness and readaptation to 1-G an overview. *Exp. Brain Res.*, 64: 291–298.

20

Three-Axis Approaches to Ocular Motor Control: A Role for the Cerebellum

Mark F. Walker, Heimo Steffen, and David S. Zee

20.1   Introduction

It is a privilege for us to make a contribution to this volume honoring Professor Ian Howard. Following from one of the many areas of research in which he has made an important contribution, we will discuss some of our recent findings about how the brain controls eye torsion, that is, the movements that rotate the globe around its line of sight. Our focus will be on the contribution of the cerebellum to the control of eye movements, and on eye torsion in particular. The cerebellum plays a pivotal role in the generation of eye movements of all types, both in their immediate, on-line control and in their long-term adaptive calibration. Not surprisingly, we find that cerebellar lesions lead to disturbances of torsion, and more generally, to disturbances in generating eye movements that rotate the globe around the correct three-dimensional axis (horizontal, vertical, and torsion), in response to a particular visual or vestibular stimulus.

20.2   Perceptual Disturbances Related to Abnormalities of Torsion

Foveate animals have explicit ocular motor requirements for best vision. Images of objects of interest must be brought to the fovea, where visual acuity is highest, and kept there, relatively still, for a long enough period of time so that the brain can interpret what is happening in the visual scene. The map of the visual world upon the retina is, of course, two-dimensional, and hence to move the fovea to its center we rotate the globe around two axes (horizontal and vertical) that are orthogonal. But there is a third degree of freedom that allows for torsion — eye movements that rotate the globe around an axis that is roughly parallel to the line of sight. Torsion neither moves images away from the fovea nor increases, by much, motion of images that are already on the fovea. In this sense, a change in the orientation of the globe produced by torsion has little affect on foveal vision. But there are other reasons to have a mechanism that controls the torsional orientation of the eyes. Changes in torsion will affect the perception of objects with images that lie eccentric to the fovea. Such information from the peripheral retina, for example, contributes to the perception of the position of our head (actually the orbits) with respect to the veridical visual upright. Patients with an acute unilateral loss of labyrinthine function show torsion of the eyes such that the top pole rotates toward the side of the lesion (Curthoys et al., 1991).[1] When asked to align the position of a vertical bar in an otherwise dark room to earth-vertical, patients tilt the image of the bar such that its top is rotated toward the side of the lesion (Curthoys and Wade, 1995).

Torsion also becomes important for optimal binocular visual function. The difference in the torsional positions of the two eyes — the angle of cyclovergence — determines the torsional disparity between the images of an object on the two eyes. Torsional disparity contributes to our perception of the shape, orientation, and location of objects in depth relative to the position of our heads. In normal subjects the angle of cyclovergence is tightly controlled (Van Rijn et al., 1994). The absolute torsional position of the eyes may fluctuate, but they do so together. As an example of a clinical disturbance that produces abnormal torsional alignment of the eyes, consider patients with a unilateral paralysis of the superior oblique muscle (Lindblom et al., 1997). In this case the palsied eye is relatively extorted (top pole rotated outward). Images of upright objects may appear slanted, with the top of the object seemingly closer than it really is. There also may be torsional diplopia, such that when viewing a horizontal bar, for example, the two images will be slanted with respect to each other, with the apparent intersection of the lines pointing toward the side of the weak, relatively extorted eye.

During rotation of the head around its naso-occipital (roll) axis the eyes must rotate around an axis parallel to that of the head to minimize motion of images in the retinal periphery. If the eyes are directed roughly straight ahead in the orbit, this compensatory motion of the eyes is also along the line of sight, and hence equivalent to eye torsion. Certainly some degree of "torsional slip" is tolerated naturally since even for normal subjects, the amplitude of "compensatory" eye movements in response to roll motion of the head is considerably less than that of the head (Tweed et al., 1994). The amount of compensation for roll movements of the head also varies with the nearness of the object of interest. For near objects the need for optimal foveal function — fine stereopsis and fusion — begin to

---

[1] Actually, with an imbalance (physiological or pathological) in the vestibular inputs that rotate the head around the roll (naso-occipital) axis, the consequent rotation of the eye is a "counterroll" around a head-fixed, nasal-occipital axis. Hence, if the eyes are eccentric in the orbit, the compensatory counterroll will not be a torsional rotation around the line of sight. The caveat, of course, is that one must be clear about which coordinate frame — head fixed or eye fixed – one is using when discussing torsional rotation of the globe.

dominate over the need for stabilization in the retinal periphery (Misslisch et al., 2001).

20.3   Listing's Law and the Cerebellum

One of the more controversial issues about the control of eye torsion is related to the mechanisms that underlie Donders' and Listing's laws. Donders' law states that during steady fixation, the torsional orientation of the eyes is specified by the horizontal and vertical position of the eyes, and is independent of how the eyes reached that position. Ocular torsion is fixed for a given eccentric eye position. Listing's law describes a way in which the eyes can rotate from one position to another in order to implement the torsion predicted by Donders' law. Listing's law dictates a primary position of the eyes, and if this position is chosen as the reference position, all other positions can be reached from this position by rotation around an axis that lies in a plane (Listing's plane) that is perpendicular to primary position. This arrangement satisfies Donders' law. An important mathematical relationship between angular velocity and eye position for Listing's law to be obeyed is the "half-angle rule." For eye trajectories that do not pass through primary position, the angular velocity vector tilts by half the angle of the orthogonal eye position eccentricity (Haslwanter, 1995).

We recently reported the results of three-axis recordings in a large group of patients with degenerative cerebellar disease and no evidence for neurological disturbances outside the cerebellum (Straumann et al., 2000). Here we summarize some of those findings relevant to Listing's law. All patients had typical cerebellar ocular motor findings, including downbeat and horizontal gaze-evoked nystagmus, and impaired smooth pursuit. We asked if these patients also had torsional drift of the eyes — which they did — and what was its cause. Could the drift be interpreted as a violation of Listing's law, or was it due to some other cause, for example, a torsional vestibular bias, or an impaired torsional gaze-holding network (torsional integrator)? We first performed a conventional Listing's analysis during eye fixations by plotting torsional eye position as a function of horizontal and vertical eye position and then seeing how well the data could be fit to a plane, as is called for by Listing's law. Figure 20.1 shows the analysis from one patient. The data were rotated from a coordinate system defined by the coil frame to the Listing's coordinate system (Bergamin et al., 2001). The slow drifts of downbeat and horizontal gaze-evoked nystagmus are evident in the plot of horizontal vs. vertical position (as seen from the front of the subject) (Figure 20.1, top row, left panel).

From the middle (torsional-horizontal, x-z projection) and lower panels (torsional-vertical, x-y projection) of Figure 20.1 it can be seen that the width of Listing's plane in the patient was larger than that in the normal subject. This so-called thickness of Listing's plane (as reflected in the mean values of the standard deviations of torsional eye position from the plane) was significantly higher,

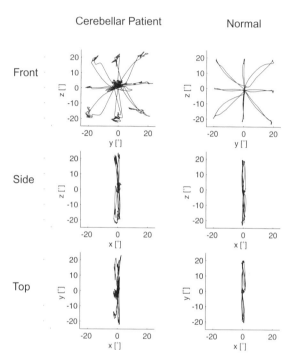

FIGURE 20.1. Attempted fixations of a central and eight eccentric targets on a tangent screen. x: torsional, y: vertical, z: horizontal components of rotation vectors in degrees. Data are rotated into Listing's coordinates. Signs of base vector directions according to the right-hand rule.(for horizontal, positive is leftward; for vertical, positive is downward; and for torsion, positive is clockwise). Patient data is in left panels and normal subject in right panels. Upper panels: front view (y-z-plane). Middle panels: side view (x-z-plane). Lower panels: top view (x-y-plane). Note the slightly increased thickness of Listing's planes in the cerebellar patient. Modified with permission from Straumann et al. (2000), *J. Neurophysiol.*, 83: 1125–1140.

by about 25%, in the group of 18 patients (0.95 deg ±0.37), than in the group of 25 normal comparison subjects (0.75 deg ±0.22). The slopes of the planar fits, which are related to the orientation of Listing's plane relative to the straight ahead reference position and are described by horizontal and vertical primary position, were not significantly different between the two groups. The difference between subjects and patients did not seem to be related to the shape or "twist" of a surface, when a second-order fit was used to compare the two groups.

This analysis, based on the positions of the eye during attempted fixation, suggests that Listing's law is less valid in patients with cerebellar disease. But this conclusion is not on firm ground because the eyes were constantly moving and it was therefore impossible to determine a true Listing's plane, which would have required steady fixation. Accordingly, to look further into how the cerebellum might contribute to the control of the torsional position of the eyes, we examined trajectories of eye drift at the angular velocity level, since Listing's law makes predictions about the axis around which the eye is rotating when it is moving. We first examined how well each of the drift components (horizontal, vertical, and torsional) correlated with one another, and then used the 'half-angle rule' to see if Listing's law was obeyed.

The main results of this analysis were as follows. The torsional drift was independent of the upward drift of downbeat nystagmus, that is, it was not simply the result of a constant cross coupling of torsion with vertical, as one might expect if the torsional drift were vestibular in origin, for example. The torsional drift was related to horizontal eye position; for the right eye, intorsion when the eye was in abduction and extorsion when in adduction; for the left eye, vice versa. But torsion did not correlate with horizontal drift velocity in the same way among all patients, which argues against a simple relationship between torsional and horizontal drift. Finally, by comparing the torsional drift associated with horizontal eccentricities in up and in down gaze positions, and using the half-angle rule, we could show that the drift still was not what one would find if Listing's law were obeyed (Figure 20.2).

Further evidence for non-Listing's law behavior in cerebellar patients comes from the study of horizontal pursuit eye movements at different vertical eccentricities. Again, because of the half-angle rule, the horizontal angular velocity vector should tilt during pursuit according to the degree of vertical eccentricity. Figure 20.3 shows horizontal and torsional angular velocity during horizontal pursuit at two vertical eccentricities for a normal subject. One can see that the phase between the torsional and horizontal angular velocity reverses on up and down gaze, indicating that the angular velocity vector is tilting in the opposite direction during horizontal pursuit in up and in down gaze. The results of this type of analysis in a group of cerebellar patients and in normal subjects are summarized in Figures 20.4 and 20.5. We calculated the tilt of the angular velocity vector (arctangent of the ratio of torsional and horizontal velocity) at the various vertical eccentricities, and then computed the relationship of the tilt to vertical eye position (calling this number the tilt slope). For Listing's behaviour the tilt slope should be 0.5. Note that the values for the normal subjects cluster tightly around 0.65,

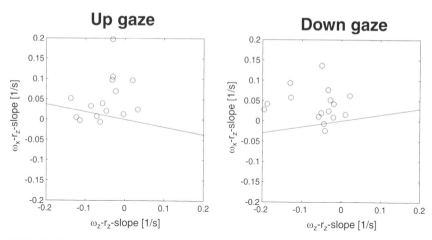

FIGURE 20.2. Cerebellar patients, ω_z-r_z-slopes vs. ω_x-r_z-slopes, where ω is angular velocity and r is angular position. The solid lines indicate where the data points must lie at an upward gaze eccentricity of 21.2 deg (left panel) and a downward gaze eccentricity of 16.8 deg (right panel) to obey Listing's law. Modified with permission from Straumann et al. (2000). *J. Neurophysiology*, 83: 1125–1140.

while there was considerable spread in the values of the patients, which ranged from about 0.3 to 1.2. In sum, using an angular velocity analysis of the smooth tracking by our cerebellar patients, we showed non-Listing's law behavior during smooth pursuit. These data are compatible with the hypothesis that there are abnormalities in the control of Listing's law in cerebellar patients

We also performed a similar analysis on the slow phases of optokinetic nystagmus (OKN) in response to a full-field rotating stimulus. To induce OKN, subjects were rotated in a vestibular chair in darkness at a constant velocity, and when the vestibular response died away, the lights were turned on as the subjects continued to rotate. Subjects were then asked to change their vertical eye position at various times while still rotating. As shown in Figure 20.5, in normal subjects, the tilt slope was about 0.25, a value close to the tilt expected from a rotational vestibular stimulus (Fetter et al., 1994). Note that this value was different from values during smooth pursuit, and is another piece of evidence that smooth pursuit and OKN have, at least in part, separate premotor circuitry. This distinction between smooth pursuit and OKN parallels the distinction between the translational and rotational VOR (Miles, 1997). Like pursuit, the translational VOR roughly obeys Listing's law, and like OKN, the rotational VOR does not (Angelaki et al., 2000; Walker et al., 2000).

In our cerebellar patients we found that the tilt slopes for OKN had a much wider spread of values than normals, ranging from about -0.2 to 0.6. Furthermore, the tilt slopes for pursuit and OKN were not well correlated in many patients, though they were in the normal subjects (Figure 20.6). Whether or not this

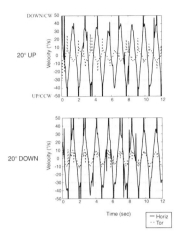

FIGURE 20.3. Data from a normal subject during horizontal smooth pursuit at up and down 20 deg. Horizontal (solid line) and torsional (dashed line) components (in head coordinates) of angular eye velocity are shown. Note the reversal of the phase of torsional velocity with vertical eye position. Data here are presented using the right-hand rule.

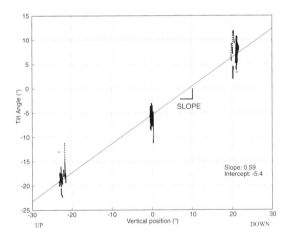

FIGURE 20.4. Data from one normal subject during horizontal smooth pursuit (0.7 Hz). Vertical eye position is represented on the x-axis and the instantaneous tilt angle (arc-tangent of the ratio of torsional and horizontal velocity) on the y-axis. The straight line represents a least-squares linear regression. The intercept of this regression is the tilt intercept (tilt angle at zero vertical eye position). The slope of the regression is the tilt slope, the variation of the tilt angle with vertical eye position. Data here are presented using the right-hand rule.

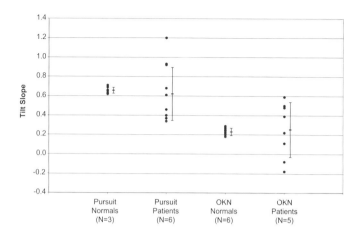

FIGURE 20.5. Tilt slopes for all normal subjects and cerebellar patients for smooth pursuit and for OKN. The values for the normal subjects cluster tightly; those for the patients cluster over a much wider range.

dissociation between pursuit and OKN indicates that separate structures within the cerebellum control the axis around which the eye rotates for the different sub-types of eye movements remains to be proven.

What might be the role of the cerebellum, then, in the elaboration of Listing's law? Recently, Demer and colleagues have shown the importance of orbital "pulleys" — connective tissue sleeves through which the eye muscle tendons pass — in determining eye torsion (Clark et al., 2000). The position of these pulleys, like the trochlea itself, determines the pulling direction of the eye muscles on the globe, and hence the degree of torsion associated with a given horizontal and vertical gaze position. The pulleys then could be responsible for the amount of torsion that is imparted to the globe during eye fixations, and hence allow List-ing's law to be satisfied during fixation. Furthermore, the pulleys are surrounded by smooth muscles that are innervated by fibres carried in the ocular motor nerves themselves (Demer et al., 2000). Hence, the position of the pulleys could be mod-ified by changes in the innervation to the smooth muscles that surround them. Thus, if the positions of the pulleys could be modified "on-line" they also might help account for the fact that some types of eye movements obey Listing's law (e.g., smooth pursuit, saccades and the translational VOR) while others do not (e.g., the angular VOR and optokinetic nystagmus when stimulated by rotation of the head (or visual scene) around the roll axis). The orientation of Listing's plane also is modulated according to the state of vergence, even when the eye it-self does not change its orbital position, as is the case for the viewing eye during accommodative vergence (Steffen et al., 2000; Kapoula et al., 1999). It remains to be seen whether these changes in Listing's behaviour can be attributed solely to changes in the position of pulleys or to changes in the innervation of the cyclover-

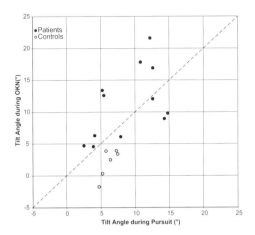

FIGURE 20.6. Comparison of tilt slopes for pursuit and OKN in normal subjects and cerebellar patients. Each point represents the responses of a single subject to pursuit and OKN in one direction (either right or left). Note how closely the values for the normal subjects cluster and how widely scattered are the values for many of the cerebellar patients.

tical muscles that directly impart torsion to the globe. In either case it is also not known if the cerebellum is responsible for these immediate, online adjustments in Listing's law.

In the long term, too, it seems unlikely that Listing's law could be implemented (or not implemented, during head rotation, for example) to its high degree of specificity without appropriate adjustments in innervation, either to the pulleys themselves or to the muscles where they attach to the globe. Consequently, we predict there must be a central mechanism that monitors torsional eye position as a function of gaze eccentricity, and, in the longterm adjusts innervation to the smooth muscles of the pulleys, to the eye muscles that insert on the globe, or to both, to ensure that Listing's law is obeyed or not obeyed, according to circumstances. The cerebellum may be the structure that performs this function. This specific hypothesis is similar to other ideas about cerebellar ocular motor function. For example, the cerebellum assures that the pulse, slide and step of innervation to the ocular muscles is matched correctly according to the mechanical properties of orbital tissues so that there is no unwanted drift immediately following each saccade (Optican et al., 1986).

These considerations, of course, imply that Listing's law is mutable. We have recently tested this idea experimentally in four normal subjects who wore, for 72 hours, a vertically displacing prism (Steffen et al., 2002) that produced a left over right disparity of 7–11 diopters, which they could fuse. The orientation of Listing's plane of each eye was measured, under monocular viewing conditions, before and after the period of exposure to the prism, We found a significant shift of the relative orientation of the vertical primary positions (right eye minus left

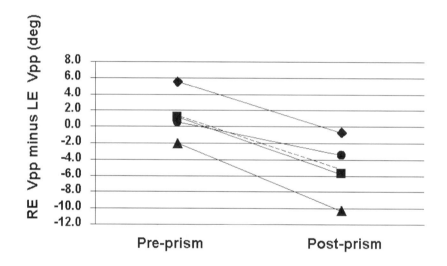

FIGURE 20.7. Relative orientation of the vertical primary positions ((vPP_{diff}) = vertical primary position (vPP) of the right eye minus vPP of the left eye), measured during monocular viewing, in unadapted and adapted states in four subjects. Reprinted from *Invest. Opthal. Vis. Sci.*, 43: 668-672, 2002, Steffen, H., et al., Changes in Listing's plane following sustained vertical fusion, with permission of the Association for Research in Vision and Ophthalmology.

eye) of 6.3 deg ± 1.7 (Figure 20.7). In other words, prolonged fusion of a vertical disparity leads to a shift in the orientation of Listing's plane, even when measured under monocular viewing conditions. Whether or not the cerebellum mediates this change too, remains to be discovered.

20.4   A Labyrinthine Coordinate Scheme for Smooth Pursuit: Torsion During Vertical Pursuit

Thus far we have attempted to infer the role of the cerebellum in controlling the torsional position of the eye during fixation. One can also ask if the cerebellum contributes to the control of torsional velocity in the slow-phase response to visual and vestibular stimuli. Recall that for optimal visuomotor function, the three-dimensional axis around which the eye rotates must match that of the visual or vestibular stimulus. We studied three patients with isolated cerebellar lesions (cavernous angiomas) in the region of the middle cerebellar peduncle close to the fourth ventricle (FitzGibbon et al., 1996). Each showed a direction-changing torsional nystagmus during vertical smooth pursuit and during vertical VOR cancellation when fixing upon a target moving with the head, but not during vertical saccades or the vertical VOR in darkness. During upward smooth tracking, the upper poles of the eyes rotated toward the side of the lesion, and during downward smooth tracking toward the side opposite the lesion. The slow-phase velocity of the torsional eye movement was proportional to that of the vertical component. Torsional eye movements were also present during vertical tracking produced with an eccentric afterimage, indicating that actual motion of images on the retina was not necessary for the torsion to appear.

We asked if an abnormality in Listing's or Donders' laws might explain this abnormal torsion. This seemed unlikely because there was no abnormal torsion during saccades, and the speed and the direction of torsion were related to eye velocity and not to eye position. Alternatively, we proposed that the abnormal torsion during vertical pursuit could be interpreted by considering the nature of signal processing associated with generating pure vertical slow phases of the vestibulo-ocular reflex (VOR). In particular, we hypothesized that pursuit signals are encoded in coordinate frame similar to that of the vestibular responses from the labyrinth. Recall that each vertical semicircular canal responds optimally to a mixture of roll and pitch motion of the head, and that stimulation of an individual vertical semicircular canal produces a mixed vertical-torsional slow-phase eye movement. The vertical direction is determined by whether the anterior (producing upward slow phases) or posterior (producing downward slow phases) canal is stimulated. The direction of torsion, however, is the same for stimulation of the anterior and posterior canals in the same labyrinth; the top poles of the eyes always rotate away from the side of stimulation.

Hence, to produce a pure vertical slow phase, for example, in response to a pure pitch movement of the head, the anterior or posterior canals on both sides of the head must be stimulated simultaneously, so that the vertical components add and the torsional components cancel. We propose here that an analogous type of signal processing, requiring cancellation of oppositely directed torsional slow phases, must occur for pure vertical pursuit. Indeed, such an organization for visual tracking is suggested from physiological studies of another visual following system — optokinetic nystagmus (OKN) — which is driven by full-field stimuli and works in concert with the angular VOR. In rabbits, for example, visual information for OKN is encoded in a labyrinthine coordinate system (Van der Steen et al., 1994). Sensitivity to vertical and torsional motion is combined on neurons carrying information for OKN in the same way that information about roll and pitch motion of the head is combined on primary vestibular afferents. A cancellation of torsion then becomes necessary to generate pure vertical visual tracking. The hypothesis for our patients, then, assumes that pursuit movements are generated using some of the same circuitry that underlies the generation of full-field visual-following responses such as OKN. Indeed, there is some evidence that pursuit signals within the primate flocculus might be encoded in such a canal framework (Krauzlis and Lisberger, 1996).

To explain the specific pattern of abnormal vertical pursuit shown by our patients, we assume there is a partial loss of pursuit signals that are carried in a labyrinthine coordinate scheme. Because of the consistent location of the lesions in these patients we suggest that the middle cerebellar peduncle carries visual information (probably relaying information from the pontine nuclei to the vestibulocerebellum and dorsal vermis) encoded in anterior SCC coordinates and that interruption of this pathway leads to torsional nystagmus during vertical pursuit. This would lead to the pattern of torsional nystagmus during upward and downward tracking that was observed in these patients. Whether the lack of cancellation of torsional signals occurs in the cerebellum per se, or in more downstream, outflow pursuit pathways such as the vestibular nuclei, is not settled. There are other examples in the cerebellum in which the influences upon anterior and upon posterior semicircular canal pathways are disparate. The flocculus, for example, has inhibitory projections to the vestibular nuclei that mediate anterior but not posterior canal responses. This dichotomy has been suggested as the cause for the frequent finding of down beating nystagmus in cerebellar patients (Zee et al., 1981; Baloh and Yee 1989; Böhmer and Straumann, 1998).

20.5   Inappropriate Torsional Responses to Vestibular Stimulation: Cross-Coupling in the VOR

We have previously reported that cerebellar patients frequently have inappropriate vertical slow-phase responses in response to yaw axis (horizontal) head rotation (Walker and Zee, 1999). This can also occur during responses to brief, high-acceleration, head thrusts, which are presumably comprised of relatively high-frequency components, as well as with more sustained, relatively low-frequency

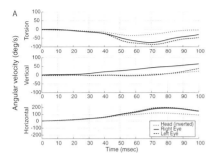
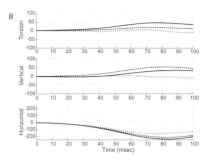

FIGURE 20.8. Binocular dual axis scleral search coil recordings in a patient with cerebellar degeneration. Angular velocity components for both eyes and the head (inverted) are shown for head impulses in both horizontal directions. Note the increased horizontal gain, the upward vertical cross-coupling with horizontal head rotations in either direction, and the torsional cross-coupling that changes direction with the direction of horizontal head rotation. The responses are also disconjugate. (a) Rightward head rotation. (b) Leftward head rotation. Right, up, and clockwise are positive here.

vestibular responses, such as during constant-velocity head rotations in the dark. Recently, we have looked at the torsional as well as the vertical responses to horizontal head thrusts in cerebellar patients, and some have shown a response compatible with a disturbance in the control of anterior semicircular canal pathways. For example, Figure 20.8 shows the response to horizontal head thrusts from a patient with cerebellar degeneration. The main features include

1. vertical cross-coupling that was always upward for both horizontal directions of rotation;

2. torsional cross-coupling that changed direction; clockwise for rightward head thrusts and counterclockwise for leftward head thrusts; and

3. the cross-coupling was disconjugate; torsion was greater in the ipsilateral eye and vertical greater in the contralateral eye. This last response is a dynamic skewing of the eyes during head rotation.

A hypothesis to explain this abnormal pattern of vestibular response is that the cerebellar lesion has led to a dynamic release of inhibition upon vestibular pathways that carry information from the anterior semicircular canals. A second assumption is that excitation of a semicircular canal is a more effective stimulus than inhibition (Ewald's second law), so that the excited anterior semicircular canal (which is in the opposite labyrinth relative to the direction of head rotation) contributes more to the response than the inhibited anterior semicircular canal on the other side. In this way, one might expect an upward slow phase to be associated with horizontal head rotation in either direction. In addition, the disconjugate pattern of torsion and vertical motion could reflect the fact that the primary excitatory connections of the anterior semicircular canal pathways are to the ipsilateral

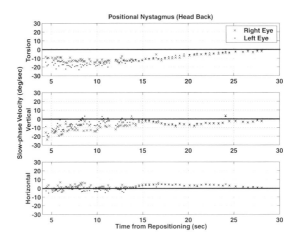

FIGURE 20.9. Binocular dual axis scleral search coil recordings in a patient with a lesion of the nodulus. Note the positional nystagmus just after the head is placed in about 50 deg of extension. There were strong vertical and torsional components. Also nnote that initially the nystagmus is disconjugate. Data here are presented using the right-hand rule.

superior rectus and contralateral inferior oblique muscles, hence leading to relatively more vertical rotation in one eye and torsion in the other. There are, of course, other explanations for this pattern of abnormal vestibular response, but our hypothesis would be another example in which an alteration of activity in anterior semicircular pathways is suggested as a cause of a vestibular disturbance in cerebellar disease. Finally, the disconjugacy of the abnormal responses implicates the cerebellum in the maintenance of dynamic eye alignment during high acceleration, high-frequency head rotation. Although the exact structures within the cerebellum responsible for control of the axis of eye rotation in response to head thrusts are not known, it is attractive to speculate that the flocculus and paraflocculus might be involved.

Recently we had the opportunity to examine a patient with a relatively isolated lesion of the cerebellar nodulus. She had two main ocular motor findings. First, there was a head-shaking induced nystagmus in which rapid, back-and-forth horizontal head rotation led to a post-head-shaking torsional nystagmus. Also, when the head was quickly pitched back to a supine position, the patient developed a mixed pattern of vertical and torsional nystagmus which was somewhat dissociated between the two eyes, more torsion in one, more vertical in the other. (Figure 20.9) In contrast, her responses to horizontal head thrusts were quite normal. These findings further implicate the cerebellar nodulus in the control of low-frequency vestibular-ocular responses, including the control of the axis around which the eye is rotating relative to the stimulus (Angelaki and Hess, 1995; Sheliga et al., 1999; Radtke et al., 2001; Wiest et al., 1999; Wearne et al., 1998). Finally, they also suggest a role for the nodulus in the control of the conjugacy of vestibulo-ocular responses (Zee, 1996).

In sum, we have presented a series of studies in patients with cerebellar disturbances that implicate the cerebellum in the control of eye torsion. We suggest that the cerebellum is important for the control of torsion during fixation, and hence in the elaboration of Listing's law. During movements of the eyes, too, the cerebellum appears important for assuring that the eye rotates around the correct three-dimensional axis in response to visual and vestibular stimuli. We propose a common organizational scheme for phylogenetically old vestibular responses driven by labyrinthine stimulation, and more recent, foveally driven, smooth pursuit tracking. We emphasize that our clinics can be laboratories in which we not only devise better ways to diagnose and treat patients, but also learn more about how the brain functions.

Acknowledgements

NIH grants RO1 EY0-1849 and K32EY-0400, German Research Foundation DFG Ste 86o/2-1, and the Arnold-Chiari foundation. Ronald Tusa referred one of the patients.

References

Angelaki, D. E. and Hess, B. J. (1995). Inertial representation of angular motion in the vestibular system of rhesus monkeys. II. Otolith-controlled transformation that depends on an intact cerebellar nodulus. *J. Neurophysiol.*, 73: 1729–1751.

Angelaki, D. E., McHenry, M. Q. and Hess, B. J. M. (2000). Primate translational vestibulocular reflexes. I. High-frequency dynamics and three-dimensional properties during lateral motion. *J. Neurophysiol.*, 83: 1637–1647.

Baloh, R. W. and Yee, R. D. (1989). Spontaneous vertical nystagmus. *Rev. Neurol. (Paris)*, 145: 527–532.

Bergamin, O., Zee, D. S., Roberts, D. C., Landau, K., Lasker, A. G. and Straumann, D. (2001). Three-dimensional Hess screen test with binocular dual search coils in a three-field magnetic system. *Invest. Ophthal. Vis. Sci.* 42: 660–667.

Böhmer, A. and Straumann, D. (1998). Pathomechanism of downbeat nystagmus: a simple hypothesis. *Neurosci. Lett.*, 250: 127–130.

Clark, R. A., Miller, J. M. and Demer, J. L. (2000). Three-dimensional location of human rectus pulleys by path inflections in secondary gaze positions. *Invest. Ophthal. Vis. Sci.*, 41: 3787–3797.

Curthoys, I. S., Dai, M. J. and Halmagyi, M. (1991). Human ocular torsional position before and after unilateral vestibular neurectomy. *Exp. Brain Res.*, 85: 218–225.

Curthoys, I. S. and Wade, S. W. (1995). Ocular torsion position and the perception of visual orientation. *Acta Oto - Laryngol.*, 520 (Suppl.): 298–300.

Demer, J. L., Oh, S. Y. and Poukens, V. (2000). Evidence for active control of rectus extraocular muscle pulleys. *Invest. Ophthal. Vis. Sci.*, 41: 1280–1290.

Fetter, M., Tweed, D., Misslisch, H., and Koenig, E. (1994). Three-dimensional human eye movements are organized differently for the different oculomotor subsystems. *Neuro-ophthal.*, 14: 147–152.

FitzGibbon, E. J., Calvert, P. C., Dieterich, M. D., Brandt, T. and Zee, D. S. (1996). Torsional nystagmus during vertical pursuit. *J. Neuroophthalmol.*, 16: 79–90.

Haslwanter, T. (1995). Mathematics of three-dimensional eye rotations. *Vis. Res.*, 35: 1727–1739.

Kapoula, Z., Bernotas, M., and Haslwanter, T. (1999). Listing's plane rotation with convergence: role of disparity, accommodation, and depth perception. *Exp. Brain Res.*, 126: 175–186.

Krauzlis, R. J. and Lisberger, S. G. (1996). Directional organization of eye movement and visual signals in the floccular lobe of the monkey cerebellum. *Exp. Brain Res.*, 109: 289–302.

Lindblom, B., Westheimer, G. and Hoyt, W. F. (1997). Torsional diplopia and its perceptual consequences. *Neuro-ophthal.*, 18: 105–110.

Miles, F. A. (1997). Visual stabilization of the eyes in primates. *Current Opin. in Neurobiol.*, 7: 867–871.

Misslisch, H., Tweed, D. and Hess, B. J. M. (2001). Stereopsis outweighs gravity in the control of the eyes. *J. Neurosci.*, 21: 414–418.

Optican, L. M., Zee, D. S. and Miles, F. A. (1986). Floccular lesions abolish adaptive control of post-saccadic drift in primates. *Exp. Brain Res.*, 64: 596–598.

Radtke, A., Bronstein, A. M., Gresty, M. A., Faldon, M., Taylor, W., Stevens, J. M. and Rudge, P. (2001). Paroxysmal alternating nystagmus and skew deviation after partial destruction of the uvula. *J. Neurol. Neurosurg. Psych.*, 70: 790–793.

Sheliga, B. M., Yakushin, S. B., Silvers, A., Raphan, T. and Cohen, B. (1999). Control of spatial orientation of the angular vestibulo-ocular reflex by the nodulus and uvula of the vestibulocerebellum. *Ann. NY Acad. Sci.*, 871: 94–122.

Steffen, H., Walker, M. F. and Zee, D. S. (2000). Rotation of Listing's plane with convergence: independence from eye position. *Invest. Ophthal. Vis. Sci.*, 41: 715-721.

Steffen, H., Walker, M. F. and Zee, D. S. (2002). Changes in Listing's plane following sustained vertical fusion. *Invest. Ophthal. Vis. Sci.*, 43: 668–672.

Straumann, D., Zee, D. S. and Solomon, D. (2000). Three-dimensional kinematics of ocular drift in humans with cerebellar atrophy. *J. Neurophysiol.*, 83: 1125–1140.

Tweed, D., Sievering, D., Misslisch, H., Fetter, M., Zee, D. S. and Koenig, E. (1994). Rotational kinematics of the human vestibuloocular reflex. I. Gain matrices. *J. Neurophysiol.*, 72: 2467–2479.

Van der Steen, J., Simpson, J. I. and Tan, J. (1994). Functional and anatomic organization of three-dimensional eye movements in rabbit cerebellar flocculus. *J. Neurophysiol.*, 72: 31–46.

van Rijn, L. J., Van der Steen, J. and Collewijn, H. (1994). Instability of ocular torsion during fixation: Cyclovergence is more stable than cycloversion. *Vis. Res.*, 34: 1077–1087.

Walker, M. F. and Zee, D. S. (1999). Directional abnormalities of vestibular and optokinetic responses in cerebellar disease. *Ann. NY Acad. Sci.*, 871: 205–220.

Walker, M. F., Zee, D. S., Shelhamer, M., Roberts, D. C. and Lasker, A. G. (2000). Variation of eye velocity axis with vertical eye position during horizontal pursuit, interaural translation, and yaw rotation in normal humans. *Soc. Neurosci. Abstr.*, 26: 1718.

Wearne, S., Raphan, T. and Cohen, B. (1998). Control of spatial orientation of the angular vestibuloocular reflex by the nodulus and uvula. *J Neurophysiol.*, 79: 2690–2715.

Wiest, G., Deecke, L., Trattnig, S. and Mueller, C. (1999). Abolished tilt suppression of the vestibulo- ocular reflex caused by a selective uvulo-nodular lesion. *Neurol.*, 52: 417–419.

Zee, D. S., (1996). Considerations on the mechanisms of alternating skew deviation in patients with cerebellar lesions. *J. Vestibular Res.*, 6: 1–7.

Zee, D. S., Yamazaki, A., Butler, P. H. and Gücer, G. (1981). Effects of ablation of the flocculus and paraflocculus on eye movements in primate. *J. Neurophysiol.*, 46: 878–899.

Author Index

Subject Index